Biological Invasions and Animal Behaviour

How does behaviour affect biological invasions? Can it explain why some animals are such successful invaders?

With contributions from experts in the field and covering a broad range of animals, this book examines the role of behaviour in biological invasions from the point of view of both invaders and native species. The chapters cover theoretical aspects, relevant behaviours and well-documented case studies, showing that behaviour is critical to the success, and ecological and socio-economic impact, of invasive species. Its insights suggest methods to prevent and mitigate those impacts, and offer unique opportunities to understand the adaptive role of behaviour.

Offering a comprehensive overview of current understanding of the subject, the book is intended for biological invasion researchers and behavioural ecologists, as well as ecologists and evolutionary biologists interested in how organisms deal with anthropogenic environmental changes such as climate change and habitat loss.

Judith S. Weis is Professor Emerita in Biological Sciences at Rutgers University, New Jersey, USA, and serves on advisory committees for federal agencies and the UN. Her research focuses on salt marshes, fish, crabs and stresses in estuaries, including pollution and invasive species.

Daniel Sol is a National Spanish Research Council (CSIC) Scientist at the Centre for Ecological Research and Forestry Applications (CREAF) in Catalonia, Spain. His research focuses on the causes and consequences of animal responses to environmental changes.

Biological Invasions and Animal Behaviour

Edited by

JUDITH S. WEIS
Rutgers University, New Jersey, USA

and

DANIEL SOL
National Spanish Research Council (CSIC), Spain

CAMBRIDGE
UNIVERSITY PRESS

University Printing House, Cambridge CB2 8BS, United Kingdom

One Liberty Plaza, 20th Floor, New York, NY 10006, USA

477 Williamstown Road, Port Melbourne, VIC 3207, Australia

314-321, 3rd Floor, Plot 3, Splendor Forum, Jasola District Centre, New Delhi - 110025, India

79 Anson Road, #06-04/06, Singapore 079906

Cambridge University Press is part of the University of Cambridge.

It furthers the University's mission by disseminating knowledge in the pursuit of education, learning and research at the highest international levels of excellence.

www.cambridge.org
Information on this title: www.cambridge.org/9781107434714

© Cambridge University Press 2016

First published 2016
First paperback edition 2019

A catalogue record for this publication is available from the British Library

Library of Congress Cataloging in Publication data
Names: Weis, Judith S., 1941–, editor. | Sol, Daniel, 1967–, editor.
Title: Biological invasions and animal behaviour / edited by Judith S. Weis,
Rutgers University, New Jersey, USA, and Daniel Sol, National Spanish
Research Council (CSIC), Spain.
Description: New York : Cambridge University Press, 2016. | Includes index.
Identifiers: LCCN 2016019011 | ISBN 9781107077775
Subjects: LCSH: Introduced animals–Behavior. | Introduced animals–Ecology. |
Biological invasions.
Classification: LCC QL86 .B56 2016 | DDC 591.5–dc23
LC record available at https://lccn.loc.gov/2016019011

ISBN 978-1-107-07777-5 Hardback
ISBN 978-1-107-43471-4 Paperback

Contents

Preface

Ever since Ernst Mayr, behaviour has increasingly been perceived as crucial to understanding how animals invade new regions and interact with native species. Surprisingly, however, up to now there has been no book entirely devoted to discussing this subject. The present book is an attempt to fill the gap. By integrating a variety of topics, approaches and study systems, the book presents a broad, up-to-date overview of the mechanisms by which behaviour affects biological invasions. Although in spirit the book is based on basic research, many findings reported throughout the chapters also have obvious conservation applications to prevent and mitigate the impact of invaders.

The genesis of this book is strange, in that the two people most responsible for the idea of developing such a book are no longer involved with it. Back in early 2013, Suzanne Albrecht, who at that time was a Senior Manager at John Wiley publishers, and Mark Hauber of Hunter College of the City University of New York organized an on-line conference on the topic of introduced species and behaviour. Suzanne and Mark had worked together on the journal *Ethology*, had previously done an on-line conference together on kin recognition, and felt that to do something more applied was a natural follow-up. A number of contributors to this volume participated in the on-line conference, along with one of the co-editors (JSW). During the conference, pre-recorded talks were available to be viewed on specific days, and questions and answers were available to all through on-line access. Following the successful on-line conference, a number of participants expressed interest in developing a book about invasive species from a behavioural ecologist's perspective. At that time, Mark was willing to be a co-editor. JSW agreed to be a co-editor too, but we felt we wanted another co-editor. Suzanne proposed DS as third editor, and the three of us began contacting additional potential chapter authors. In the fall of 2013, before we had submitted a formal book proposal to Wiley, two major changes took place: Suzanne departed Wiley and Mark decided for professional reasons not to continue as a co-editor, since he had become editor-in-chief of *The Auk: Ornithological Advances* and Acting Associate Provost for Research at Hunter College, and no longer could dedicate enough time to the project. This left the two of us feeling rather stranded. However, Suzanne had written to her friend Martin Griffiths, a Commissioning Editor for Life Sciences at Cambridge University Press, about the project and he contacted us about the possibility of publishing the book with Cambridge, which we decided to do. Then in the fall of 2014, we were told by Cambridge University Press that Martin Griffiths had left. Despite all the changes and bumps in the road, we are very pleased with how the book has turned out. We are very appreciative of the work of

Suzanne Albrecht and Mark Hauber for the initial idea for the on-line symposium and book, grateful to Martin Griffiths for getting us through the approval process at Cambridge University Press and grateful to Tim Hyland for editorial assistance in the last stages of the publication process.

Finally, we are in debt to all the authors who contributed to the book. We feel lucky to have been able to join such a bunch of excellent scientists. Our biggest thanks go to all of them. The book also benefited tremendously from many colleagues who accurately reviewed the chapters and we thank all them for their effort.

Judith Weis and Daniel Sol

Contributors

Mark A. Albins
Auburn University, USA

Laura Aquiloni
University of Florence, USA

Ignasi Bartomeus
Uppsala University, Sweden

Amber J. Brace
University of South Florida, USA

Grzegorz Buczkowski
Purdue University, USA

Martina Carrete
Estación Biológica de Doñana, Spain

Phillip Cassey
The University of Adelaide, Australia

David G. Chapple
Monash University, Australia

Julien Cote
University of Toulouse, France

Amy E. Deacon
University of St Andrews, UK

Marie Diquelou
University of Newcastle, Australia

I. Federspiel
University of Vienna, Austria

Jochen Fründ
University of Guelph, Canada and University of California, USA

Stephanie S. Gervasi
University of South Florida, USA

Andrea S. Griffin
Newcastle University, Australia

Tomáš Grim
Palacky University, Czech Republic

Edwin D. Grosholz
University of California, USA

D. Guez
University of Newcastle, Australia

Leon Hohl
The University of Adelaide, Australia

Steven A. Juliano
Illinois State University, USA

Holly J. Kilvitis
University of South Florida, USA

Sonia Kleindorfer
Flinders University, Australia

F. Lermite
University of Newcastle, Australia

L. Philip Lounibos
University of Florida, USA

Anne E. Magurran
University of St Andrews, UK

Lynn B. Martin
University of South Florida, USA

Joan Maspons
CREAF (Centre for Ecological Research and Forestry Applications), Spain

Katharina J. Peters
Flinders University, Australia

Ben L. Phillips
James Cook University, Australia

Jennifer S. Rehage
Florida International University, USA

Andrew Sih
University of California, Davis, USA

Jules Silverman
North Carolina State University, USA

Daniel Sol
Centre for Ecological Research and Applied Forestries, Spain

Bård G. Stokke
Norwegian University of Science and Technology, Norway

Andrew V. Suarez
University of Illinois at Urbana-Champaign, USA

Frank J. Sulloway
University of California, USA

José L. Tella
Estación Biológica de Doñana, Spain

Elena Tricarico
University Firenze, Italy

Judith S. Weis
Rutgers University, USA

Elizabeth H. Wells
California Department of Water Resources, USA

Neal M. Williams
University of California, USA

Bob B.M. Wong
Monash University, Australia

1 Introduction

Andrew V. Suarez and Phillip Cassey

Human modification of the Earth has turned ecologists and evolutionary biologists into harbingers of global change and stewards of the planet's remaining diversity. Global change takes many forms including habitat loss and fragmentation, biological invasions, emerging infectious diseases, harvesting/exploiting natural resources at unsustainable rates, pollution and climate change. Moreover, these mechanisms of change are interrelated, each increases the impact of the others. If we do not take immediate measures to reduce the impact of these processes, we will face a new era of mass extinction – the Anthropocene – where global diversity will be replaced by a subset of species whose distribution and persistence is aided predominantly by human transport and activity.

The interdisciplinary field of conservation biology has been recognized as a discipline for over 35 years (Soulé and Wilcox, 1980; Soulé, 1985). However, for its first decade, the role of behaviour in the field was largely restricted to issues related to captive breeding and reintroduction efforts. Empirical and theoretical research in conservation largely focused on the genetics of small populations, viability analyses, community ecology, responses to habitat fragmentation, and measures of diversity and its loss. Over the last 20 years, a number of books and edited volumes have been published that have attempted to fill in the gap created by the omission of behavioural perspectives in conservation (Sutherland, 1996; Clemmons and Buchholz, 1997; Caro, 1998; Gosling and Sutherland, 2000; Festa-Bianchet and Apollonio, 2003; Candolin and Wong, 2011). However, these publications spread their focus across the many stochastic and deterministic processes responsible for species' decline and subsequently often only scratched the surface of how behaviour can be used to manage and mitigate the loss of diversity.

Biological invasions are now recognized as a major form of human-induced global change (Vitousek et al., 1996). The implications of biological invasions have been discussed for almost 200 years (since Darwin) and Charles Elton's (1958) seminal contribution laid the foundation for the study of invasive species to be appreciated as its own integrative discipline. As with conservation biology generally, the field of invasion biology has benefited immensely by adding the perspective of behavioural research (Holway and Suarez, 1999; Sih et al., 2010; Chapple et al., 2012). Behaviour is the mechanistic link that determines how species interact with each other and their environment and is

Biological Invasions and Animal Behaviour, eds J.S. Weis and D. Sol. Published by Cambridge University Press. © Cambridge University Press 2016.

therefore essential for understanding why some species succeed in new environments and what impacts they will have once established.

In this edited volume, Daniel Sol and Judith Weis bring together a diverse group of scientists to examine how a careful study of behavioural mechanisms can guide efforts to understand the processes associated with biological invasions including their establishment, spread and impact. These chapters draw from diverse perspectives, ecosystems and taxa, including fish, birds and a variety of terrestrial and aquatic arthropods.

Many chapters examine the role of behavioural flexibility as a mechanism for success across the various stages of invasions. Amy Deakin and Anne Magurran examine how behavioural flexibility facilitates poeciliids to establish and spread in new environments. Jennifer Rehage, Julien Cote and Andrew Sih compare individual variation in behaviour (e.g. personality) between native and invaded habitats in mosquitofish. They also examine how personality traits can promote success through different steps of the invasion pathway (introduction, establishment, spread, impact on community). Andrea Griffin and colleagues examine behavioural flexibility from the perspective of learning and cognition in an introduced bird, the Indian myna. Finally, Ben Phillips addresses the issue of how different behavioural types within populations influence dispersal rates and can potentially cause accelerating rates of spread post establishment.

Invasions provide a powerful unintended experimental system for examining novel species interactions including predator–prey and host–parasite coevolution. Ted Grosholz and Elizabeth Wells, for example, discuss predator–prey recognition with novel assemblies of predators and prey of crabs and whelks, and examine the roles of novel weapons and enemy release in these interactions. Judith Weis reviews the invasion of arthropods in aquatic systems to examine how behavioural mechanisms may allow a novel predator to escape detection by naïve prey, and how novel prey may be avoided by resident predators. She also examines the defensive strategies of introduced prey and the characteristics that may make introduced species particularly effective predators. Andrea Griffin and colleagues use experiments to provide evidence of the importance of learning in avoiding novel predators.

Predation is not the only negative species interaction caused by introductions. Chapters by Jules Silverman and Grzegorz Buczkowski on ants and by Elena Tricarico and Laura Aquiloni on crayfish examine how behavioural traits may influence not only predation but also inter- and intraspecific competition. Invasions also offer unique opportunities to study host–parasite relationships. Steven Juliano and Phillip Lounibos examine the role of invasive mosquitoes as novel vectors of disease. They discuss how both adult and larval behaviours influence interspecific interactions such as predation and competition with other mosquitoes. Sonia Kleindorfer and colleagues question whether behavioural studies can guide efforts to save Darwin's finches threatened by an introduced parasitic fly, which causes high nestling mortality. Tomáš Grim and Bård Stokke use introduced birds to examine the coevolution of brood parasite–host behaviour in novel ecological settings.

Not all species interactions are negative and novel mutualisms may also arise due to species introductions. New links in a mutualism network (e.g. pollination) can have both positive and negative impacts on existing constituent members. Ignasi Bartomeus and colleagues write about how introduced plants affect the behaviour of pollinators and the complex impact that this can have in pollinators' webs.

Behavioural processes undoubtedly act synergistically with other mechanisms to determine why some species are successful invaders while others are not. Lynn Martin and colleagues address how physiological and behavioural traits combine to mediate species' invasiveness. Specifically they examine how hormone regulation might influence invasions via plasticity, e.g. modifying a phenotype to match a new environment before evolution has time to act. Mark Albins examines the range of life history and behavioural traits that have facilitated the rapid invasion of introduced lionfish (*Pterois* spp.) in the Atlantic. In particular, the high population growth rates of lionfish are likely influenced by fast growth and early maturation, high fecundity, high fertilization success, high survival of eggs and larvae, and a high capacity for those propagules to travel long distances. These traits are all backed-up by an impressive tolerance to a variety of environmental conditions. Daniel Sol and Joan Maspons ask how examining behaviour together with life history traits can assist our understanding of how species persist in novel environments.

Finally, many authors use behavioural insights to provide key management recommendations for their specific systems and for directing future research on invasion biology more generally. The role of behaviour in facilitating invasions has been overlooked, often, in the study of global biological change. In many cases this has been due to our scientific obsession with quantifying and interpreting the risk-based measures of interspecific invasion success. The result has been that we know much less about the intraspecific traits that influence biological invasions throughout the transitions of the invasion pathway. The chapters in this book on 'Biological Invasions and Animal Behaviour' go a long way towards balancing this ledger.

References

Candolin, U. and Wong, B. (eds) (2011). *Behavioral Responses to a Changing World*. Oxford, UK: Oxford University Press.

Caro, T. (ed.) (1998). *Behavioral Ecology and Conservation Biology*. Oxford, UK: Oxford University Press.

Chapple, D.G., Simmonds, S.M. and Wong, B.M. (2012). Can behavioral and personality traits influence the success of unintentional species introductions? *Trends in Ecology and Evolution*, 27, 57–64.

Clemmons, J.R. and Buchholz, R. (eds) (1997). *Behavioral Approaches to Conservation in the Wild*. New York, NY: Cambridge University Press.

Elton, C. S. (1958). *The Ecology of Invasions by Animals and Plants*. London: Methuen.

Festa-Bianchet, M. and Apollonio, M. (eds) (2003). *Animal Behavior and Wildlife Conservation.*, Washington, DC: Island Press.

Gosling, L.M. and Sutherland, W.J. (eds) (2000). Behaviour and Conservation. New York, NY: Cambridge University Press.

Holway, D.A. and Suarez, A.V. (1999). Animal behavior: an essential component of invasion biology. *Trends in Ecology and Evolution*, 14, 328–330.

Sih, A., Bolnik, D.I., Luttbeg, B., *et al.* (2010). Predator-prey naiveté, antipredator behavior, and the ecology of predator invasions. *Oikos* 119:610–621.

Soulé, M.E. (1985). What is conservation biology? *BioScience* 35:727–734.

Soulé, M.E. and Wilcox, B. (eds) (1980). *Conservation Biology: An Evolutionary–Ecological Perspective.* Sunderland, MA: Sinauer Associates.

Sutherland, W.J. (1996). *From Individual Behaviour to Population Ecology. Oxford Series in Ecology and Evolution.* Oxford, UK: Oxford University Press.

Vitousek, P.M., D'Antonio, C.M., Loope, L.L. and Westbrooks, R. (1996). Biological invasions as global environmental change. *American Scientist*, 84, 469–478.

Part I

Behaviour and the Invasion Process

2 The Role of Behavioural Variation across Different Stages of the Introduction Process

David G. Chapple and Bob B.M. Wong

Human activities are responsible for the movement of individuals from thousands of different species to new regions each day, some of which are deliberate while others are not (Elton, 1958; Davis, 2009; Richardson, 2011; Lockwood *et al.*, 2013). For those that become invasive, establishment and spread are the final stages of what, in many cases, would have already been a long and treacherous journey (Blackburn *et al.*, 2011; Chapple *et al.*, 2012a). Indeed, to be successful, the invader would have had to overcome a series of successive hurdles and challenges to reach and establish in their new home in what is essentially the culmination of a multi-stage process (i.e. transport, introduction, establishment, spread) (Blackburn *et al.*, 2011; Chapple *et al.*, 2012a; Lockwood *et al.*, 2013). An inability to negotiate any one of these stages or barriers results in the ultimate failure of the invasion (Blackburn *et al.*, 2011; Chapple *et al.*, 2012a). This 'filtering' process is important because it is expected to reduce propagule size (i.e. propagule pressure), which in turn can affect both establishment and spread.

While a range of species-level (e.g. life history, habitat generalist, diet) or abiotic traits (climatic match been source and introduced regions) have been shown to be predictors of introduction success within particular taxonomic groups (Kolar and Lodge, 2001; Colautti *et al.*, 2006; Hayes and Barry, 2008), propagule pressure is often the primary determinant of establishment and invasion success in many animals (Lockwood *et al.*, 2005; Simberloff, 2009; Blackburn *et al.*, 2015). Behaviour may contribute to determining propagule pressure. This is because it affects the likelihood of extinction by demographic stochasticity and Allee effects (Taylor and Hastings, 2005; Drake and Lodge, 2006; Tobin *et al.*, 2011). Since behaviour mediates how animals interact with their environment, it should also influence the propensity for individuals to be transported (via either deliberate or unintentional means) and their ability to transition through each stage of the introduction process (Holway and Suarez, 1999; Chapple *et al.*, 2012a, b; Carrete *et al.*, 2012). Indeed, in several taxa, the inclusion of behavioural traits improves the predictions of establishment success (Sol *et al.*, 2002, 2008; Suarez *et al.*, 2005; Blackburn *et al.*, 2009). While researchers have long acknowledged the role of behaviour in the success of biological invasions (Holway and Suarez, 1999; Sol and Lefebvre, 2000), we currently know much less about how behavioural variation – both between and within species – might assist a species' progress through each stage

Biological Invasions and Animal Behaviour, eds J.S. Weis and D. Sol. Published by Cambridge University Press. © Cambridge University Press 2016.

of the introduction process (Chapple *et al.*, 2012a, b; Carrete *et al.*, 2012). Accordingly, the goal of our chapter is to highlight the role of behavioural variation throughout the invasion process and, in particular, to identify the knowledge gaps that could help further our understanding of how this variation might contribute to the success or failure of biological invasions.

Why Behavioural Variation Matters

Variation in animal behaviour has been a source of fascination and wonder since the time of Aristotle. Heritable behavioural differences are not only widespread throughout the animal kingdom, but maintained over time (Wolf and Weissing, 2012). Once considered a nuisance in behavioural studies, considerable research effort is now being devoted to understanding the ecological and evolutionary consequences of behavioural variation. In this respect, different behaviours are often correlated in different contexts (jointly called *behavioural syndromes*) and individuals of a species may also differ in their precise combinations of behavioural tendencies (the notion of a *behavioural type*; Sih *et al.*, 2004a, b; Bell, 2007). Aggressive individuals might, for example, also be more exploratory, and bolder in their anti-predator behaviours, compared to those that are less aggressive (Wolf and Weissing, 2012). An individual's behavioural type has a direct impact on its fitness, and the behavioural types present within a population or species (and the relative prevalence of each) may influence a range of ecological and evolutionary processes (e.g. population dynamics, life history, distribution and abundance; Dall *et al.*, 2012; Sih *et al.*, 2012; Wolf and Weissing, 2012).

The role of behavioural variation in the success or failure of species introductions has been largely overlooked. However, recent reviews (Chapple *et al.*, 2012a; Sih *et al.*, 2012) predict that greater interspecific behavioural variation may be important in the success or failure of potential invaders. To invade successfully, individuals of a species must navigate their way through all four stages of the introduction process (Figure 2.1). Each stage poses its own set of challenges, where behavioural variation can be critical (see Priority 3 below). Comparative studies have implied that the behaviour of successful invaders distinguishes them from that of unsuccessful invaders; but these have tended to use indirect measures of behaviour like brain size rather than directly considering behavioural syndromes (Sol *et al.*, 2002, 2005, 2008; Réale *et al.*, 2007; Amiel *et al.*, 2011). In this regard, the most proficient invader should be expected to exhibit a high dispersal tendency (through natural or human-mediated means), high foraging activity and boldness in novel environments, along with social tendencies appropriate for avoiding Allee effects (loss of fitness associated with diminished population density) (Tobin *et al.*, 2011; Chapple *et al.*, 2012a; Sih *et al.*, 2012; Table 2.1). Surprisingly, while previous studies have compared invasive and non-invasive species, and found differences in single behaviour categories (e.g. exploratory behaviour, Chapple *et al.*, 2011; dispersal tendency, Rehage and Sih, 2004; feeding rate, Rehage *et al.*, 2005a), there has been no investigation of how whole suites of behaviours associated with success across the entire introduction process might differ between invading and non-invading species.

Table 2.1 Behaviours that may influence success of species introductions at different stages of the introduction process (adapted from Chapple et al., 2012a)

| Behavioural trait[a] | Transport | | Introduction | Establishment | Spread | References |
	Uptake	Transit				
Actively hide/seek shelter	+[b]	+[b]	+/−[b]	−	+[b]	Aubry et al., 2006; Toy and Newfield, 2010; Chapple et al., 2011; Witmer et al., 2014
Activity[6]	+	−	+	+	+[b]	Cote et al., 2010; Cromie and Chapple, 2012; Brodin and Drotz, 2014; Monceau et al., 2014; Truhlar and Aldridge, 2015
Anti-predator behaviour[1]	?	?	?	+/−	+/−	Rehage et al., 2005b; Pintor et al., 2008; Polo-Cavia et al., 2008; Weis, 2010; Wright et al., 2010; Cisterne et al., 2014
Anti-parasite behaviour	?	?	?	+/−	+/−	Hughes and Cremer, 2007
Attraction to/tolerance of human-occupied environments	+	?	+	+[b]	+/−	Holway and Suarez, 1999; Sol et al., 2002; Suarez et al., 2008; Short and Petren, 2008
Boldness[1,4,5,6]	+	−	+	+	+[b]	Short and Petren, 2008; Pintor et al., 2008; Cote et al., 2010, 2011; Brodin and Drotz, 2014; Gonzalez-Bernal et al., 2014; Monceau et al., 2014
Dispersal tendency[3,4,5]	+	−	+	+/−[b]	+[b]	Rehage and Sih, 2004; Barbaresi et al., 2004; Bubb et al., 2006; Phillips et al., 2006; Duckworth and Kruuk, 2009; Blackburn et al., 2009; Weis, 2010; Cote et al., 2010, 2011; Llewelyn et al., 2010; Berthouly-Salazar et al., 2012; McGrammachan and Lester, 2012; Knop et al., 2013
Exploratory behaviour[4]	+[b]	−	+	+/−[b]	+[b]	Martin and Fitzgerald, 2005; Cote et al., 2010; Russell et al., 2010; Wright et al., 2010; Chapple et al., 2011; Henry et al., 2013; Carvahlo et al., 2013; Hui and Pinter-Wollman, 2014; Monceau et al., 2014; Truhlar and Aldridge, 2015
Foraging behaviour and flexibility[1,2]	+	?	+	+[b]	+	Rehage and Sih, 2004; Martin and Fitzgerald, 2005; Rehage et al., 2005a; Pintor et al., 2008, 2009; Blackburn et al., 2009; Weis, 2010; Wright et al., 2010; Alexander et al., 2014; Liebl and Martin, 2014; Mueller et al., 2014

(cont.)

Table 2.1 (*cont.*)

Behavioural trait[a]	Transport		Introduction	Establishment	Spread	References
	Uptake	Transit				
Habitat preferences and flexibility	+	+	+	+[b]	+[b]	Cure et al., 2014; Lee and Gelembiuk, 2008; Suarez et al., 2008; Blackburn et al., 2009; Weis, 2010; Wright et al., 2010; Cromie and Chapple, 2012
Intraspecific aggression[1,2,4]	−	−	−	−[b]	+/−[b]	Holway and Suarez, 1999; Tsutsui et al., 2000; Holway et al., 2002; Suarez et al., 2002; Tsutsui and Suarez, 2003; Abbott et al., 2007; Ugelvig et al., 2008; Pintor et al., 2008, 2009; Sagata and Lester, 2009; Weis, 2010; Wright et al., 2010
Interspecific aggression[3]	?	?	+	+[b]	+[b]	Holway and Suarez, 1999; Usio et al., 2001; Barbaresi et al., 2004; Heinze et al., 2006; Duckworth and Badyaev, 2007; Rowles and O'Dowd, 2007; Duckworth and Kruuk, 2009; Carpintero and Reyes-Lopez, 2008; Sagata and Lester, 2009; Weis, 2010; Grangier and Lester, 2012
Mate choice and mating behaviour	?	?	?	+/−[b]	+/−[b]	Liu et al., 2007; Blackburn et al., 2009; Luan et al., 2013
Nesting/oviposition behaviour	+[b]	?	?	+[b]	+[b]	Holway et al., 2002; Tsutsui and Suarez, 2003; Sol et al., 2002; Suarez et al., 2005; Heinze et al., 2006; Suarez et al., 2008
Parental care	+	+	?	+	+	
Social tendency[4,5]	+	+/−	+	+[b]	+/−[b]	Tsutsui et al., 2000; Cote et al., 2010, 2011; Carvalho et al., 2013; Truhlar and Aldridge, 2015
Species recognition	?	?	+	+	+	Heavener et al., 2014
Thermoregulatory behaviour and flexibility	+	+	+	+	+	Lee and Gelembiuk, 2008; Polo-Cavia et al., 2009, 2010; Cromie and Chapple, 2012

These behaviours may have either a positive or negative impact on success. The reference column indicates the studies that have examined the role of the behaviour during the introduction process. Please note that because Holway and Suarez (1999) have already provided an excellent summary at the time of their review, only studies since that review have been included here.

[a] The numbers indicate which behaviours have been linked in a behavioural syndrome in invasive species: [1] = Pintor et al., 2008, [2] = Pintor et al., 2009, [3] = Duckworth and Kruuk, 2009, [4] = Cote et al., 2010, [5] = Cote et al., 2011, [6] = Brodin and Drotz, 2014, [7] = Carvalho et al., 2013).

[b] At least one empirical study has indicated that this behaviour is associated with success at this stage in the introduction process.

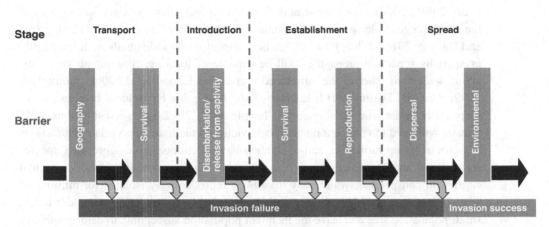

Figure 2.1 Outline of the invasion process for species introductions (adapted from Blackburn *et al.*, 2011, and Chapple *et al.*, 2012a) – a sequence of four stages (transport, introduction, establishment and spread) through which individuals must pass to become successful invaders. Each stage has one or more barriers to overcome before the next stage is accessible. Invasive species are those that successfully navigate through all such barriers, and spread into the recipient region.

In the next section of our chapter, we focus on outlining six key research priorities to improve our understanding of how behavioural variation might influence success across the introduction process. We build upon the framework presented in Chapple *et al.* (2012a) and Sih *et al.* (2012), and discuss the recent contributions to this field.

Research Priorities for Enhancing our Understanding of the Role of Behavioural Variation Across the Introduction Process

1.   A Greater Focus on the Earlier Stages of the Introduction Process

The initial stages of the introduction process still represent a major gap in our knowledge of how behaviour, and inter-individual behavioural variation, impacts the success of biological invasions. As far as we are aware, no empirical behavioural studies have examined the pre-establishment stages (transport, introduction) since the review of Chapple *et al.* (2012a) highlighted the dearth of knowledge in the early stages of the introduction process. Behavioural research has continued to focus on the post-establishment phase of the introduction process, and particularly the spread stage (Table 2.1; see Priority 4 below).

In addition, there remains limited information on how populations initially establish in new non-native environments. Behavioural syndromes can influence a wide range of ecological and evolutionary processes within populations (e.g. Sih *et al.*, 2012), and are likely to play an important role during the initial establishment of incipient populations in non-native regions. In animals, most introduction propagules comprise a single individual or small groups (Mack *et al.*, 2000; Gill *et al.*, 2001; Ward *et al.*, 2006; Chapple *et al.*, 2013b, 2016), which arrive at different times and often from multiple regions of the native range (Kolbe *et al.*, 2004; Simberloff, 2009; Chapple *et al.*, 2013b, 2016; Rius and Darling, 2014). While it has always been assumed that individuals from temporally or spatially separated propagules will be capable of locating, recognizing and interacting with each other in the introduced region (Lockwood *et al.*, 2005; Simberloff, 2009; Rius and Darling, 2014; Edelaar *et al.*, 2015), this has seldom been examined empirically in the wild or laboratory (Chapple *et al.*, 2012a). In such situations, individuals will need to employ a range of behaviours to facilitate appropriate interactions (e.g. social group formation, anti-predator behaviour, intraspecific aggression, species recognition) and reproductive activities (e.g. mate choice, nesting behaviour, parental care) with conspecifics, which may have been reproductively isolated for millions of years (Chapple *et al.*, 2013a, b; Rius and Darling, 2014). Failure to do so could lead to a small population size and leave the incipient population susceptible to demographic or environmental stochasticity and Allee effects (Taylor and Hastings, 2005; Tobin *et al.*, 2011).

2.   An Increased Focus on Unintentional Introductions

Chapple *et al.* (2012a) highlighted that most investigations of invasion success have focused on deliberate introductions (e.g. Kolar and Lodge, 2001; Hayes and Barry, 2008). This bias has most likely been due to the greater ease in studying deliberate introductions, as researchers generally have access to documentation on the source region/population, transport pathway and method, number of individuals released and the timing of introduction (e.g. Blackburn *et al.*, 2011, 2015; Cassey *et al.*, 2004; Kraus, 2009). As a consequence, we have a limited understanding of the behaviours that assist invaders to transition through the transport and introduction phase of the introduction process (Chapple *et al.*, 2012a), which is where most unintentional introductions fail (Kolar and Lodge, 2001; Puth and Post, 2005; Ward *et al.*, 2006). Carrete *et al.* (2012) point out that the pre-establishment phase can also be important in the case of deliberate species introductions, the key difference being that the selection on individuals in the initial transport phase is the result of direct human action in deliberate introductions, and indirectly from human action in unintentional introductions. While the distinction between deliberate and unintentional introductions is evident in the initial stages of the introduction process (transport, introduction), depending on the filtering process prior to establishment, both introduction types could be similar during the post-introduction phases (establishment, and spread). Indeed, unintentional events (e.g. escape of individuals from captivity or confinement, or human-assisted spread across the invaded range) often occur during deliberate species introductions (White and Shine, 2009; Blackburn *et al.*, 2011; Lockwood *et al.*, 2013) and, similarly, there may be the intentional spread of unintentionally introduced species (e.g. pet trade) in the invaded range (Meenken, 2012). Thus, the delineation between deliberate and unintentional introductions becomes blurred as the introduction progresses.

We are not aware of any study since the review of Chapple *et al.* (2012a) that has directly examined the role of behaviour in the transport or introduction stage in either deliberately or unintentionally introduced species. Nevertheless, it is worthwhile considering the role behaviour (and inter-individual behavioural variation) might play in early stages of both deliberate and unintentional species introductions.

Unintentional Introductions

The species that will have the greatest 'opportunity' for human-assisted dispersal are those that occur in high densities in human-occupied environments and have widespread distributions that overlap with multiple transport hubs (Floerl and Inglis, 2005; Hulme, 2009; Tingley *et al.*, 2010; Toy and Newfield, 2010). The 'uptake' of individuals into transport vectors may occur via either passive (e.g. individuals in ballast water, or individuals residing in or sheltering in valuable commodities (e.g. fresh produce, timber, soil)) or active means (e.g. individuals with high activity or exploratory behaviour entering freight and cargo in storage warehouses or wharves) (Chapple *et al.*, 2012a). However, once ensnared within freight or cargo, the stowaways will need to avoid detection, survive transit and arrive at the destination in good health if they are to disembark and begin establishing (Chapple *et al.*, 2012a).

The invasive delicate skink (*Lampropholis delicata*), the only Australian lizard species that has successfully invaded outside of Australia (Kraus, 2009; Chapple *et al.*, 2013a), provides some insight into how behavioural differences among species may influence their propensity for human-assisted dispersal and unintentional species introductions. Compared to the non-invasive garden skink (*L. guichenoti*) – which is a largely sympatric species in its native range and, hence, has a similar 'opportunity' for human-assisted transportation – the delicate skink exhibits higher activity and exploratory behaviour and a greater tendency to actively hide in sheltered areas (Chapple *et al.*, 2011; Cromie and Chapple, 2012; Bezzina *et al.*, 2014). This may contribute to the delicate skink's greater propensity for getting into freight and cargo (~6 times greater), while also allowing it to avoid biosecurity checks at the border (Chapple *et al.*, 2011, 2013b). As a result, the delicate skink has been accidentally introduced (with subsequent, successful establishment) to Lord Howe Island on five separate occasions, once to the Hawaiian Islands (with six subsequent human-assisted colonizations among islands within the archipelago) and once to New Zealand (with frequent long-distance jump dispersal within the country) (Chapple *et al.*, 2013a, b, 2014; Moule *et al.*, 2015; Tingley *et al.*, 2016).

Deliberate Introductions

As we alluded to earlier, pre-establishment selective filters are also an inherent component of intentional species introductions (reviewed in Carrete *et al.*, 2012). Animals are generally collected from the wild (or bred in captivity) before being deliberately transported to a new location, yet particular collection methods may be biased with regard to the behavioural types that they sample from the population (i.e. trappability may differ between active, bold individuals and inactive, shy individuals; Biro and Dingemanse, 2009; Garamszegi *et al.*, 2009; Carter *et al.*, 2012; Biro, 2013). An individual's behaviour may influence its capacity to adapt to captivity, and survive subsequent transportation and the quarantine period in the new location (Teixeira *et al.*, 2007; Carrete *et al.*, 2012). Importantly, behaviour may play a role in which individuals escape (e.g. active or bold individuals; Minderman *et al.*, 2009), or are released from captivity (e.g. aggressive or shy individuals) into the wild in the introduced range (e.g. Carrete *et al.*, 2012). Thus, there are sequential selective processes in play during transport (collection, captivity, transit) and introduction (quarantine, escape/release from captivity) that filter out which individuals are likely to reach and establish in a non-native location. Here, researchers may wish to focus particular attention on behaviours that are likely to be important but have generally been overlooked or understudied, such as aggression, or the avoidance of predators and parasites (see also Table 2.1).

3.   Behavioural Syndromes Versus Behavioural Flexibility

When considering the role of behavioural variation in invasion success, it is crucial to understand not only the role of behavioural syndromes, but also the importance of behavioural flexibility. The existence of a behavioural syndrome, by its very nature, implies that an individual should have limited behavioural flexibility once its behavioural type is set, and that any shift in behaviour should take time to occur (Sih *et al.*, 2004a, b). However, a substantial body of literature suggest that behavioural plasticity is often critical in allowing individuals to respond quickly to altered environmental conditions (Van Buskirk, 2012; Wong and Candolin, 2015), including those that might be encountered when animals are introduced into new areas (Mayr, 1965; Sol, 2003; Phillips and Suarez, 2012).

Phenotypic plasticity describes the ability of a particular genotype to produce different phenotypes as a function of the environment (Thibert-Plante and Hendry, 2011). In the context of behaviour, plasticity can allow animals to adjust rapidly to the conditions of its immediate (new) environment, and to allow it to thrive (Van Buskirk, 2012). A lack of plasticity might therefore be expected to contribute to the failure of species in new areas. Here, the plasticity of traits that have evolved under past environmental conditions will be important in determining a species fate outside its natural range, and whether it will become a successful invader or not.

In a well-known study investigating the invasion success of introduced birds, Sol *et al.* (2002) showed that species with relatively larger brains – a proxy for behavioural flexibility – had a higher probability of establishing themselves when introduced outside their native range. Similar patterns have since been reported in invasive amphibians (Amiel *et al.*, 2011) and mammals (Sol *et al.*, 2008). More recently, plasticity in habitat use – a trait that is often associated with flexibility in foraging behaviour – is believed to have facilitated the successful invasion of Pacific red lionfish (*Pterois volitans*) into the Atlantic (Cure *et al.*, 2014). Behavioural plasticity can also be important in competitive interactions, with a study of invasive wasps (*Vespula vulgaris*) showing adjustment of interference behaviour when wasps are competing with native ants over food (Grangier and Leister, 2014). Such examples underscore the importance of adaptive plastic responses in mediating the success of species invasions.

It is important to realize, however, that not all plastic behavioural responses are adaptive (Wong and Candolin, 2015) and that this can have important repercussions for the fate of newly introduced species. One way that plastic behavioural responses can be detrimental is if novel environmental conditions cause animals to make maladaptive behavioural decisions (so-called evolutionary traps) (Robertson *et al.*, 2013). Hence, mechanisms that act as a limit on behavioural flexibility – such as the existence of a behavioural syndrome or behavioural type – could conceivably prevent species from engaging in behaviours that may otherwise be detrimental to their chances of succeeding in new areas. On the other hand, from an evolutionary perspective, genetic correlations that link different behaviours together into behavioural types, could potentially constrain adaptation to new environments (Walsh and Blows, 2009). Thus, although researchers acknowledge that both behavioural syndromes and behavioural flexibility can be important for understanding invasion success, more work is clearly needed to disentangle their relative contributions to biological invasions.

4. Examining a Greater Diversity of Behavioural Traits Across Multiple Stages of the Introduction Process

Despite the important role that behaviour may play in biological invasions, much of the research attention continues to focus on only a handful of behavioural traits. For instance, there have been numerous studies documenting the importance of dispersal behaviour, foraging and aggression (both intra- and interspecific) in invasion success within species, but comparatively less attention has been given to other ecologically relevant behaviours, such as mate choice or parental care (Table 2.1). Redressing this imbalance will be important in advancing our understanding of the role of behaviour in invasion ecology. For instance, interspecific differences in reproductive behaviours, such as male courtship and mate guarding, is believed to have been key to the widespread invasion of the whitefly (*Bemisia tabaica*) and its subsequent displacement of native competitors in its introduced range (Liu *et al.*, 2007; Luan *et al.*, 2013).

Most studies relating behaviour to invasion success also have a tendency to compare differences in behaviour at the distributional extremes of the invasive range – either by looking at variation in behaviour between the invasion front and more established populations, or between the native and introduced range (reviewed in Hudina *et al.*, 2014). By contrast, as we have already pointed out (Priority 2, above), fewer studies have focused on specific stages of the introduction process – and even fewer have considered behaviours across multiple stages (Chapple *et al.*, 2012a). Since factors that contribute to an animal's ability to successfully traverse through each stage of the introduction process can differ from stage to stage, it is important, therefore, to consider the role of behaviours across multiple stages of the introduction process. Indeed, while certain behaviours may operate in a complementary fashion across the different stages, others could have counteractive effects. As an example, being exploratory may be beneficial in helping a species to establish and spread once it reaches its new destination, but it can also increase the chances of detection during transit (Chapple *et al.*, 2011). Similarly, due to the existence of behavioural types, individuals that exhibit the strongest dispersal tendencies could also be less socially inclined (e.g. Cote *et al.*, 2010), thereby increasing the risk of Allee effects during establishment and spread. Moreover, it is important to realize that the value of particular behaviours could also differ depending on the environmental context (Sol *et al.*, 2011).

5.   Evidence for Selective Filters During the Introduction Process

Behavioural variation can have important consequences for species introductions (both intentional and unintentional). The introduction process acts as a sequential selective filter (Chapple *et al.*, 2012a), so cohorts of invaders that undergo repeated founder effects may exhibit less behavioural variation than their source population. This prediction remains to be tested empirically, although, in some invaders, behaviour has been compared between native and introduced populations or long-established and range-edge invasive populations (as outlined previously in Priority 4). These studies indicate that individual variation may become increasingly eroded, leading to populations becoming increasingly more dispersive (e.g. cane toads, *Rhinella marina*, Lindstrom *et al.*, 2013; bluebirds, *Sialia*, Duckworth and Badyaev, 2007), active (e.g. rusty crayfish, *Orconectes rusticus*, Pintor and Sih, 2009), bold (e.g. house sparrows, *Passer domesticus*, Liebl and Martin, 2014), or aggressive (e.g. signal crayfish, *Pacifastacus leniusculus*, Pintor *et al.*, 2008; bluebirds, *Sialia*, Duckworth and Badyaev, 2007) as the introduction progresses (but see Lopez *et al.*, 2012, African jewelfish, *Hemichromis letourneuxi*; Truhlar and Aldridge, 2015, amphipod, *Dikerogammarus villosus*). Although the concept of selective filters during the introduction process requires behaviours related to invasion success to have a genetic basis, this has rarely been examined. In one of the few studies, Mueller *et al.* (2014) provided evidence for this by showing that variation in the *DRD4* gene is related to the response to novelty in the yellow-crowned bishop (*Euplectes afer*) in its introduced range in Europe.

The progressive erosion of behavioural variation throughout the introduction process (e.g. the invasive wasp, *Vespa velutina* in Europe; Monceau *et al.*, 2014) may act to slow or even stall the invasion. The impact of reduced behavioural variation may be particularly pronounced in situations where different behavioural types (or traits in general) are favoured in different stages of the invasion process. Indeed, empirical studies (mosquito fish: positive correlation between boldness, activity level, exploratory behaviour and sociability, Cote *et al.*, 2011; Argentine ants: exploratory behaviour and nest site selection, Hui *et al.*, 2014) and theoretical models (Fogarty *et al.*, 2011) suggest that the spread of invasive species is enhanced when there is a mix of behavioural types in a population. In one possible course of events, asocial individuals may have greater propensity for dispersal and colonize new habitats, but are unlikely to have sufficient densities to establish self-sustaining populations (Fogarty *et al.*, 2011). Social individuals that are also present in the population will eventually move into patches previously colonized by the asocial individuals (Fogarty *et al.*, 2011). The resulting increase in population density drives the asocial individuals to disperse further (Fogarty *et al.*, 2011). In such a scenario, range expansion involving repeated cycles of dispersal, colonization and population growth demands a mix of behavioural types in the population (Sih *et al.*, 2012; Hui *et al.*, 2014).

This creates an intriguing conundrum. Invasion success might be enhanced through a mix of behavioural types within a population (Cote *et al.*, 2011; Fogarty *et al.*, 2011), but at the same time, the introduction process is predicted to decrease behavioural variation (Chapple *et al.*, 2012a). This represents a key puzzle in the field: how one species can maintain the mix of divergent traits required for negotiating the many barriers encountered en route to a successful invasion. Empirical research to resolve this puzzle should be a priority.

6.   The Importance of Behavioural Traits in an Invasion Syndrome

As we have already discussed, the idea that suites of behaviours may be correlated can have important implications in explaining whether individuals are able to successfully transition through each and every stage of the invasion process. In this regard, correlated behaviours that may facilitate success across multiple stages of the invasion process could represent an 'invasive syndrome' (Sih *et al.*, 2004b), with growing empirical support for the existence of such syndromes in taxa as diverse as crustaceans (e.g. Pintor *et al.*, 2008; Pintor and Sih, 2009; Brodin and Drotz, 2014), fish (Cote *et al.*, 2010, 2011) and birds (Duckworth and Kruuk, 2009; Carvahlo *et al.*, 2013). The existence of invasive syndromes could thus help to explain why some species are repeatedly successful invaders, whereas others are not (Mack *et al.*, 2000; Gill *et al.*, 2001; Sih *et al.*, 2004; Kraus, 2009). However, it is important to appreciate that invasion syndromes may include more than simply behavioural traits. As Sol and Maspons (2015) recently highlighted, behaviour is part of a broader complex of adaptive traits (e.g. life history, physiology, morphology) that can mediate the response of individuals to new environments. For instance, an animal's life history – i.e. the way that individuals allocate time and resources to reproduction, growth and survival – is closely tied to behaviour, and can also influence population dynamics (e.g. invasive birds, Sol *et al.*, 2012; grey squirrels, *Sciurus carolinensis*, Goldstein *et al.*, 2015). Thus, greater research effort should be invested in exploring the interplay between behaviour and other traits, and their relative importance, in contributing towards an invasion syndrome (Hudina *et al.*, 2014; Sol and Maspons, 2015).

Summary and Conclusions

It is clear that biological invasions, whether the result of deliberate or accidental introductions, are the product of a multi-stage process, which require the invader to successfully negotiate a suite of sequential, selective filters. Studies have shown that behaviour, and more specifically behavioural variation, can play a leading role in facilitating success during establishment and spread. However, this chapter highlights that numerous gaps in our understanding remain. In particular, researchers need to place a greater emphasis on unintentional species introductions, as the initial selective filters differ from those associated with deliberate introductions. More generally, irrespective of how animals reach non-native areas, there needs to be a greater focus not only on the initial stages of the introduction, but also how behaviours may facilitate or hamper success from stage to stage – topics that have largely been neglected so far. Here, it would be important to consider a broader range of behaviours and, in doing so, we need to understand not only the role of behavioural syndromes, but also behavioural flexibility. Evidence suggests that both are likely to be important, but more work is needed to disentangle their relative contributions, especially given the prediction that the introduction process, by acting as a selective filter, is expected to erode variation as the invasion progresses. Lastly, a consideration of a so-called 'invasion syndrome' needs to broaden the focus to examine how behaviour interacts with other traits (e.g. life history, physiology, morphology). Addressing these knowledge gaps will provide us with a fuller understanding of the role of behaviour, and behavioural variation, in invasion success throughout the introduction process.

Acknowledgements

DGC was supported by funding from the Australian Research Council (DP0771913), Hermon Slade Foundation (HSF09/02) and the National Geographic Society (8085–06 and 8952–11).

References

Abbott, K.L., Greaves, S.N.J., Ritchie, P.A., *et al.* (2007). Behaviourally and genetically distinct populations of an invasive ant provides insight into invasion history and impacts on a tropical ant community. *Biological Invasions*, 9, 453–463.

Alexander, M.E., Dick, J.T.A., Weyl, O.L.F., *et al.* (2014). Existing and emerging high impact invasive species are characterized by higher functional responses than native. *Biology Letters*, 10, 20130946.

Amiel, J.J., Tingley, R. and Shine, R. (2011). Smart moves: effects of relative brain size on establishment success of invasive amphibians and reptiles. *PLoS ONE*, 6, e18277.

Aubry, S., Labaune, C., Magnin, F., *et al.* (2006). Active and passive dispersal of an invading land snail in Mediterranean France. *Journal of Animal Ecology*, 75, 802–813.

Barbaresi, S., Santini, G., Tricarico, E., *et al.* (2004). Ranging behaviour of the invasive crayfish, *Procambarus clarkii* (Girard). *Journal of Natural History*, 38, 2821–2832.

Bell, A.M. (2007). Future directions in behavioural syndromes research. *Proceedings of the Royal Society B*, 274, 755–761.

Berthouly-Salazar, C., van Rensburg, B.J., Le Roux, J.J., *et al.* (2012). Spatial sorting drives morphological variation in the invasive bird, *Acridotheres tristis*. *PLoS ONE*, 7, e38145.

Bezzina, C.N., Amiel, J.J. and Shine, R. (2014). Does invasion success reflect superior cognitive ability? A case study of two congeneric lizard species (*Lampropholis*, Scincidae). *PLoS ONE*, 9, e86271.

Biro, P.A. (2013). Are most samples of animals systematically biased? Consistent individual trait differences bias samples despite random sampling. *Oecologia*, 171, 339–345.

Biro, P.A. and Dingemanse, N.J. (2009). Sampling bias resulting from animal personality. *Trends in Ecology and Evolution*, 24, 66–67.

Blackburn, T.M., Cassey, P. and Lockwood, J.L. (2009). The role of species traits in the establishment success of exotic birds. *Global Change Biology*, 15, 2852–2860.

Blackburn, T.M., Pyšek, P., Bacher, S., *et al.* (2011). A proposed unified framework for biological invasions. *Trends in Ecology and Evolution*, 26, 333–339.

Blackburn, T.M., Lockwood, J.L. and Cassey, P. (2015). The influence of numbers on invasion success. *Molecular Ecology*, 24, 1942–1953.

Brodin, T. and Drotz, M.K. (2014). Individual variation in dispersal associated behavioral traits of the invasive Chinese mitten crab (*Eriocheir sinensis*, H. Milne Edwards, 1854) during initial invasion of Lake Vanern, Sweden. *Current Zoology*, 60, 410–416.

Bubb, D.H., Thom, T.J. and Lucas, M.C. (2006). Movement, dispersal and refuge use of co-occurring introduced and native crayfish. *Freshwater Biology*, 51, 1359–1368.

Carpintero, S. and Reyes-Lopez, J. (2008). The role of competitive dominance in the invasive ability of the Argentine ant (*Linepithema humile*). *Biological Invasions*, 10, 25–35.

Carrete, M., Edelaar, P., Blas, J., *et al.* (2012). Don't neglect pre-establishment individual selection in deliberate introductions. *Trends in Ecology and Evolution*, 27, 67–68.

Carter, A.J., Heinsohn, R., Goldizen, A.W., *et al.* (2012). Boldness, trappability and sampling bias in wild lizards. *Animal Behaviour*, 83, 1051–1058.

Carvalho, C.F., Leitao, A.V., Funghi, C., *et al.* (2013). Personality traits are related to ecology across a biological invasion. *Behavioral Ecology*, 24, 1081–1091.

Cassey, P., Blackburn, T.M., Sol, D., *et al.* (2004). Global patterns of introduction effort and establishment success in birds. *Proceedings of the Royal Society B*, 271, S405–S408.

Chapple, D.G., Simmonds, S.M. and Wong, B.B.M. (2011). Know when to run, know when to hide: can behavioral differences explain the divergent invasion success of two sympatric lizards? *Ecology and Evolution*, 1, 278–289.

Chapple, D.G., Simmonds, S.M. and Wong, B.B.M. (2012a). Can behavioral and personality traits influence the success of unintentional species introductions? *Trends in Ecology and Evolution*, 27, 57–64.

Chapple, D.G., Simmonds, S.M. and Wong, B.B.M. (2012b). Intraspecific behavioral variation is important in both deliberate and unintentional species introductions: response to Carrete *et al.* *Trends in Ecology and Evolution*, 27, 68–69.

Chapple, D.G., Miller, K.A., Kraus, F., *et al.* (2013a). Divergent introduction histories among invasive populations of the delicate skink (*Lampropholis delicata*): has the importance of genetic admixture in the success of biological invasions been overemphasized? *Diversity and Distributions*, 19, 134–146.

Chapple, D.G., Whitaker, A.H., Chapple, S.N.J., *et al.* (2013b). Biosecurity interceptions of an invasive lizard: origin of stowaways and human-assisted spread within New Zealand. *Evolutionary Applications*, 6, 324–339.

Chapple, D.G., Miller, K.A., Chaplin, K., *et al.* (2014). Biology of the invasive delicate skink (*Lampropholis delicata*) on Lord Howe Island. *Australian Journal of Zoology*, 62, 498–506.

Chapple, D.G., Knegtmans, K., Kikillus H., van Winkel, D. (2016). Biosecurity of exotic reptiles and amphibians in New Zealand: building upon Tony Whitaker's legacy. *Journal of the Royal Society of New Zealand*, 46, 66–84.

Cisterne, A., Vanderduys, E.P., Pike, D.A., *et al.* (2014). Wary invaders and clever natives: sympatric house geckos show disparate responses to predator scent. *Behavioral Ecology*, 25, 604–611.

Colautti, R.I., Girgorovich, I.A. and MacIsaac, H.J. (2006). Propagule pressure: a null model for biological invasions. *Biological Invasions*, 8, 1023–1037.

Cote, J., Fogarty, S., Weinersmith, K., *et al.* (2010). Personality traits and dispersal tendency in the invasive mosquitofish (*Gambusia affinis*). *Proceedings of the Royal Society B*, 277, 1571–1579.

Cote, J., Fogarty, S., Brodin, T., *et al.* (2011). Personality-dependent dispersal in the invasive mosquitofish: group composition matters. *Proceedings of the Royal Society B*, 278, 1670–1678.

Cromie, G.L. and Chapple, D.G. (2012). Impact of tail loss on the behaviour and locomotor performance of two sympatric *Lampropholis* skink species. *PLoS One*, 7, e34732.

Cure, K., McIlwain, J.L. and Hixon, M.A. (2014). Habitat plasticity in native Pacific red lionfish *Pterois volitans* facilitates successful invasion of the Atlantic. *Marine Ecology Progress Series*, 506, 243–253.

Dall, S.R.X., Bell, A.M., Bolnick, D.I., *et al.* (2012). An evolutionary ecology of individual differences. *Ecology Letters*, 15, 1189–1198.

Davis, M.A. (2009). *Invasion Biology*. Oxford: Oxford University Press.

Drake, J.M. and Lodge, D.M. (2006). Allee effects, propagule pressure and the probability of establishment: risk analysis for biological invasions. *Biological Invasions*, 8, 365–375.

Duckworth, R.A. and Badyaev, A.V. (2007). Coupling of dispersal and aggression facilitates the rapid range expansion of a passerine bird. *Proceedings of the National Academy of Sciences, USA*, 104, 15017–15022.

Duckworth, R.A. and Kruuk, L.E.B. (2009). Evolution of genetic integration between dispersal and colonization ability in a bird. *Evolution*, 63, 968–977.

Edelaar, P., Roques, S., Hobson, E.A., Gonvalves da Silva, A., *et al.* (2015). Shared genetic diversity across the global invasive range of the monk parakeet suggests a common restricted geographic origin and the possibility of convergent selection. *Molecular Ecology*, 24, 2164–2176.

Elton, C.S. (1958). *The Ecology of Invasions by Animals and Plants*. Chicago, IL: University of Chicago Press.

Floerl, O. and Inglis, G.J. (2005). Starting the invasion pathway: the interaction between source populations and human transport vectors. *Biological Invasions*, 7, 589–606.

Fogarty, S., Cote, J. and Sih, A. (2011). Social personality polymorphism and the spread of invasive species: a model. *American Naturalist*, 177, 273–287.

Garamszegi, L.Z., Eens, M. and Török, J. (2009). Behavioural syndromes and trappability in free-living collared flycatchers, *Ficedula albicollis*. *Animal Behaviour*, 77, 803–812.

Gill, B.J., Bejakovich, D. and Whitaker, A.H. (2001). Records of foreign reptiles and amphibians accidentally imported to New Zealand. *New Zealand Journal of Zoology*, 28, 351–359.

Goldstein, E.A., Butler, F. and Lawton, C. (2015). Frontier population dynamics of an invasive squirrel species: do introduced populations function differently than those in the native range? *Biological Invasions*, 17, 1181–1197.

González-Bernal, E., Brown, G.P. and Shine, R. (2014). Invasive cane toads: social facilitation depends upon an individual's personality. *PLoS One*, 9(7), e102880.

Grangier, J. and Lester, P.J. (2012). Behavioral plasticity mediates asymmetric competition between invasive wasps and native ants. *Communicative and Integrative Biology*, 5, 127–129.

Hayes, K.R. and Barry, S.C. (2008). Are there any consistent predictors of invasion success? *Biological Invasions*, 10, 483–506.

Heavener, S.J., Carthey, A.J.R. and Banks, P.B. (2014). Competitive naiveté between a highly successful invader and a functionally similar native species. *Oecologia*, 175, 73–84.

Heinze, J., Cremer, S., Eckl, N., *et al.* (2006). Stealthy invaders: the biology of *Cardiocondyla* tramp ants. *Insectes Sociaux*, 53, 1–7.

Henry, P.Y., Salgado, C.L., Muñoz, F.P., *et al.* (2013). Birds introduced to new areas show rest disorders. *Biology Letters*, 9, 20130463.

Holway, D.A. and Suarez, A.V. (1999). Animal behavior: an essential component of invasion biology. *Trends in Ecology and Evolution*, 14, 328–330.

Holway, D.A., Lach, L., Suarez, A.V., *et al.* (2002). The causes and consequences of ant invasions. *Annual Review of Ecology and Systematics*, 33, 181–233.

Hudina, S., Hock, K. and Zganec, K. (2014). The role of aggression in range expansion and biological invasions. *Current Zoology*, 60, 401–409.

Hughes, D.P. and Cremer, S. (2007). Plasticity in antiparasite behaviours and its suggested role in invasion biology. *Animal Behaviour*, 74, 1593–1599.

Hui, A. and Pinter-Wollman, N. (2014). Individual variation in exploratory behaviour improves speed and accuracy of collective nest selection by Argentine ants. *Animal Behaviour*, 93, 261–266.

Hulme, P.E. (2009). Trade, transport and trouble: managing invasive species pathways in an era of globalization. *Journal of Applied Ecology*, 46, 10–18.

Knop, E., Rindlisbacher, N., Ryser, S., *et al.* (2013). Locomotor activity of two sympatric slugs: implications for the invasion success of terrestrial invertebrates. *Ecosphere*, 4, 92.

Kolar, C.S. and Lodge, D.M. (2001). Progress in invasion biology: predicting invaders. *Trends in Ecology and Evolution*, 16, 199–204.

Kolbe, J.J., Glor, R.E., Schettino, L.R., *et al.* (2004). Genetic variation increases during biological invasion by a Cuban lizard. *Nature*, 431, 177–181.

Kraus, F. (2009). *Alien Reptiles and Amphibians*. Berlin: Springer.

Lee, C.E. and Gelembiuk, G.E. (2008). Evolutionary origins of invasive populations. *Evolutionary Applications*, 1, 427–448.

Liebl, A.L. and Martin, L.B. (2014). Living on the edge: range edge birds consume novel foods sooner than established ones. *Behavioral Ecology*, 25, 1089–1096.

Lindstrom, T., Brown, G.P., Sisson, S.A., *et al.* (2013). Rapid shifts in dispersal behavior on an expanding range edge. *Proceedings of the National Academy of Sciences, USA*, 110, 13452–13456.

Liu, S.S., De Barro, P.J., Xu, J., *et al.* (2007). Asymmetric mating interactions drive widespread invasion and displacement in a whitefly. *Science*, 318, 1769–1772.

Llewelyn, J., Phillips, B.L., Alford, R.A., *et al.* (2010). Locomotor performance in an invasive species: cane toads from the invasion front have greater endurance, but not speed, compared to conspecifics from a long-colonised area. *Oecologia*, 162, 343–348.

Lockwood, J.L., Cassey, P. and Blackburn, T. (2005). The role of propagule pressure in explaining species invasions. *Trends in Ecology and Evolution*, 20, 223–228.

Lockwood, J.L., Hoopes, M.F. and Marchetti, M.P. (2013). *Invasion Ecology*, 2nd edn. Oxford, UK: Wiley-Blackwell.

Lopez, D.P., Jungman, A.A. and Rehage, J.S. (2012). Nonnative African jewelfish are more fit but not bolder at the invasion front: a trait comparison across an Everglades range expansion. *Biological Invasions*, 14, 2159–2174.

Luan, J.B., De Barro, P.J., Ruan, Y.M. and Liu, S.S. (2013). Distinct behavioural strategies underlying asymmetric mating interactions between invasive and indigenous whiteflies. *Entomologia Experimentalis et Applicata*, 146, 186–194.

Mack, R.N., Simberloff, D., Lonsdale, W.M., *et al.* (2000). Biotic invasions: causes, epidemiology, global consequences and control. *Ecological Applications*, 10, 689–710.

Martin, L.B. and Fitzgerald, L. (2005). A taste for novelty in invading house sparrows, *Passer domesticus*. *Behavioral Ecology*, 16, 702–707.

Mayr, E. (1965). The nature of colonising birds. In *The Genetics of Colonizing Species*, ed. Baker, H.G. and Stebbins, G.L. New York: Academic Press, pp. 29–43.

McGrannachan, C.M. and Lester, P.J. (2012). Temperature and starvation effects on food exploitation by Argentine ants and native ants in New Zealand. *Journal of Applied Entomology*, 137, 550–559.

Meenken, D. (2012). *Pet Biosecurity in New Zealand: Current State of the Domestic Pet Trade System and Options Going Forward*. Wellington, New Zealand: Ministry for Primary Industries.

Minderman, J., Reid, J.M., Evans, P.G.H., *et al.* (2009). Personality traits in wild starlings: exploration behaviour and environmental sensitivity. *Behavioral Ecology*, 20, 830–837.

Monceau, K., Moreau, J., Poidatz, J., *et al.* (2014). Behavioural syndrome in a native and invasive hymenoptera species. *Insect Science*, 10.1111/1744–7917.12140.

Moule, H., Chaplin, K., Bray, R.D., *et al.* (2015). A matter of time: temporal variation in the introduction history and population genetic structuring of an invasive lizard. *Current Zoology*, 61, 456–464.

Mueller, J.C., Edelaar, P., Carrete, M., *et al.* (2014). Behaviour-related DRD4 polymorphism in invasive bird populations. *Molecular Ecology*, 23, 2876–2885.

Phillips, B.L. and Suarez, A.V. (2012). The role of behavioural variation in the invasion of new areas. In *Behavioural Responses to a Changing World: Mechanisms and Consequences*, ed. Candolin, U. and Wong, B.B.M. Oxford: Oxford University Press, pp. 190–200.

Phillips, B.L., Brown, G.P., Webb, J.K., *et al.* (2006). Invasion and the evolution of speed in toads. *Nature*, 439, 803.

Pintor, L.M. and Sih, A. (2009). Differences in growth and foraging behavior of native and introduced populations in an invasive crayfish. *Biological Invasions*, 11, 1895–1902.

Pintor, L.M., Sih, A. and Bauer, M.L. (2008). Differences in aggression, activity and boldness between native and introduced populations of an invasive crayfish. *Oikos*, 117, 1629–1636.

Pintor, L.M., Sih, A. and Kerby, J.L. (2009). Behavioral correlations provide a mechanism for explaining high invader densities and increased impacts on native prey. *Ecology*, 90, 581–587.

Polo-Cavia, N., Lopez, P. and Martin, J. (2008). Interspecific differences in responses to predation risk may confer competitive advantages to invasive freshwater turtle species. *Biological Invasions*, 11, 1755–1765.

Polo-Cavia, N., Lopez, P. and Martin, J. (2009). Interspecific differences in heat exchange rates may affect competition between introduced and native freshwater turtles. *Ethology*, 114, 115–123.

Polo-Cavia, N., Lopez, P. and Martin, J. (2010). Competitive interactions during basking between native and invasive freshwater turtle species. *Biological Invasions*, 12, 2141–2152.

Puth, L.M. and Post, D.M. (2005). Studying invasion: have we missed the boat? *Ecology Letters*, 8, 715–721.

Réale, D., Reader, S.M., Sol, D., *et al.* (2007). Integrating animal temperament within ecology and evolution. *Biological Reviews*, 82, 291–318.

Rehage, J.S. and Sih, A. (2004). Dispersal behavior, boldness, and the link to invasiveness: a comparison of four *Gambusia* species. *Biological Invasions*, 6, 379–391.

Rehage, J.S., Barnett, B.K. and Sih, A. (2005a). Foraging behaviour and invasiveness: do invasive *Gambusia* exhibit higher feeding rates and broader diets than their noninvasive relatives? *Ecology of Freshwater Fish*, 14, 352–360.

Rehage, J.S., Barnett, B.K. and Sih, A. (2005b). Behavioral responses to a novel predator and competitor of invasive mosquitofish and their non-invasive relatives (*Gambusia* sp.). *Behavioral Ecology and Sociobiology*, 57, 256–266.

Richardson, D.M. (ed.) (2011). *Fifty Years of Invasion Ecology: The Legacy of Charles Elton*. Oxford, UK: Wiley-Blackwell.

Rius, M. and Darling, J.A. (2014). How important is intraspecific genetic admixture to the success of colonising populations? *Trends in Ecology and Evolution*, 29, 233–242.

Robertson, B.A., Rehage, J.S. and Sih, A. (2013). Ecological novelty and the emergence of evolutionary traps. *Trends in Ecology and Evolution*, 28, 552–560.

Rowles, A.D. and O'Dowd, D.J. (2007). Interference competition by Argentine ants displaces native ants: implications for biotic resistance to invasion. *Biological Invasions*, 9, 73–85.

Russell, J.C., McMorland, A.J.C. and MacKay, J.W.B. (2010). Exploratory behaviour of colonizing rats in novel environments. *Animal Behaviour*, 79, 159–164.

Sagata, K. and Lester, P.J. (2009). Behavioural plasticity associated with propagule size, sources and the invasion success of the Argentine ant, *Linepithema humile. Journal of Applied Ecology*, 46, 19–27.

Short, K.H. and Petren, K. (2008). Boldness underlies foraging success of invasive *Lepidodactylus lugubris* geckos in the human landscape. *Animal Behaviour*, 76, 429–437.

Sih, A., Bell, A. and Johnson, J.C. (2004a). Behavioral syndromes: an ecological and evolutionary overview. *Trends in Ecology and Evolution*, 19, 372–378.

Sih, A., Bell, A.M., Johnson, J.C. and Ziemba, R.E. (2004b). Behavioral syndromes: an integrative overview. *The Quarterly Review of Biology*, 79, 241–277.

Sih, A., Cote, J., Evans, M., Fogarty, S. and Pruitt, J. (2012). Ecological implications of behavioural syndromes. *Ecology Letters*, 15, 278–289.

Simberloff, D. (2009). The role of propagule pressure in biological invasions. *Annual Review of Ecology, Evolution and Systematics*, 40, 81–102.

Sol, D. (2003). Behavioural flexibility: a neglected issue in the ecological and evolutionary literature. In *Animal Innovation*, ed. Reader, S.M. and Laland, K.N. Oxford, UK: Oxford University Press, pp. 63–82.

Sol, D. and Lefebvre, L. (2000). Behavioural flexibility predicts invasion success in birds introduced to New Zealand. *Oikos*, 90, 599–605.

Sol, D. and Maspons, J.M. (2015). Integrating behavior into life history theory: a comment on Wong and Candolin. *Behavioral Ecology*, 26, 677–678.

Sol, D., Timmermans, S. and Lefebvre, L. (2002). Behavioural flexibility and invasion success in birds. *Animal Behaviour*, 63, 495–502.

Sol, D., Duncan, R.P., Blackburn, T.M., *et al.* (2005). Big brains, enhanced cognition, and response of birds to novel environments. *Proceedings of the National Academy of Sciences, USA*, 102, 5460–5465.

Sol, D., Bacher, S., Reader, S.M., *et al.* (2008). Brain size predicts the success of mammal species introduced to novel environments. *American Naturalist*, 172, S63–S71.

Sol, D., Griffin, A.S., Bartomeus, I., *et al.* (2011). Exploring or avoiding novel food resources? The novelty conflict in an invasive bird. *PLoS One*, 6, e19535.

Sol, D., Maspons, J., Vall-llosera, M., *et al.* (2012). Unraveling the life history of successful invaders. *Science*, 337, 580–583.

Suarez, A.V., Holway, D.A., Liange, D., *et al.* (2002). Spatiotemporal patterns of intraspecific aggression in the invasive Argentine ant. *Animal Behaviour*, 64, 697–708.

Suarez, A.V., Holway, D.A. and Ward, P.S. (2005). The role of opportunity in the unintentional introduction of nonnative ants. *Proceedings of the National Academy of Sciences, USA*, 102, 17032–17035.

Suarez, A.V., Holway, D.A. and Tsutsui, N.D. (2008). Genetics and behavior of a colonizing species: the invasive Argentine ant. *American Naturalist*, 172, S72–S84.

Taylor, C.M. and Hastings, A. (2005). Allee effects in biological invasions. *Ecology Letters*, 8, 895–908.

Teixeria, C.P., Schetini de Azevado, C., Mendi, M., *et al.* (2007). Revisiting translocation and reintroduction programmes: the importance of considering stress. *Animal Behaviour*, 73, 1–13.

Thibert-Plante, X. and Hendry, A.P. (2011). The consequences of phenotypic plasticity for ecological speciation. *Journal of Evolutionary Biology*, 24, 326–342.

Tingley, R., Romagosa, C.M., Kraus, F., *et al.* (2010). The frog filter: amphibian introduction bias driven by taxonomy, body size and biogeography. *Global Ecology and Biogeography*, 19, 496–503.

Tingley, R., Thompson, M.B., Hartley, S. and Chapple, D.G. (2016). Patterns of niche filling and expansion across the invaded ranges of an Australian lizard. *Ecography*, 39, 270–280.

Tobin, P.C., Berec, L. and Liebhold, A.M. (2011). Exploiting Allee effects for managing biological invasions. *Ecology Letters*, 14, 615–624.

Toy, S.J. and Newfield, M.J. (2010). The accidental introduction of invasive animals as hitchhikers through inanimate pathways: a New Zealand perspective. *Revue Scientifique et Technique-Office International des Epizooties*, 29, 123–133.

Truhlar, A.M. and Aldridge, D.C. (2015). Differences in the behavioural traits between two potentially invasive amphipods, *Dikerogammarus villosus* and *Gammarus pulex*. *Biological Invasions*, 17, 1569–1579.

Tsutsui, N.D. and Suarez, A.V. (2003). The colony structure and population biology of invasive ants. *Conservation Biology*, 17, 48–58.

Tsutsui, N.D., Suarez, A.V., Holway, D.A., *et al.* (2000). Reduced genetic variation and the success of an invasive species. *Proceedings of the National Academy of Sciences, USA*, 97, 5948–5953.

Ugelvig, L.V., Drijfhout, F.P., Kronauer, D.J.C., *et al.* (2008). The introduction history of invasive garden ants in Europe: integrating genetic, chemical and behavioural approaches. *BMC Biology*, 6, 11.

Usio, N., Konishi, M. and Nakano, S. (2001). Species displacement between an introduced and a 'vulnerable' crayfish: the role of aggressive interactions and shelter competition. *Biological Invasions*, 3, 179–185.

Van Buskirk, J. (2012). Behavioural plasticity and environmental change. In *Behavioural Responses to a Changing World: Mechanisms and Consequences*, ed. Candolin, U. and Wong, B.B.M. Oxford: Oxford University Press, pp. 145–158.

Walsh, B. and Blows, M.W. (2009). Abundant genetic variation plus strong selection = multivariate genetic constraints: A geometric view of adaptation. *Annual Reviews of Ecology, Evolution and Systematics*, 40, 41–59.

Ward, D.F., Beggs, J.R., Clout, M.N., *et al.* (2006). The diversity and origin of exotic ants arriving in New Zealand via human-mediated dispersal. *Diversity and Distributions*, 12, 601–609.

Weis, J.S. (2010). The role of behavior in the success of invasive crustaceans. *Marine and Freshwater Behaviour and Physiology*, 43, 83–98.

White, A.W. and Shine, R. (2009). The extra-limital spread of an invasive species via 'stowaway' dispersal: toad to nowhere? *Animal Conservation*, 12, 38–45.

Witmer, G.W. Snow, N.P., Moulton, R.S., *et al.* (2014). Responses by wild house mice (*Mus musculus*) to various stimuli in a novel environment. *Applied Animal Behaviour Science*, 159, 99–106.

Wolf, M. and Weissing, F.J. (2012). Animal personalities: consequences for ecology and evolution. *Trends in Ecology and Evolution*, 27, 452–461.

Wong, B.B.M. and Candolin, U. (2015). Behavioral responses to changing environments. *Behavioral Ecology*, 26, 665–673.

Wright, T.F., Eberhard, J.R., Hobson, E.A., *et al.* (2010). Behavioral flexibility and species invasions: the adaptive flexibility hypothesis. *Ethology Ecology and Evolution*, 22, 393–404.

3 Invading New Environments: A Mechanistic Framework Linking Motor Diversity and Cognition to Establishment Success

Andrea S. Griffin, D. Guez, I. Federspiel, Marie Diquelou and F. Lermite

Introduction

To invade a new environment and become established all animals need to solve the same set of problems: First, they need to detect new resources and investigate them. Second, they need to develop the skills to exploit them and, third, store adequate information to be able to identify and handle them in the future. Fourth, they need to detect new predators and avoid them. Admittedly, the amplitude of these challenges will most likely vary with the degree of adaptive match between the invader and the new environment (Duncan *et al.*, 2003; Sol, 2007). Invaders arriving from environments similar to their new surroundings benefit from their existing learned and evolutionary knowledge because the cues that signal resources in their new environment and the way new items need to be handled will bear some similarity to the cues they experienced and the skills they deployed in their original environment. The challenges are greatest for those invaders arriving from different environments because their existing knowledge will not apply to their new circumstances. But, overall, whatever their level of adaptive match, invaders confronted with unfamiliar surroundings and unfamiliar circumstances all face these problems to some extent. The question we address here is which behavioural and cognitive mechanisms assist alien animals in solving these challenges.

In recent years, literature-based, macro-ecological research examining the life history traits associated with invasion success has revealed that avian (Sol *et al.*, 2005), mammalian (Sol *et al.*, 2008), reptilian and amphibian (Amiel *et al.*, 2011) taxa with larger brains relative to their body size are more likely to become established when introduced to new environments than taxa with smaller relative brains. Setting aside methodological concerns with whole brain size measurements (Healy and Rowe, 2007), this finding has been taken to indicate that brain size contributes to successful passage through the initial stages of the invasion process. The existence of a correlation between brain size and invasion success does not in itself explain why larger brains might be useful during establishment, however. Sol *et al.* (2005) proposed independently from an earlier

Biological Invasions and Animal Behaviour, eds J.S. Weis and D. Sol. Published by Cambridge University Press. © Cambridge University Press 2016.

suggestion by Allman *et al.* (1993) and Deaner *et al.* (2002) that the large brain of invaders might equip them with a greater capacity to gather, store and integrate environmental information and tested this idea by correlating cross-taxon variation in brain size and invasion success with cross-taxon variation in the frequency of anecdotal reports of novel feeding behaviours, a measure referred to as 'behavioural flexibility' (Sol *et al.*, 2005). A three-way positive relationship between relative brain volume, behavioural flexibility and establishment success has been taken as supporting evidence that greater information processing afforded by larger brains is key to invasion success (Sol *et al.*, 2005).

Inherent to this conclusion, however, is the underlying assumption that larger volumes, rather than re-organization, higher cell densities, or higher conduction velocities, for example (Roth and Dicke, 2005; Smaers and Soligo, 2013), underpin greater information processing capacity. To date, this notion remains controversial (Healy and Rowe, 2007). There is a dire need to quantify cognition in invasive species experimentally to complement the correlational macro-ecological comparative approach. A good place to start is to articulate a set of a priori predictions regarding the specific cognitive mechanisms that might assist invaders. The nature of these processes will point us to the experimental methodologies needed to measure them unequivocally.

In this chapter, we focus our discussion on the establishment phase of the invasion process (Blackburn *et al.*, 2011). We discuss the information processing mechanisms that might facilitate survival and reproduction of the first few generations of invaders in a new environment. We argue that resolving the four sets of problems faced by these pioneers relies upon the ability to express a variety of motor actions coupled with a combination of information processing mechanisms pertaining to learning, memory and inhibitory control. We explain how these mechanisms are measured and illustrate them drawing particularly on our recent work on the behaviour and cognition of the common (Indian) myna (*Acridotheres tristis*; recently re-classified as *Sturnus tristis* by Christidis and Boles, 2008), a highly successful worldwide ecological invader and ideal model avian system for exploring the role of cognition in invasion success. We aim to provide a mechanistic framework that will direct future empirical work on the role of specific motor and cognitive processes in invasiveness. But first, we introduce the reader to the study of cognition.

What Is Cognition?

Cognition can be defined broadly as the acquisition, processing, storage and use of information (Dukas, 2004; Shettleworth, 2010). In essence, the term refers to a collection of unobservable processes, which emerge from the integrated activity of neural networks within the brain and result in behavioural expression. Cognition encompasses a large variety of abilities, some of which are multidimensional, meaning they can be further delimited into subsets of abilities (Table 3.1). For example, behavioural inhibition involves both the ability to inhibit ongoing behaviour and the ability to ignore previously learned information (Table 3.1). The taxonomic distribution of distinct

Table 3.1 Examples and definitions of cognitive abilities. We provide some examples of cognitive traits and their cognitive content. S stands for stimulus. Cues are typically arbitrary stimuli with little biological significance, such as a light or a simple tone

Cognitive ability	Cognitive content
Pavlovian learning	Animals learn a predictive relationship between two S
Operant learning	Animals learn a predictive relationship between a motor action and an outcome
Discrimination learning	Animals learn that one S (S+) predicts an outcome while another (S−) predicts the absence of an outcome
Habituation	Animals learn that S is irrelevant
Generalization	Animals respond to novel (i.e. not experienced during training) S that share sensory features with a learned S
Contextual learning	Animals learn to avoid/approach a place in which they have experienced an aversive/appetitive S
Memory duration	The time interval between training and test is extended
Spatial learning and memory	Landmark learning: An animal is trained to find a reward specified by a landmark configuration
	Path integration: An animal must return to its departure point based on an ongoing calculation of direction and distance travelled
Inhibitory control[a]	Go/no go: In a discrimination learning task in which S+ and S− are presented sequentially, an animal learns that S+ requires it to perform a response (e.g. peck a key), whereas S− requires it to inhibit a response (withhold from pecking a key)
	Reversal learning: Animal is trained to criterion on a discrimination learning task, at which point the significance of the cues is reversed such that the animal must inhibit previously learned (successful) behaviour
	Stop signal: An animal must inhibit an ongoing motor action in response to an acoustic cue
	Delay discounting: Animal must delay responding in order to obtain a more valuable reward
Social learning[a]	Local/S enhancement: A social cue directs an animal's attention to a particular S or location (local enhancement) or category of S (S enhancement)
	Social facilitation: An animal increases its frequency of a behaviour after witnessing others perform that behaviour
	Observational conditioning: Animals learn a predictive relationship between a cue and a social S

[a] This is a multifaceted cognitive ability and examples of different tests measure different dimensions.

cognitive abilities varies, with some abilities appearing in only a small range of taxa (e.g. theory of mind) and others showing a far broader distribution (e.g. Pavlovian learning) (Shettleworth, 2010).

As cognitive abilities are not directly manifested phenotypically, their measurement is achieved by quantifying a change in behaviour. To date, most animal cognition researchers have been interested in identifying and quantifying what cognitive ability(ies) cause observable changes in behaviour. This contrasts with the aim of behavioural ecologists, who are interested in determining the adaptive significance of a trait. For example, the behavioural ecologist might want to quantify the costs and benefits of social dominance, whereas the animal cognition researcher would want to know whether an individual that knows that individual A is dominant over individual B and that B is dominant over C, can infer that A is dominant over C, a cognitive ability known as transitive inference (Guez and Audley, 2013). Evidence for this cognitive ability is gathered by using experimental designs that disentangle behavioural responses attributable to transitive inference from those attributable to associative learning (Guez and Audley, 2013). We turn now to predicting the role of specific cognitive abilities in establishing in new environments, while acknowledging that the range of cognitive processes that might assist invaders might be potentially far greater than those discussed here.

Detecting and Investigating New Resources

Alien species arriving in a new environment lack the up-to-date knowledge necessary to locate and recognize resources key to survival and reproduction. This information gap includes local knowledge about food and nesting materials, the location and suitability of sheltering opportunities, such as roosts and burrows, and knowledge about local dangers, including predators, poisons, and climatic exposure. Motivation and willingness to approach and interact with novel stimuli are critical to persisting under such circumstances because, without them, pre-existing knowledge cannot be updated and new knowledge cannot be created. Approach and investigation of novel objects are behaviours that are included under the concept of neophilia (Greenberg and Mettke-Hofmann, 2001), the experimental measurement of which usually involves presenting a satiated individual with an unfamiliar object and measuring the number of times it approaches it and/or makes contact with it per unit time (reviewed by Griffin and Guez, 2014). As approach is elicited by a novel object per se, and not by any extrinsic reward (e.g. food), neophilia tests are taken to reflect an attraction to intrinsically rewarding novelty (Greenberg and Mettke-Hofmann, 2001).

Despite its relevance to invading and persisting in novel environments, and to changing environmental conditions more generally, there has been little work to date investigating the mechanisms, function and ecological correlates of neophilia (Mettke-Hofmann, 2007; Mettke-Hofmann et al., 2009). A small body of research suggests that expression of the trait varies considerably across and within species and that this variation is tightly linked to ecology. For example, parrot species that inhabit complex habitats and islands are quick to approach novel objects, while those species with diets involving seeds and/or flowers are slower (Mettke-Hofmann et al., 2002).

Approaching and interacting with novel objects entails risks, however. Consequently, there is a need for a protective behavioural mechanism that dampens an individual's neophilic propensity (Greenberg and Mettke-Hofmann, 2001). Neophobia – the avoidance of novelty – most likely fulfils this function (Greenberg and Mettke-Hofmann, 2001). Just like neophilia, research on mechanisms of neophobia is scant and assumptions about its function and ecological correlates remain mostly theoretical (Greenberg and Mettke-Hofmann, 2001; but see Mettke-Hofmann et al., 2002; Brown et al., 2013; Miranda et al., 2013; Candler and Bernal, 2014). As the measurement of neophobia typically involves quantifying the latency with which a food-deprived individual is willing to feed from a familiar food dish in the presence of a novel object, low neophobia is generally taken to reflect a generally low fear of novelty, or alternatively a higher propensity to overcome novelty-evoked fear when necessity is high ('risk-taking') (Griffin and Guez, 2014).

Although one might intuitively imagine neophilia and neophobia as inversely related, research to date suggests that the two traits can vary independently (Mettke-Hofmann *et al.*, 2002, 2005; Biondi *et al.*, 2010; Miranda *et al.*, 2013), meaning an animal can be both highly neophobic and highly neophilic (Greenberg and Mettke-Hofmann, 2001). For example, it might explore novel objects extensively, but only once it has overcome its initial neophobic response. One way for selection to maximize the benefits of exploring novelty, while minimizing associated risks, might be to couple relatively hard-wired levels of attraction to novelty with environmentally tailored, plastic levels of novelty avoidance. In this way, for example, individuals reared in environments where risks associated with investigating novelty are low (e.g. low predation risk) might exhibit low avoidance of novelty and high interest in novelty, whereas those reared in environments where risks are high (e.g. high predation) might exhibit high levels of neophobia. Developmentally plastic neophobia is supported by the finding that many species reared in captivity are far less neophobic than their wild counterparts, but both captive- and wild-reared individuals investigate novel objects extensively (Schuppli *et al.*, 2014). In addition to being developmentally plastic, neophobia might be continuously updated throughout an animal's lifetime as a function of experience, as suggested by the finding that wild orang-utans (*Pongo pygmaeus*) show high neophobia of objects placed within their environment, but given enough time, show high levels of approach and investigation towards them (Schuppli *et al.*, 2014).

Neophilia and neophobia have been considered most often under the umbrella of emotional responses (Greenberg and Mettke-Hofmann, 2001; Sol *et al.*, 2012b), presumably because the phenotypical manifestation of these traits, approach versus avoidance, seems most related to a fear response. This focus has overshadowed its necessary cognitive underpinnings, however (Suchail *et al.*, 2000). Indeed, recognizing novelty within a familiar setting, a proviso of responding to it, necessarily requires a comparison between current environmental information and stored environmental information (Suchail *et al.*, 2000).

Continuous updating and effective integration of neophilia and neophobia throughout an animal's lifetime relies upon learning and memory. Neophobia requires animals to have learnt to recognize the characteristics of their familiar environment, a familiar backdrop against which they recognize novel objects. To adjust neophobia as a function of experience, the animal must further learn and store information about the properties of novel objects (e.g. its harmlessness versus usefulness) over successive encounters. Habituation (Table 3.1), a learning process that allows neophobia to decrease across repeatedly harmless encounters, paves the way for approach and investigation (Schuppli *et al.*, 2014). Finally, once the object has been investigated, information about the value of the object must be placed in memory, ensuring that the object is either returned to or ignored in the future.

These considerations lead us to predict that successful establishment in novel environments will correlate positively with high levels of neophilia and high, but developmentally plastic and lifelong adjustable (via learning and memory), levels of neophobia

(Sol *et al.*, 2011). Hence, invasive species should display levels of neophobia that vary across populations as a function of local environmental risk. Within populations, individuals should be capable of continuously updating their levels of neophobia based on experience acquired as they discover and explore their new environments. Finally, the rate at which neophobia is adjusted should vary across populations, occurring more quickly in safe environments and less quickly in risky environments. In this way, invaders will minimize the risks associated with exploring new environments while maximizing the benefits of discovering novel resources.

One of only three bird species to be listed in the top '100 World's Worst Invasive Alien species' by the International Union for the Conservation of Nature (IUCN; Lowe *et al.*, 2000), the common myna provides an ideal system in which to test some of these predictions. So, far our research has focused on neophobia rather than neophilia and on comparing this trait across individuals, populations and species. Although there is a rapidly growing body of published research on various aspects of novelty responses, our work constitutes, to our knowledge, the most extensive multilevel investigation of this trait in a single system.

In tests, where a novel object is placed next to a familiar food dish, wild-caught, captive-held mynas delay feeding significantly relative to when the novel object is absent (Sol *et al.*, 2011, 2012b; Griffin *et al.*, 2013, 2014; Griffin and Diquelou, 2015). Neophobia responses vary across individuals and are consistent across time (Sol *et al.*, 2012b), a finding that extends previous work demonstrating repeatable inter-individual variation in this behaviour (Boogert *et al.*, 2006; Seok An *et al.*, 2011; Bókony *et al.*, 2012; Tebbich *et al.*, 2012). Mynas significantly decrease their latency to feed in the presence of a novel object across four successive encounters with a novel object, indicating that neophobia is rapidly adjusted as a function of ongoing experience (Figure 3.1). In addition, neophobia varies across populations as found in a small number of other species (Martin and Fitzgerald, 2005; Bókony *et al.*, 2012; Brown *et al.*, 2013; Miranda *et al.*, 2013; Candler and Bernal, 2014). Mynas captured in highly urbanized environments show lower neophobia responses than individuals captured in more urban habitats (Sol *et al.*, 2011, 2012a). Ongoing work is testing the prediction that populations differ not only in mean levels of neophobia but also in the slope of the habituation gradient. Finally, a recent cross-species comparison of mynas and the native sympatric Australian noisy miner (*Manorina melanocephala*) has revealed that neophobia is significantly higher in the invasive myna than in the native noisy miner (Griffin and Diquelou, 2015). More extensive comparisons of invasive and non-invasive avian species along the lines of those on migratory and non-migratory parrots (Mettke-Hofmann *et al.*, 2002, 2005, 2009) are now needed to ascertain to what extent the neophobia responses of mynas and other invaders are systematically higher than those of species with which they cohabit.

Exploiting Newly Discovered Resources

In some cases, approach and investigation of newly discovered resources is not sufficient to exploit them effectively. Some will require additional manipulation. Consequently,

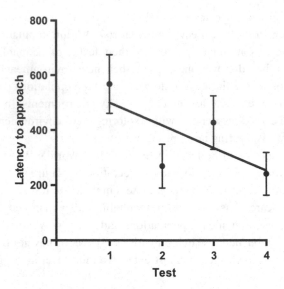

Figure 3.1 Changes in mean ± SEM neophobia across repeated presentations of a novel stimulus in common mynas ($N = 39$). Four successive presentations of the same novel object were separated by 24 h. On each day, neophobia was calculated by subtracting the latency to feed from a familiar food dish in a no-object baseline test from the latency to feed in the presence of a novel object. Best linear fit ($y = -82.22 + 581.5$); slope differs significantly from zero ($F_{(1, 154)} = 4.016, p < 0.05$) (Griffin, unpublished data).

the second challenge for an invader is to deploy a suitable handling technique and to remember which handling technique was successful so it can be used in the future. This brings us to how motor diversity and a cognitive ability known as operant learning can assist invaders in establishing in new environments.

Motor Diversity

Klopfer (1967) was the first to put forward the concept of motor stereotypy to refer to the tendency to produce only a narrow range of motor actions to accomplish a given act. He suggested that motor stereotypy might reflect an inability to adjust movements to changes in substrate. We have elaborated on this concept further, suggesting that an animal that is more diverse in its motor behaviour should be able to perform a greater number of distinct motor actions and express them with more even relative frequencies. Collectively, we have referred to these two patterns of motor behaviour expression (number and evenness) as 'motor diversity' (for calculations of motor diversity, see Griffin and Diquelou, 2015).

In cases where accessing a newly discovered resource requires it to be manipulated, invaders need to discover a motor action that allows the novel resource to be handled successfully. For example, a new food might be located under leaves that need to be lifted or surrounded by a shell that needs to be removed. To solve this problem, animals can apply a motor action that already exists in their motor repertoire to this novel context or invent a novel motor action, whereby novel motor actions represent variations of existing motor actions. In both cases, the likelihood that a suitable motor action emerges increases as a function of the number of motor actions the animal already has within its repertoire (Griffin *et al.*, 2013, 2014; Griffin and Diquelou, 2015; Diquelou *et al.*, 2015). Assuming that the more often a motor action is repeated the more likely it is to be successful, the more even relative frequencies of distinct motor actions will also facilitate solution finding (Griffin *et al.*, 2013, 2014; Griffin and Diquelou, 2015; Diquelou *et al.*, 2015). As a consequence, motor diverse animals should be better equipped to handle novel resources than those with a more limited motor output. This prediction can be tested by presenting animals with puzzle boxes that need to be manipulated to access the food, an experimental assay referred to as innovative problem solving (see Griffin and Guez, 2014 for a review). Work in this field is yielding accumulating evidence in birds and mammals that individuals and species that express a more diverse range of motor behaviours display a greater ability to solve such technical foraging problems (Overington *et al.*, 2011; Thornton and Samson, 2012; Benson-Amram and Holekamp, 2012; Mangalam and Singh, 2013; Griffin *et al.*, 2014; Griffin and Diquelou, 2015; Diquelou *et al.*, 2015). We predict that species that successfully establish in new environments will show greater diversity in their motor behaviour than species that do not become established and as a consequence will have a higher probability of, and be faster at, solving innovative foraging tasks.

Although previous work has established the importance of motor diversity to innovative problem solving, our work is unique in so far that it combines both intra- and interspecies analyses. A within-species analysis has confirmed that individual common mynas with a greater number of distinct motor actions and more even relative frequencies of expression outperform mynas with fewer motor actions and more skewed relative frequencies when tested on technical foraging problems (Griffin *et al.*, 2014). A comparative analysis of innovative problem solving has revealed that mynas outperform the Australian native noisy miner (*Manorina melanocephala*) on technical foraging tasks and that a higher motor diversity in mynas accounts for this performance difference (Griffin and Diquelou, 2015). A recent cross-species comparison of innovative problem solving in seven urbanized avian species has revealed that motor diversity consistently enhances innovative foraging (Diquelou *et al.*, 2015). The most frequent innovator and motor diverse species was the Australian raven (*Corvus coronoides*), a member of the Corvidae family and true crow. Common mynas innovated less frequently than ravens, but more frequently than three of four other native Australian species with which they were compared, as did European starlings (*Sturnus vulgaris*) (Diquelou *et al.*, 2015). Finally, it is possible that motor diversity contributes to explaining higher problem-solving abilities of urbanized mynas relative to suburban mynas (Sol *et al.*, 2011), but a comparative analysis of motor diversity in these populations remains to be conducted. More extensive cross-species comparisons are now needed to determine whether high motor diversity is a common feature of species capable of establishing in new areas.

Operant Learning

Operant (instrumental) learning involves storing to memory an association between a behaviour and an outcome (Thorndike, 1898). Behaviours followed by desirable

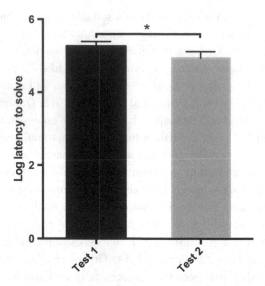

Figure 3.2 Innovative problem-solving performance in common mynas on two successive innovation opportunities. Mean ± SEM latency (logged) to solve a technical problem-solving task on two consecutive presentations 24 h apart ($N = 23$). Latencies decreased significantly from first to second solving (paired *t*-test, $t_{22} = 2.112$, $p < 0.05$) (Griffin, unpublished data).

outcomes are expressed more frequently, while those followed by adverse, undesirable outcomes are expressed less frequently. For example, a behaviour that yields a food reward will be repeated subsequently, whereas a behaviour that fails to yield food will not. Although the scientific literature pertaining to operant conditioning is largely constrained to testing common laboratory species, such as rats and pigeons, in environmentally and socially deprived environments, there are many examples of natural behaviour where operant learning is likely to be involved. For example, in many predator species, mothers bring live prey to their offspring, thus creating opportunities for them to practice and improve their capture and killing techniques, presumably because effective techniques are retained and ineffective ones discarded (Thornton and Samson, 2012).

The Role of Operant Learning in *Remembering* an Effective Motor Action

For an invader that has discovered a suitable handling technique for exploiting a novel resource, foraging more efficiently in the future requires that the effective motor action be stored to memory so that it can repeated on subsequent encounters with the resource. Operant learning will serve exactly this purpose. Over the course of repeated interactions with a given resource, the probability of displaying ineffective actions will decrease while the expression of effective ones will increase.

To our knowledge, only one study has examined the role of operant learning in the retention of newly discovered motor actions (Thornton and Samson, 2012). Experiments in mynas have shown consistently that the latency with which mynas solve a technical foraging problem after first contacting it decreases the second time they solve the task (Sol *et al.*, 2012b; Figure 3.2). Further analyses are needed to determine whether shorter solving latencies are a consequence of higher motivation due to previous reward (which might make an individual cycle through its motor repertoire more quickly for example) or whether they reflect greater frequencies of effective motor actions relative to ineffective ones as a consequence of operant learning. To date, there have been no cross-species comparisons of operant learning but this will represent an interesting avenue for future research.

The Role of Operant Learning in *Discovering* an Effective Motor Action

So far, we have only considered the problem of remembering a newly discovered handling technique so that it can be performed again in the future, but operant learning might also operate *during* the discovery of an effective motor action. Considering an animal that cycles through its motor repertoire searching for a successful handling technique, discovering an effective motor action is the result of a random process influenced only by the number of attempts made, the number of motor actions in the animal's repertoire and their relative frequency of expression. Neither operant learning, nor any other cognitive process (e.g. causal inference) is implicated in discovering a solution.

Another possibility involves honing in on actions that lead to partial solutions by gradually expressing them more often relative to others. Here, the animal selects the suitable motor action by performing a gradually closer and closer approximation of the final motor action. In this case, learning does not occur via a motor action–reward pairing, but via a pairing of a motor action and secondary cues that function as indirect cues for reward delivery. These secondary cues could be recognized largely independently from experience or learned through their own pairing with reward delivery. For example, animals might learn that lifting or moving a leaf enables the capture of prey. In this case, the movement of the leaf predicts reward delivery (see next section on Pavlovian learning). The learned significance of movement cues means they could then be used as a proxy for the reward, a process known as second-order conditioning, and generalized to future innovation contexts (Rescorla, 2014). Overington *et al.* (2011) have shown that carib grackles (*Quiscalus lugubris*) discover the solution to a foraging problem faster when they have access to movement cues than when these cues are blocked. To demonstrate unambiguously that operant learning is involved will require further demonstrating that motor actions that induce movement increase in frequency relative to ones that do not. Although we have not yet determined whether operant learning is involved in the discovery of novel foraging techniques in mynas, common mynas express significantly more effective techniques (i.e. ones that can potentially solve) relative to their levels of persistence when attempting to solve an innovative foraging task compared to the native Australian noisy miner (Griffin and Diquelou, 2015), suggesting mynas might learn more during their interactions with the task than the native species.

In sum, operant learning is central both to discovering and remembering a new handling technique. We predict that the capacity for operant learning will be enhanced in species that have successfully established in new environments. Specifically, invaders will modify their motor actions in response to environmental contingencies more often and more quickly than species that have not succeeded in becoming established. Further work is needed to test this prediction.

Inhibitory Control

To allow the expression of novel motor actions or the use of old ones under new circumstances, it is of central importance that animals are capable of inhibiting motor actions that are no longer successful. Defined as the 'ability to suppress the processing or expression of information that would disrupt the efficient completion of the goal at hand', inhibitory control allows a resistance to interference from irrelevant stimuli and the suppression of previously rewarded, but currently inappropriate, actions (Table 3.1) (Dempster, 1992; Björklund and Harnishfeger, 1995). Inhibitory control seems to be executed in the prefrontal cortex in mammals and the nidopallium in birds (Lissek *et al.*, 2002).

In *response* inhibition, previously rewarded behaviours are inhibited in favour of trying out novel solutions to problems. For example, wild meerkats solve an innovative foraging task significantly more quickly the second time they are presented with it, a decrease in solving latency attributable to decreased time spent manipulating non-functional components of a technical foraging problem (i.e. ones that do not allow the food to be accessed). This finding suggests that meerkats are able to inhibit their attempts to access the food through ineffective means (Thornton and Samson, 2012). Response inhibition should be a particularly important cognitive ability for invaders of novel environments, where new food sources have to be explored and accessed. This is because transferring known motor actions to novel contexts might not always work. Inhibiting previously rewarded motor actions in favour of trying out new solutions within the motor repertoire is key to finding solutions to novel problems. Hence, we predict that species successful at establishing in new environments will succeed particularly well in tasks investigating inhibitory control.

This prediction has led us to begin measuring inhibitory control in mynas using two distinct approaches. In a first study, mynas were given access to a technical foraging task in which two possible solving techniques could be used. Once mynas had discovered one of these, birds were trained until they rapidly solved the task using this technique. At this point, the use of this technique was blocked, such that the mynas had to discover the alternative opening technique. Upon succeeding, this technique was in turn trained, and then blocked while making the first one available again. Figure 3.3 depicts the latencies with which the birds discovered the initial two solving techniques for the first time, as well as the latencies with which they successfully alternated between the two techniques over the course of four such reversal tests. One can see that once mynas had discovered the two techniques for the first time, they learnt to switch quickly between the two possible techniques when one was blocked (Figure 3.3).

In a second study, mynas were trained to retrieve food from inside an opaque tube, which was rotated on each presentation, a task known as the 'tube task' (Sandel *et al.*, 2009). All mynas learned to retrieve the food from inside the tube within a maximum of two presentations. During subsequent test trials, mynas were presented with food inside an identical tube, but this time the tube was translucid rather than opaque. Individuals that have high inhibitory control are predicted to inhibit pecking at the food through the wall of the tube, and instead to perform the detour and access the food through its

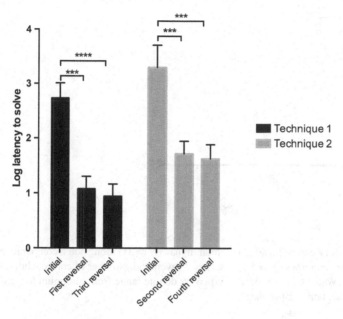

Figure 3.3 Acquisition and reversal of motor performance in mynas. Mean ± SEM latency (logged) to solve a puzzle box using a previously learnt alternative foraging technique when a first technique is blocked ($N = 14$). Two-way repeated measure analysis of variance (ANOVA): time $p < 0.0001$, $F_{(2,26)} = 32.51$; technique $p < 0.05$, $F_{(1,13)} = 5.13$, time by technique interaction $p > 0.05$. **** $p < 0.0001$; *** $p < 0.001$ (Griffin, unpublished data).

open end (Sandel *et al.*, 2009). On ten successive test trials, two mynas made six correct choices, detouring the tube to access food through its open end, and 29 mynas made eight or more correct choices (Figure 3.4). All but two birds made the correct choice on the very first presentation of the translucid tube. This level of performance is no different from that of adult carrion crows tested on the same task (Figure 3.4).

The above experimental findings using the tube task should be interpreted with caution. It is possible that individuals that execute the detour and retrieve the food during test trials with the translucid tube are more inclined to perform a memorized motor sequence (i.e. the one they learnt during baseline trials with an opaque tube), ignoring or overlooking the food visible inside the translucid tube. Although more experimental work is needed to address this caveat, our current measurements of inhibitory control in the invasive myna suggest that their performance is on a par with that of a true crow, considered to be the evolutionary pinnacle of avian intelligence (Cnotka *et al.*, 2008; Seed *et al.*, 2009).

Learning Predictors of New Resources and Predators

Motivation and willingness to approach and investigate novelty, motor diversity and inhibitory control to generate effective handling techniques, and operant learning to

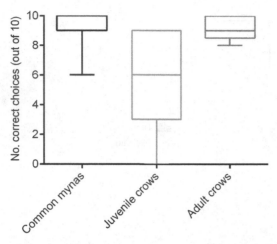

Figure 3.4 Inhibitory control in adult common mynas ($N = 31$) alongside a comparative data collection in carrion crows ($N = 5$ adults, $N = 7$ juveniles). Number of correct choices made out of 10 (Kruskal–Wallis test, $p = 0.015$). Whiskers denote range from minimum to maximum value (Federspiel, unpublished data).

place those skills in memory are insufficient to increase future behavioural efficiency unless they are coupled with the ability to recognize and re-locate recently discovered resources in the future. Without this, investigation and skill development would need to be repeated each time the resource is encountered. To become established, invaders also need to be able to respond adaptively to novel predators they might encounter. These final challenges facing invaders bring our discussion to a cognitive ability known as Pavlovian learning (Table 3.1).

Pavlovian Learning

In Pavlovian learning, also known as classical conditioning, an initially neutral stimulus (conditioned stimulus, CS; e.g. a light) is presented repeatedly together with a biologically significant event, which evokes a spontaneous response (unconditioned stimulus, US; e.g. food). As a result, animals learn a stimulus–stimulus association between the CS and US (Table 3.1). Learning of this association is reflected in that the CS acquires the ability to evoke a response that is related to the response evoked by the US (e.g. both CS and US elicit foraging behaviour). It is generally accepted that learning occurs when appearance of the CS predicts, or signals, the subsequent occurrence of the US. Hence, the function or adaptive significance of Pavlovian learning is to allow organisms to recognize predictive signals and to prepare themselves for biologically significant events. For instance, American crows (*Corvus brachyrhynchos*) acquire aggressive responses to humans (CS) by whom they have been previously handled (US) (Marzluff *et al.*, 2010). Although the Pavlovian learning mechanism is taxonomically widespread, it is now well-established that, at least in some species, this basic stimulus–stimulus learning mechanism can mediate learning of sophisticated types of information, including temporal information and causal reasoning (Arcediano *et al.*, 2003; Blaisdell *et al.*, 2006).

The stimulus–stimulus association mechanism underpinning Pavlovian learning allows animals to learn two types of information particularly relevant to coping with new habitats and new circumstances (Griffin, 2010). First, Pavlovian learning allows for the physical attributes of a new resource to be learnt, such as its smell and/or its visual features so that it can be recognized in the future. For example, domestic chicks (*Gallus gallus*) that taste a coloured unpalatable grain will subsequently avoid similarly coloured food items (Marples and Roper, 1997). Second, Pavlovian learning allows for animals to learn the location (CS) in which they experienced biologically important events (US), a type of Pavlovian learning known as contextual conditioning (Shettleworth, 2010, table 1). For example, a rodent will avoid returning to a chamber in which it received a foot shock (Fanselow, 2000).

Another aspect of stimulus–stimulus association learning particularly relevant to invading new environments is that Pavlovian learning of associations can be triggered by social USs, such as the communication signals of conspecifics or heterospecifics. For instance, young white-tailed ptarmigan chicks (*Lagopus leucurus*) learn to forage on foods high in protein (CS), which they have associated with their mother's food calls (US; Allen and Clarke, 2005) and rats learn to recognize the smell of a new food (CS) they detect on the breath of a conspecific (US; Galef, 1996). In the predation domain, social conditioning of anti-predator behaviour towards novel predators is taxonomically widespread, presumably because it is less risky than learning through direct experience with predators (see Griffin, 2004 for a review). Prior to establishment when populations are small and surroundings are new, the ability to use the behaviour of other individuals to learn about local conditions is likely to facilitate rapid transfer of information across individuals (Wright *et al.*, 2010).

It is not known whether invasive species outperform non-invasive species in tests of Pavlovian learning. But one might imagine that selection for dealing with novel environments could act to increase the range of stimuli that can be associated, both in terms of CSs and USs, increase the speed with which associations are made and/or the complexity with which different types of information acquired via operant and Pavlovian learning can interact (Blaisdell *et al.*, 2006). Both work from our lab and others is providing increasing support for the idea that mynas readily learn a diverse range of associations. There are published reports of acquired bait avoidance (Feare, 2010) and learning of armed myna shooters, as well as their whereabouts (Dhami and Nagle, 2009). Work in our lab has focused on exploring social transmission of information about novel predators (Griffin, 2008; Griffin and Boyce, 2009; Griffin *et al.*, 2010). As predicted, mynas learn readily from social information. A series of studies has revealed that mynas increase their anti-predator vigilance in response to a novel predator after just one pairing of the stimulus and the alarm response of a social companion (Griffin, 2008, 2009). The species is also capable of socially acquired place avoidance (Griffin and Boyce, 2009; Griffin *et al.*, 2010; Griffin and Haythorpe, 2011). Mynas trained to forage in a particular location behave significantly more cautiously on subsequent visits to the foraging location after they have observed a conspecific trapped in that area and expressing an alarm response towards a human trying to net them (Griffin and Boyce, 2009). Acquisition was not merely the result of a straightforward place (CS)–social alarm (US) association, however. Observer mynas only became more vigilant if they were given visual access to information about the cause of the demonstrator myna's alarm response (human capture) (Griffin and Haythorpe, 2011), a mechanism that presumably safeguards against learning erroneous associations between places and social alarm responses (Laland, 2004). Together, operant and Pavlovian learning form general purpose learning mechanisms, which can produce an enormous range of flexible and adaptive behaviours (Shettleworth, 2010). Moderation of acquisition by contextual information (i.e. the presence or absence of a cause) provides a glimpse of the sophistication that associative learning mechanisms might reach in invaders.

Motor and Cognitive Abilities: Correlations with Neural Volumes

Having presented the specific behavioural and cognitive abilities we predict should facilitate passage through the early stages of the invasion process (establishment), we now touch on the evidence that these abilities are linked to brain size.

Although motor diversity does not fall under the umbrella of cognition, there is a body of work suggesting that motor abilities are positively associated with brain space. Starting with the organization of the primary motor cortex, it has long been known that the amount of brain matter devoted to any particular body part represents the amount of control that the primary motor cortex has over that body part (Penfield and Rasmussen, 1950). Larger amounts of brain matter are associated with an increase in the degree of precision of movement that body part can achieve (Penfield and Rasmussen, 1950). More recently, larger motor repertoires, defined as the number of behaviours in published ethograms (logged) and obtained for 24 mammalian species, as well as the number of muscle types (logged), computed across eight mammalian orders, have both been found to increase with increasing encephalization (Changizi, 2003). There is evidence that the primary motor cortex motor is developmentally highly plastic in terms of its organization (Sanes and Donoghue, 2000). This plasticity might extend to the amount of brain matter devoted to a given set of movements. For example, comparative studies of professional and amateur musicians suggest that the premotor and motor areas show significantly increased amounts of grey matter volume in the former than in the latter group (Gaser and Schlaug, 2003). There is similar evidence for developmental plasticity of motor areas in non-humans. In canaries, the RA, a brain nucleus involved in the production of bird song, expands and shrinks seasonally as the birds learn new and different song repertoires on successive years (Nottebohm, 1981), suggesting that the amount of devoted neural tissue increases and decreases along with the number of different songs sung. Hence, it is possible that selection for an increased capacity to deal with novel environments might lead to increased motor diversity with an associated increase in motor cortices. Although this relationship supports the idea of a brain size–invasion link, it raises the possibility that increases in motor abilities rather than cognitive abilities underpin this relationship.

To our knowledge, only one study has examined the relationship between response inhibition and neural volumes. In a very recent cross-taxon analysis of 36 species of birds and mammals, variation in brain size (absolute and relative) has been found to explain the greatest proportion of variation in cross-taxon variation in self-control (MacLean *et al.*, 2014). Worthy of note is that in the same study, an analysis of the predictors of self-control in primates revealed that, along with brain size, the breadth of food consumed was a significant predictor of response inhibition (MacLean *et al.*, 2014). As it is possible that motor diversity is linked to diet breadth (Klopfer, 1967; Griffin, 2016; Griffin and Guez, 2016), this finding suggests a link between ecology, motor diversity, inhibition and neural volumes.

Turning to operant and Pavlovian learning, a recent study in guppies (*Poecilia reticulata*) has found that among a group of individuals selected for larger relative brain size, large-brained females learned more quickly to approach a discriminatory cue in a foraging task than large-brained males and a group of individuals from a non-selected genetic line (smaller brained females and males) (Kotrschal *et al.*, 2013a, but see Healy and Rowe, 2013, and Kotrschal *et al.*, 2013b). Similarly, in rodents, such as mice, spatial discrimination learning, speed and reversal have also been associated with greater brain size (Elias, 1970). Even in the modest honeybee (*Apis mellifera*), brain volume is correlated with faster olfactory conditioning of the proboscis extension reflex (Gronenberg and Couvillon, 2010). Together, these studies suggest that both performance on tests of operant and Pavlovian learning increase with relative and/or absolute brain size. Whether these volumetric measurements are caused by more neurons, more synaptic connections between neurons and/or more neuronal support cells (e.g. glial cells) is poorly understood. Although there is still much work to be done to address stark critics of cognition–brain space relationships (Healy and Rowe, 2007, 2013), this body of work is in line with the finding from the large-scale comparative literature that brain size facilitates invasion success (Sol *et al.*, 2005), but suggests more explicitly that specific cognitive abilities, namely those that we have predicted to assist invaders in dealing with new environments and new circumstances, are positively associated with brain space.

Conclusions

A relationship between brain size, cognition and invasion success has been the topic of much macro-ecological work and much discussion. However, cognition is often used in its broadest sense with no explicit reference to the particular processes that might assist invaders in establishing in new environments. We have put forward motor diversity and several cognitive abilities, including operant and Pavlovian learning and inhibitory control as specific cognitive processes that might assist the first few generations of animals released into new environments in coping with novel habitats and novel circumstances. We have illustrated these processes using our recent work in one of the most successful worldwide invasive species, the common myna. We predict that these cognitive abilities should be enhanced in those species that have succeeded in becoming established relative to those that have not. We also predict that these abilities should be enhanced within species, in individuals on the invasion front relative to individuals from long-established populations (Phillips and Suarez, 2012; Chapple *et al.*, 2012). We suggest that a more focused experimental comparative approach in which researchers apply standardized tests known to measure specific cognitive abilities will provide a much needed complement to literature-based comparative methodologies to ascertain whether cognition facilitates invasion success.

References

Allen, T. and Clarke, J. (2005). Social learning of food preferences by white-tailed ptarmigan chicks. *Animal Behaviour*, 70, 305–310.

Allman, J., McLaughlin, T. and Hakeem, A. (1993). Brain weight and life-span in primate species. *Proceedings of the National Academy of Sciences, USA*, 90, 118–122.

Amiel, J.J., Tingley, R. and Shine, R. (2011). Smart moves: effects of relative brain size on establishment success of invasive amphibians and reptiles. *PLoS ONE*, 6, e18277.

Arcediano, F., Escobar, M. and Miller, R.R. (2003). Temporal integration and temporal backward associations in human and nonhuman subjects. *Learning and Behavior*, 31, 242–256.

Benson-Amram, S. and Holekamp, K.E. (2012). Innovative problem solving by wild spotted hyenas. *Proceedings of the Royal Society of London, Series B*, 279, doi: 10.1098/rspb.(2012). 1450.

Biondi, L.M., Bó, M.S. and Vassallo, A.I. (2010). Inter-individual and age differences in exploration, neophobia and problem-solving ability in a neotropical raptor (*Milvago chimango*). *Animal Cognition*, 13, 701–710.

Björklund, D.F. and Harnishfeger, K.K. (1995). The evolution of inhibition mechanisms and their role in human cognition and behavior. In *Interference and Inhibition in Cognition*, ed. Dempster, F.N. and Brainerd, C.J., pp. 142–169. San Diego, CA: Academic Press.

Blackburn, T.M., Pyšek, P., Bacher, S., *et al.* (2011). A proposed unified framework for biological invasions. *Trends in Ecology and Evolution*, 26, 333–339.

Blaisdell, A.P., Sawa, K., Leising, K.J. and Waldmann, M.R. (2006). Causal reasoning in rats. *Science*, 311, 1020–1022.

Bókony, V., Kulcsár, A., Tóth, Z. and Liker, A. (2012). Personality traits and behavioral syndromes in differently urbanized populations of house sparrows (*Passer domesticus*). *PLoS one*, 7, e36639.

Boogert, N.J., Reader, S.M. and Laland, K.N. (2006). The relation between social rank, neophobia and individual learning in starlings. *Animal Behaviour*, 72, 1229–1239.

Brown, G.E., Ferrari, M.C.O., Elvidge, C.K., Ramnarine, I. and Chivers, D.P. (2013). Phenotypically plastic neophobia: a response to variable predation risk. *Proceedings of the Royal Society of London, Series B*, 280, doi: 10.1098/rspb.(2012).2712.

Candler, S. and Bernal, X.E. (2014). Differences in neophobia between cane toads from introduced and native populations. *Behavioral Ecology*, 26, 97–104.

Changizi, M.A. (2003). Relationship between number of muscles, behavioral repertoire size, and encephalization in mammals. *Journal of Theoretical Biology*, 220, 157–168.

Chapple, D.G., Simmonds, S.M. and Wong, B.B.M. (2012). Can behavioral and personality traits influence the success of unintentional species introductions? *Trends in Ecology and Evolution*, 27, 57–64.

Christidis, L. and Boles, W. (2008). *Systematics and Taxonomy of Australian Birds*. Collingwood, Australia: CSIRO Publishing.

Cnotka, J., Güntürkün, O., Rehkämper, G., Gray, R.D. and Hunt, G.R. (2008). Extraordinary large brains in tool-using New Caledonian crows (*Corvus moneduloides*). *Neuroscience Letters*, 433, 241–245.

Deaner, R.O., Barton, R.A. and van Schaik, C.P. (2002). Primate brains and life histories: renewing the connection. In *Primate Life Histories and Socioecology*, ed. Kappeler, P. M. and Pereira, M. E. Chicago, IL: The University of Chicago Press, pp. 233–265.

Dempster, F.N. (1992). The rise and fall of the inhibitory mechanism: toward a unified theory of cognitive development and aging. *Developmental Review*, 12, 45–75.

Dhami, M.K. and Nagle, B. (2009). Review of the biology and ecology of the common myna (*Acridotheres tristis*) and some implications for management of this invasive species. Report. Auckland, New Zealand: Pacific Invasives Initiatives, pp. 1–28.

Diquelou, M., Griffin, A.S. and Sol, D. (2015). Solving new foraging problems: Motor processes are key to behavioural innovation in birds. *Behavioral Ecology*, doi:10.1093/beheco/arv190.

Dukas, R. (2004). Evolutionary biology of animal cognition. *Annual Review of Ecology, Evolution, and Systematics*, 35, 347–374.

Duncan, R.P., Blackburn, T.M. and Sol, D. (2003). The ecology of bird introductions. *Annual Review of Ecology, Evolution, and Systematics*, 34, 71–98.

Elias, M.F. (1970). Spatial discrimination reversal learning for mice genetically selected for differing brain size: a supplementary report. *Perceptual and Motor Skills*, 30, 239–245.

Fanselow, M.S. (2000). Contextual fear, Gestalt memories, and the hippocampus. *Behavioural Brain Research*, 110, 73–81.

Feare, C.J. (2010). The use of Starlicide® in preliminary trials to control invasive common myna *Acridotheres tristis* populations on St Helena and Ascension islands, Atlantic Ocean. *Conservation Evidence*, 7, 52–61.

Galef, B.G.J. (1996). Social enhancement of food preferences in Norway rats: A brief review. In *Social Learning in Animals: The Roots of Culture*, ed. Heyes, C.M. and Galef, B.G.J. San Diego, CA: Academic Press, pp. 49–64.

Gaser, C. and Schlaug, G. (2003). Gray matter differences between musicians and nonmusicians. *Annals of the New York Academy of Sciences*, 999, 514–517.

Greenberg, R.S. and Mettke-Hofmann, C. (2001). Ecological aspects of neophilia and neophobia in birds. *Current Ornithology*, 16, 119–178.

Griffin, A.S. (2004). Social learning about predators: a review and prospectus. *Learning and Behavior*, 32, 131–140.

Griffin, A.S. (2008). Social learning in Indian mynahs, *Acridotheres tristis*: the role of distress calls. *Animal Behaviour*, 75, 79–89.

Griffin, A.S. (2009). Temporal limitations on social learning of novel predators by Indian mynahs, *Acridotheres tristis*. *Ethology*, 115, 287–295.

Griffin, A.S. (2010). Learning and conservation. In *Encyclopedia of Animal Behavior*, Vol. 2, ed. Breed, M.D. and Moore, J. Amsterdam: Elsevier, pp. 259–264.

Griffin, A.S. (2016). Innovativeness as an emergent property: a new alignment of comparative and experimental research on animal innovation. *Philosophical Transactions of the Royal Society B: Biological Sciences*, doi: 10.1098/rstb.2015.0544.

Griffin, A.S. and Boyce, H.M. (2009). Indian mynahs, *Acridotheres tristis*, learn about dangerous places by observing the fate of others. *Animal Behaviour*, 78, 79–84.

Griffin, A.S. and Diquelou, M. (2015). Innovative problem solving in birds: a cross-species comparison of two highly successful Passerines. *Animal Behaviour*, 100, 84–94.

Griffin, A.S. and Guez, D. (2014). Innovation and problem solving: a review of common mechanisms. *Behavioural Processes*, 109, 121–134.

Griffin, A.S. and Guez, D. (2016). Bridging the gap between cross-taxon and within-species analyses of behavioral innovations in birds: making sense of discrepant cognition–innovation relationships and the role of motor diversity. In *Advances in the Study of Behavior*, ed. Naguib, M., *et al.* New York, NY: Academic Press, pp. 1–40.

Griffin, A.S. and Haythorpe, K. (2011). Learning from watching alarmed demonstrators: does the cause of alarm matter? *Animal Behaviour*, 81, 1163–1169.

Griffin, A.S., Boyce, H.M. and MacFarlane, G.R. (2010). Social learning about places: observers may need to detect both social alarm and its cause in order to learn. *Animal Behaviour*, 79, 459–465.

Griffin, A.S., Lermite, F., Perea, M. and Guez, D. (2013). To innovate or not: contrasting effects of social groupings on safe and risky foraging in Indian mynahs. *Animal Behaviour*, 86, 1291–1300.

Griffin, A.S., Diquelou, M. and Perea, M. (2014). Innovative problem solving in birds: a key role of motor diversity. *Animal Behaviour*, 92, 221–227.

Gronenberg, W. and Couvillon, M.J. (2010). Brain composition and olfactory learning in honey bees. *Neurobiology of Learning and Memory*, 93, 435–443.

Guez, D. and Audley, C. (2013). Transitive or not: a critical appraisal of transitive inference in animals. *Ethology*, 119, 703–726.

Healy, S.D. and Rowe, C. (2007). A critique of comparative studies of brain size. *Proceedings of the Royal Society of London, Series B*, 274, 453–464.

Healy, S.D. and Rowe, C. (2013). Costs and benefits of evolving a larger brain: doubts over the evidence that large brains lead to better cognition. *Animal Behaviour*, 86, e1–e3.

Klopfer, P.H. 1967. Behavioural stereotypy in birds. *Wilson Bulletin*, 79, 290–300.

Kotrschal, A., Rogell, B., Bundsen, A., *et al.* (2013a). Artificial selection on relative brain size in the guppy reveals costs and benefits of evolving a larger brain. *Current Biology*, 23, 168–171.

Kotrschal, A., Rogell, B., Bundsen, A., *et al.* (2013b). The benefit of evolving a larger brain: big-brained guppies perform better in a cognitive task. *Animal Behaviour*, 86, e4–e6.

Laland, K.N. (2004). Social learning strategies. *Learning and Behavior*, 32, 4–14.

Lissek, S., Diekamp, B. and Güntürkün, O. (2002). Impaired learning of a color reversal task after NMDA receptor blockade in the pigeon (*Columbia livia*) associative forebrain (neostriatum caudolaterale). *Behavioral Neuroscience*, 116, 523–529.

Lowe, S., Browne, M., Boudjelas, S. and de Porter, M. (2000). *100 of the World's Worst Invasive Alien Species*. A Selection from the Global Invasive Species Database. Auckland: The Invasive Species Specialist Group (ISSG), a specialist group of the Species Survival Commission (SSC) of the World Conservation Union (IUCN).

MacLean, E.L., Hare, B., Nunn, C.L. *et al.* (2014). The evolution of self-control. *Proceedings of the National Academy of Sciences, USA*, 111, E2140–8.

Mangalam, M. and Singh, M. (2013). Flexibility in food extraction techniques in urban free-ranging bonnet macaques, *Macaca radiata*. *PLoS ONE*, 8, e85497.

Marples, N.M. and Roper, T.J. (1997). Response of domestic chicks to methyl anthranilate odour. *Animal Behaviour*, 53, 1263–1270.

Martin, L.B. and Fitzgerald, L. (2005). A taste for novelty in invading house sparrows, *Passer domesticus*. *Behavioral Ecology*, 16, 702–707.

Marzluff, J.M., Walls, J., Cornell, H.N., Withey, J.C. and Craig, D.P. (2010). Lasting recognition of threatening people by wild American crows. *Animal Behaviour*, 79, 699–707.

Mettke-Hofmann, C. (2007). Object exploration of garden and Sardinian warblers peaks in spring. *Ethology*, 113, 174–182.

Mettke-Hofmann, C., Winkler, H. and Leisler, B. (2002). The significance of ecological factors for exploration and neophobia in parrots. *Ethology*, 108, 249–272.

Mettke-Hofmann, C., Ebert, C., Schmidt, T., Steiger, S. and Stieb, S. (2005). Personality traits in resident and migratory warbler species. *Behaviour*, 142, 1357–1375.

Mettke-Hofmann, C., Lorentzen, S., Schlicht, E., Schneider, J. and Werner, F. (2009). Spatial neophilia and spatial neophobia in resident and migratory warblers (*Sylvia*). *Ethology*, 115, 482–492.

Miranda, A.C., Schielzeth, H., Sonntag, T. and Partecke, J. (2013). Urbanization and its effects on personality traits: a result of microevolution or phenotypic plasticity? *Global Change Biology*, 19, 2634–2644.

Nottebohm, F. (1981). A brain for all seasons: cyclical anatomical changes in song control nuclei of the canary brain. *Science*, 214, 1368–1370.

Overington, S.E., Cauchard, L., Côté, K.-A. and Lefebvre, L. (2011). Innovative foraging behaviour in birds: what characterizes an innovator? *Behavioural Processes*, 87, 274–285.

Penfield, W. and Rasmussen, T. (1950). *The Cerebral Cortex of Man: A Clinical Study of Localization of Function*. Oxford, UK: Macmillan Edn.

Phillips, B.L. and Suarez, S.D. (2012). The role of behavioural variation in the invasion of new areas. In *Behavioural Responses to a Changing World: Mechanisms and Consequences*, ed. Candolin, U. and Wong, B.B.M. Oxford, UK: Oxford University Press, pp. 190–200.

Rescorla, R.A. (2014). *Pavlovian Second-Order Conditioning. Studies in Associative Learning (Psychology Revivals)*. New York: Psychology Press.

Roth, G. and Dicke, U. (2005). Evolution of the brain and intelligence. *Trends in Cognitive Sciences*, 9, 250–257.

Sandel, A.A., MacLean, E.L. and Hare, B. (2009). Inhibitory control in an object retrieval task in five lemur species. *American Journal of Primatology*, 71, 80.

Sanes, J. and Donoghue, J. (2000). Plasticity and primary motor cortex. *Annual Review of Neuroscience*, 23, 393–415.

Schuppli, C., Sofia, F. and van Schaik, C.P. (2014). How sociality affects independent exploration: Evidence gathered in two populations of wild orangutans. *American Journal of Physical Anthropology*, 153, 234.

Seed, A., Emery, N. and Clayton, N. (2009). Intelligence in corvids and apes: a case of convergent evolution? *Ethology*, 115, 401–420.

Seok An, Y., Kriengwatana, B., MacDougall-Shackleton, E., Newman, A. and MacDougall-Shackleton, S. (2011). Social rank, neophobia and observational learning in black-capped chickadees. *Behaviour*, 148, 55–69.

Shettleworth, S.J. (2010). *Cognition, Evolution, and Behavior*, 2nd edn. Oxford, UK: Oxford University Press.

Smaers, J.B. and Soligo, C. (2013). Brain reorganization, not relative brain size, primarily characterizes anthropoid brain evolution. *Proceedings of the Royal Society of London, Series B*, 280, doi: 10.1098/rspb.2013.0269.

Sol, D. (2007). Do successful invaders exist? Pre-adaptations to novel environments in terrestrial vertebrates. In *Biological Invasions*, ed. Nentwig, W. Berlin: Springer, pp. 127–144.

Sol, D., Duncan, R.P., Blackburn, T.M., Cassey, P. and Lefebvre, L. (2005). Big brains, enhanced cognition, and response of birds to novel environments. *Proceedings of the National Academy of Sciences, USA*, 102, 5460–5465.

Sol, D., Bacher, S., Reader, S.M. and Lefebvre, L. (2008). Brain size predicts the success of mammal species introduced into novel environments. *The American Naturalist*, 172 Suppl., S63–71.

Sol, D., Griffin, A.S., Bartomeus, I. and Boyce, H. (2011). Exploring or avoiding novel food resources? The novelty conflict in an invasive bird. *PLoS ONE*, 6, e19535.

Sol, D., Bartomeus, I. and Griffin, A.S. (2012a). The paradox of invasion in birds: competitive superiority or ecological opportunism? *Oecologia*, 169, 553–564.

Sol, D., Griffin, A.S. and Barthomeus, I. (2012b). Consumer and motor innovation in the common myna: the role of motivation and emotional responses. *Animal Behaviour*, 83, 179–188.

Suchail, S., Guez, D. and Belzunces, L.P. (2000). Characteristics of imidacloprid toxicity in two *Apis mellifera* subspecies. *Environmental Toxicology and Chemistry*, 19, 1901–1905.

Tebbich, S., Stankewitz, S. and Teschke, I. (2012). The relationship between foraging, learning abilities and neophobia in two species of Darwin's finches. *Ethology*, 118, 135–146.

Thorndike, E.L. (1898). Animal intelligence: An experimental study of the associative processes in animals. *Psychological Monographs: General and Applied*, 2, 1125–1127.

Thornton, A. and Samson, J. (2012). Innovative problem solving in wild meerkats. *Animal Behaviour*, 83, 1459–1468.

Wright, T.F., Eberhard, J.R., Hobson, E.A., Avery, M.L. and Russello, M.A. (2010). Behavioral flexibility and species invasions: the adaptive flexibility hypothesis. *Ethology Ecology and Evolution*, 22, 393–404.

4 Invader Endocrinology: The Regulation of Behaviour in Pesky Phenotypes

Lynn B. Martin, Amber J. Brace, Holly J. Kilvitis and Stephanie S. Gervasi

Introduction

There are at least two reasons that the hormonal mechanisms underlying behavioural variation might be important in biological invasions. First, hormones, cytokines and functionally comparable molecules often orchestrate gene-by-environment interactions, the basis of most phenotypic variation (Sinervo and Calsbeek, 2003; Hau, 2007; McGlothlin and Ketterson, 2008). By focusing on hormones, the behaviours influencing invasion success may often be traceable to particular genetic, epigenetic or environmental factors (Badyaev and Uller, 2009; Martin et al., 2010b; Cohen et al., 2012; Martin and Cohen, 2014). Indeed, alterations to the regulation of key hormones, particularly during development, may be a mechanism whereby plastic modifications of behaviour (and physiology and morphology) occur before selection reinforces new mutations via accommodation (Martin et al., 2011). Hormones and associated plasticity could thus enable faster matching of traits to environments than selection on novel mutations, especially when populations are small and selection is weak, which often occur during invasions (Badyaev, 2013a).

A second reason that hormones might be important in the behaviours influencing invasions is that they are critical for homeostasis (Woods and Wilson, 2014). Although homeostasis is usually emphasized to canalize variation, hormones facilitate alterations to physiological set points and alter physiological network structure (Martin et al., 2011; Cohen et al., 2012). Such regulated adjustments enable organisms to exploit and compensate external and internal opportunities and challenges in a coordinated fashion and thus elicit behaviours most conducive to success under (often novel) conditions (Denver, 2009; Crespi et al., 2013; Lema and Kitano, 2013). In this light, invader endocrinology is particularly exciting because one might learn something general about the role of hormones in ecological and evolutionary processes (Angelier and Wingfield, 2013; Badyaev, 2014; Noble et al., 2014). Specifically, as genomes spread or are forced into new environments, one might elucidate how homeostatic systems determine the fate of individuals and populations, setting the table to revealing the relative roles of selection and plasticity in microevolution (West-Eberhard, 2003; Flatt and Heyland, 2011).

Biological Invasions and Animal Behaviour, eds J.S. Weis and D. Sol. Published by Cambridge University Press. © Cambridge University Press 2016.

In an ideal sense, the hormones mediating the behaviours of invaders could even serve as biomarkers of invasion risk (Martin *et al.*, 2010b). The idea warrants study but may be too idealistic given the regulatory complexity of most hormone–behaviour relationships.

In the present chapter, we have several aims with regards to the behavioural endocrinology of animal invaders. First, we summarize the as yet modest literature on the behavioural endocrinology of the various stages of invasions. Diverse physiological processes, such as salt (Braby and Somero, 2006) and water balance (Juliano *et al.*, 2002; Chown *et al.*, 2007), thermoregulation (Kimball *et al.*, 2004; Kolbe *et al.*, 2010) and other physiological traits (Kearney and Porter, 2009) are unquestionably important to invasion outcomes. However, the hormonal basis of behavioural variation is less studied. Second, we highlight how thinking about hormones as drivers of plastic phenotypic variation could be lucrative going forward in invasion biology. In plants, phenotypic plasticity in many traits is important to introduction success (Higgins and Richardson, 2014). In animals, hormones have extensive plastic, pleiotropic effects on behaviour, making hormones and their intimate linkages to the nervous system likely mediators of invader phenotypes (Dufty *et al.*, 2002; Ricklefs and Wikelski, 2002; Groothuis and Schwabl, 2008). As there are presently so few data to review, most of our efforts below entail speculations about the various ways that hormone regulation might influence invasions. We strongly advocate for more research in this new area.

Hormones and Behaviour among the Stages of Invasions

Some hormones have unique functions, as they appear to be directed to balance organisms' life priorities contingent on the environments that lineages experienced historically (Ricklefs and Wikelski, 2002; Martin *et al.*, 2011). However, as invasions are composed of stages, different hormones should take prominence at different points of an invasion depending on the biological processes they tend to dominate (Figure 4.1). Because invasions are diverse, it might be unreasonable to expect hormones to play consistent roles across all invasion stages and the habitats and histories of the sites invaded (i.e. natural versus anthropogenic). We might seek a consistent hormonal profile of invaders, but it is premature to claim that general patterns will be found.

To seek to reveal such generality, or at least understand the proximate basis by which invasions progress, invasions should be decomposed into stages. When humans move incipient invaders to areas outside their native ranges, invasion success is contingent on an organism (i) initially being caught from the native environment, (ii) being transported to a new environment, (iii) becoming established in the new environment, and (iv) expanding from the original site of introduction and/or mixing with local, reproductively compatible organisms already there (Colautti and MacIsaac, 2004). Natural invasions can also occur, as native or introduced populations respond to and exploit opportunities (Bradley and Altizer, 2007; Bonier, 2012). In these cases, certain individuals must be inherently predisposed to behaviours that facilitate range expansion, or they must be particularly competent in dealing with the novel challenges that arise where they reside (Chapple *et al.*, 2012). Given this complexity, we focus here only on

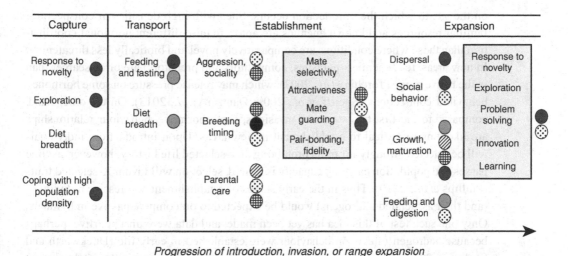

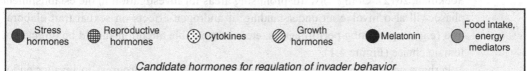

Figure 4.1 Opportunities in invader endocrinology. Putative factors mediating behavioural variation at various stages of an invasion. Tinted circles depict candidate hormones/cytokines; each one is positioned next to a relevant behaviour at different stages.

anthropogenic invasions and what roles we should expect of hormones across the stages of invasions.

Capture and Transport

The first two stages of anthropogenic invasions entail capture and transport of individuals to new areas. During these stages, glucocorticoids (GCs) and other hormones involved in metabolism, feeding and coping with novelty should be important (Figure 4.1). Individuals or species that are prone to find themselves in shipping containers, part of the pet trade or in otherwise close proximity to humans are more likely to found new populations. Further, individuals that can cope well with the low quantity or quality of food that co-occurs with unintended transport should have better chances at colonizing a new area. So far, to our knowledge, there are no data for vertebrates in these stages, although there has been some interest in the endocrinology of translocation of threatened species (Dickens *et al.*, 2009), which may be illuminating when designing studies specific to invaders.

Establishment and Expansion

The latter two stages of a range expansion are more complex than the first two. First, establishment success and the requisite hormone–behaviour links will vary contingent on the site to which the introduction occurs. Sites with high densities of conspecifics, familiar resources and known enemies are apt to require different hormonal coordination than those where conditions are comparatively novel and biotically less threatening. In new areas, too, enemies (parasites, competitors and predators) are often scarcer than in the native range (Torchin *et al.*, 2003), which may release pressure on some hormone–behaviour connections (Sheriff *et al.*, 2009; Dantzer *et al.*, 2013). On the other hand, compared to the first two stages of invasions, some hormone–behaviour relationships are likely more critical to establishment in new areas. Upon introduction, individuals will be selected strongly for traits promoting an *r*-selected life history; however, as time passes and population carrying capacity is neared, selection will favour *K*-selected traits (Phillips *et al.*, 2010). Thus in the early stages of establishment, aggressive individuals (and their associated androgens) would be expected to outcompete passive individuals. Only one such test of this idea has yet been made, and data were unsupportive, perhaps because androgen effects on behaviour were established in early life (Duckworth and Sockman, 2012). Going forward, promising areas for investigation in the establishment phase will also involve an understanding of androgen effects on sexual trait elaboration (e.g. song, mating performances, etc.); comparable attention should be directed at female choice (Figure 4.1).

Perhaps one of the areas with greatest promise for rapid progress in invader endocrinology is the coordination of breeding post-transport. This adjustment to the duration of breeding events is routinely observed in successful expansions (Partecke *et al.*, 2004; Atwell *et al.*, 2014). Also, the endocrinology of opportunistic breeding (Hahn *et al.*, 1995; Hau *et al.*, 2004), the timing of breeding generally (MacDougall-Shackleton *et al.*, 2014) and the breeding endocrinology of domesticated species, organisms selectively bred to mature rapidly and breed prolifically at high densities, are richly studied areas. Moreover, in house finches (*Carpodacus mexicanus*), range expansion success may be tied to endocrine adjustments to oogenesis (Badyaev, 2013b) although data are as yet unavailable.

The last stage of invasions, the expansion phase, is so far the most studied in terms of hormone–behaviour relationships. Figure 4.2, for example, summarizes key findings from two of the most studied avian systems. The hormone–behaviour relationships about which we know most in invasions involve GCs and their associations with boldness, response to novelty, exploratory tendencies and other traits associated with coping with adversity in new areas (Table 4.1). To date, there has been little consistency in the directionality of relationships; sometimes GCs are positively related to these behaviours and, other times, the patterns are reversed. As above, these discrepancies might have something to do with the sites being invaded (i.e. natural habitats versus cities; Crespi *et al.*, 2013) as well as distinct study designs. Few studies so far involve multiple populations of different ages; most involve two-site comparisons, which are always difficult to interpret (Garland and Adolph, 1994). In all cases, examples are correlative, and even the correlative work has been at the level of populations. Correlations between behaviours and corticosterone within individuals are presently lacking at any invasion stage. For androgens and reproductive behaviours, there is some evidence of expected individual covariation (Atwell *et al.*, 2014). As with androgens and establishment though, this absence of GC–behaviour links might be due to organizational effects of

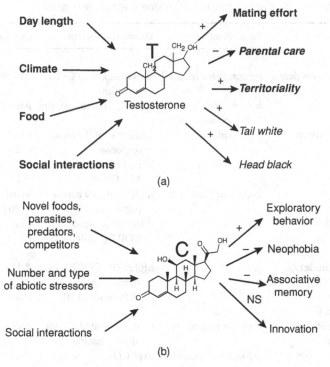

Figure 4.2 (a) Environmentally sensitive hormones can have pleiotropic effects on suites of traits, including positive (plus signs) and negative (minus sign) modulation of social behaviour and plumage, including prior experimental (boldface) and correlational (italics) results from dark-eyed juncos. Such hormonal signals may shape multi-trait responses when populations encounter novel or changing environments. Reprinted with permission from Atwell *et al.*, *Am Nat*, 2014. (b) Relationships between environmental conditions in novel range and behaviours relevant to range expansion in Kenyan house sparrows as mediated by corticosterone (C) released in response to restraint. Symbols and arrows reflect same criteria as in (a).

hormones on behaviours. If true, to determine directly how hormones affect any stage of range expansion, it might be imperative to manipulate them in early life. In side-blotched lizards (*Uta stansburiana*), home range size and activity were impacted by GC implants in adulthood (Denardo and Sinervo, 1994), but in other species, GC effects on behaviour occurred via treatment of pregnant individuals (common lizards; *Lacerta lacerta*); juvenile dispersal distance increased when mothers were treated with corticosterone (Cote *et al.*, 2006).

A final observation about Figure 4.1 and Table 4.1 is that they include cytokines as hormones, which is a non-traditional approach. Our motivation to include these elements is not to convince readers that all cytokines should be considered hormones. We simply emphasize that the regulators of the immune system warrant investigation, especially because variation in immune function affects introduction and invasion success (Lee and Klasing, 2004; Martin *et al.*, 2010b, 2015). Given the extensive effects of some cytokines, such as interleukin-1β, IL-6, TNF-α (Adelman and Martin, 2009) on sickness behaviour, we think cytokine regulation warrants particular attention (Lee *et al.*, 2005;

Table 4.1 Examples of variation in endocrine and immune regulation among vertebrate invaders

Species	Invasion context	Focal traits	Hormonal pattern	Citations
Endocrine				
Junco hyemalis	colonist vs. resident	parental care reduced	lower peak T to GnRH injection	Atwell *et al.*, 2014
	(urban/native)	bolder, more exploratory	lower CORT to restraint	Atwell *et al.*, 2012
Passer domesticus	range expansion	high exploration	smaller CORT stress response	Liebl and Martin, 2012
		low neophobia	higher hippocampal MR/GR ratio	Liebl and Martin, 2013, 2014
		endocrine plasticity	high plasticity across populations	Martin and Liebl, 2014
Passer montanus	range (altitude) expansion	n/a	different seasonal CORT profile	Li *et al.*, 2008
Zonotrichia leucophrys	range (altitude) expansion	n/a	higher BL CORT and stress response	Addis *et al.*, 2011
Sialia mexicanus	colonizers vs residents	n/a	no difference in BL T	Duckworth and Sockman, 2012
Parrots, various	wild-caught and captive	invasion propensity	CORT elevated longer in wild-caught	Cabezas *et al.*, 2013
Rhinella marina	range expansion	dessication resistance	expt CORT = more water loss and mortality	Jessop *et al.*, 2013
	range-edge animals	CORT effects on immune	positive association w/ oxidative burst	Graham *et al.*, 2012
Boiga irregularis	invasive: captive-wild,		wild: elevated BL CORT	Moore *et al.*, 2005
	novel habitat		low estradiol and progesterone	
Immune				
Passer domesticus	range expansion	coping with bacterial infection	higher TLR expression; no IL-6 pattern	Martin *et al.*, 2014
		stress–immune interactions	high covariation among individuals	Martin *et al.*, 2015
Carpodacus mexicanus	range expansion	novel infection in new range	altered splenic immune gene expression	Bonneaud *et al.*, 2011, 2012
			more pathology and inflammatory cytokines	Adelman *et al.*, 2013

Abbreviations: CORT = corticosterone; T = testosterone; GnRH = gonadotrophin releasing hormone; BL = baseline; TLR = Toll-like receptor; IL = interleukin.

Lee *et al.*, 2006; Martin *et al.*, 2010a). As is apparent from Table 4.1 though, there is no consensus yet among the few studies to have evaluated this idea.

How Hormone Regulation Impacts Invasiveness and Range Expansions

Table 4.1 highlights what little we know about hormones and invasions to date. Clearly, given the paucity of data, we have only just begun asking whether variation in organismal traits, not just demography, can influence the outcomes of invasions (Chapple *et al.*, 2012). One type of invasion about which we know quite a bit occurs in cities. As that urbanization literature has been reviewed extensively before (Evans *et al.*, 2010; Bonier, 2012; Martin and Boruta, 2014) and represents but one form of invasions, we do not discuss it here. Most other literature on invasions involves songbirds, yet few of those have included similar enough species in similar enough contexts (i.e. habitats invaded, stage of invasion) to draw conclusions. Indeed, only one system has entailed more than a few invasive vertebrate populations: the invasion of Kenya by the house sparrow (*Passer domesticus*) (Figure 4.3).

In that system, range expansion seems mediated by a constellation of physiological and behavioural adjustments. At range edges, house sparrows release high levels of glucocorticoids (GCs) in response to restraint and extensively explore new areas (Liebl and Martin, 2012), are less fearful of novelty (Liebl and Martin, 2014), have better associative memory and are less innovative (Liebl and Martin, unpublished) than birds from the site of initial introduction (Mombasa, ~1950). Whereas GC–behaviour relationships are strong at the population level, there is little indication of linkages at the individual level. Perhaps GC effects in the invasion of Kenya come predominantly via developmental effects in early life (Love *et al.*, 2013; Martin and Liebl, 2014). Selection seems to have sorted populations (based on neutral genetic markers) in spite of extensive genetic admixture (Schrey *et al.*, 2014). However, the recency of the invasion, the continued small size of many populations, and the high epigenetic variation observed within populations (and negative covariation with population genetic variation; Liebl *et al.*, 2013) suggest a role for developmental plasticity in Kenya. Developmental plasticity, mediated by GCs, could have positive or negative effects on continued range expansion. The immune systems of range-edge sparrows, for example, are quite different than birds from the core, perhaps because responses to parasites have to be adjusted in light of GC effects on behaviour (Martin *et al.*, 2014). This argument was posed to explain why genes for receptors involved in microbial surveillance (*Toll-like receptors 2* and *4*) and GC metabolism (*glucocorticoid receptor*) strongly covary among individual sparrows from multiple sites (Martin *et al.*, 2015).

New Directions in Invader Endocrinology

In addition to obvious next steps, such as (i) manipulation of hormones, experimental introductions and subsequent tracking of individuals, and (ii) comparative studies

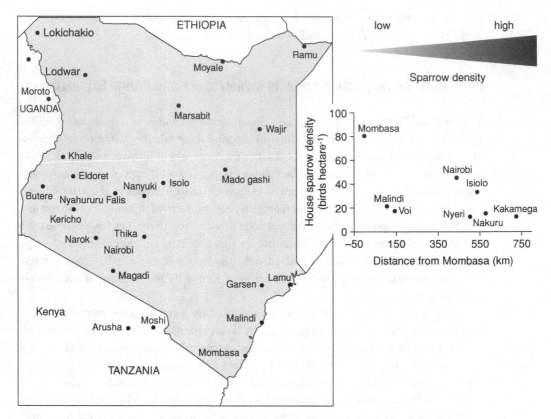

Figure 4.3 Current distribution of house sparrows in Kenya. Map depicts relative comparisons from cursory summary in summer (2010). Inset figure depicts density at several of the most intensively studied sites determined from fixed-radius point counts. Reprinted with permission from Martin *et al.*, *Proc Roy Soc B Bio*, 2014.

of hormone regulation in native and introduced species with known colonization histories, how else might our understanding of invader endocrinology progress? As we emphasized earlier, a very promising direction might entail a better appreciation of the main function of many hormones: to enable stability via change (Woods and Wilson, 2014). In dark-eyed juncos (*Junco hyemalis*), for example, the effects of gonadotropin-releasing hormone (GnRH) on testosterone (T) differed more between native and invading populations than baseline T levels (Atwell *et al.*, 2014). In Kenyan house sparrows, the oldest population was able to converge on the same GC regulatory profile as the youngest population after just a week under the same conditions (Martin and Liebl, 2014). Perhaps then, hormones mediate invasion success more via their regulatory plasticity than measurements of hormone levels in single blood samples can convey. If so, physiologically more plastic individuals should comprise more invasive populations (or species); in two invasive plants (Higgins and Richardson 2014), this idea was supported although the role of hormones was not considered. A recent study from the Kenyan house sparrow system revealed that plasticity in corticosterone regulation might have been integral to range expansion (Figure 4.4). If so, the hope for simple biomarkers of

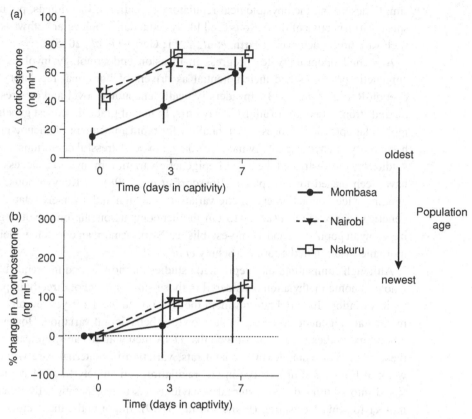

Figure 4.4 Comparisons of corticosterone response to a restraint stressor over a week of captivity in Kenyan house sparrows: (a) raw change in corticosterone (delta: 30-min post-capture – concentrations at capture); (b) % change relative to concentration at capture from the wild. Mombasa was the first colonized site (~1950) followed by Nairobi (~1980s) then Nakuru (~1990s); the latter two dates are presently speculative. Bars are means ±1 SE. Reprinted with permission from Martin and Liebl, *GCE*, 2014.

invasiveness is lost; any blood-borne signal of invasiveness will be difficult to capture in single samples.

We predict that the most successful invaders, be they individuals or species, will be those better able to up- and down-regulate circulating hormone levels to salient stimuli, adjust hormone receptor expression among tissues, or even couple and uncouple other processes (e.g. immune, growth, reproductive) to hormones (Hau *et al.*, 2008; Angelier and Wingfield, 2013). To test this hypothesis, one should be mindful of the extensive regulatory diversity involved in organismal homeostasis (Young and Crews, 1995; Lynn, 2008; Martin, 2009). In other words, what we probably most need to know about hormones and behaviour in invasions is how much variation individuals can achieve both in terms of circulating levels in response to various novel and evolutionarily unfamiliar stimuli (Ghalambor *et al.*, 2007), as well as the impacts of such variation on other traits of interest (Martin and Cohen, 2014). At the organismal level, we might benefit by simply describing the physiological regulatory potential of individuals, populations or species irrespective of the various (and likely redundant) molecular pathways by which such variation is achieved (Martin *et al.*, 2011; Gervasi *et al.*, 2015).

A second opportunity for progress in invasion endocrinology involves molecular epigenetic processes and their potential as drivers of hormonal regulatory plasticity (Ledón-Rettig *et al.*, 2013). Invaders typically encounter novel and/or stressful stimuli and strong pressures to adjust behaviours. In lieu of gradual, modest genetic change, molecular epigenetic processes might allow for rapid adjustment of phenotype and even heritability of environmentally induced change to new/stressful conditions. To date, few studies have investigated the role of epigenetic variation in invasion success, and most have been biased towards plants (Bossdorf *et al.*, 2008). In Kenyan house sparrows, genome-wide population epigenetic variation was high and inversely related to neutral genetic variation (Liebl *et al.*, 2013); in another comparison, this one involving an invading Kenyan population and a long-established North American one, a few loci exhibited more methylation in the former (Schrey *et al.*, 2012).

Although compelling, these epigenetic studies provide limited inference because the whole-genome methylation considered in those studies may not directly impact phenotypic variation. In vertebrate invaders though, no studies have yet revealed a role for molecular epigenetic processes in hormonal or behavioural variation. However, several laboratory studies have demonstrated that DNA methylation can significantly influence these traits. For example, in neonatal rats, variation in maternal care within the first week of life caused differences in GC regulation and anxiety-like behaviour that persisted into adulthood, and maternal behaviour was transgenerationally inherited from mother to female offspring (Francis *et al.*, 1999). Importantly, these environmentally induced changes were due to differential DNA methylation within the promoter region of the glucocorticoid receptor gene (Weaver *et al.*, 2004). With the advent of new technologies, used in systems such as the Kenyan house sparrow, one could probe the role of methylation for particular loci or entire transcriptomes and how these alterations impact hormonal regulation and vertebrate invasions within and across generations.

Conclusions

Understanding hormone–behaviour relationships in invaders could reveal new ways to control pests (Dell'Omo and Palmery, 2002); it is surprisingly rare that endocrinological interventions are used to stop or slow invasions (Krause *et al.*, 2014). Development of invader endocrinology might also help us manage wildlife populations with different sensitivities and responses to anthropogenic effects (Martin and Boruta, 2014). Further, more knowledge about the endocrinology of weedy species could improve our ability to bolster threatened populations through captive breeding programmes. To achieve such outcomes though, we need more controlled experiments (in spite of the challenges and ethical implications of introductions of non-natives to new areas), as almost all data now are correlational. We also need greater coverage of the early stages of invasion and work on other hormone–behaviour links besides the current favourite: GCs and behaviours directed at coping with stressors. One intriguing line of research would entail the coordination of perception centres in the brain with the hormones regulating social behaviours. Fire ants (*Solenopsis invicta*), for instance, are exceptional invaders (higher nest densities) because some colonies are less aggressive towards each other due to a lack of strong nest mate recognition/discrimination ability (Holway *et al.*, 1998). We know nothing about such brain–behaviour links in vertebrates, although the endocrinology of sociality is fairly well-studied in domesticated rodents and birds. In the end, the opportunities in invader endocrinology are many, and in some sense, we might modestly benefit from the problem we created. Invasive species have forced us to rethink how we manage natural habitats, conduct agricultural processes, and transport goods across the globe. However, they also present us with opportunities that may influence dramatically how we understand key biological processes (Bossdorf *et al.*, 2008; Ledón-Rettig *et al.*, 2013).

References

Addis, E.A., Davis, J.E., Miner, B.E. and Wingfield, J.C. (2011). Variation in circulating corticosterone levels is associated with altitudinal range expansion in a passerine bird. *Oecologia*, 167(2), 369–378.

Adelman, J. and Martin, L. (2009). Vertebrate sickness behavior: an adaptive and integrated neuroendocrine immune response. *Integrative and Comparative Biology*, 49, 202–214.

Adelman, J.S., Kirkpatrick, L., Grodio, J.L. and Hawley, D.M. (2013). House finch populations differ in early inflammatory signalling and pathogen tolerance at the peak of *Mycoplasma gallisepticum* infection. *The American Naturalist*, 181(5), 674–689.

Angelier, F. and Wingfield, J.C. (2013). Importance of the glucocorticoid stress response in a changing world: theory, hypotheses and perspectives. *General and Comparative Endocrinology*, 190, 118–128.

Atwell, J.W., Cardoso, G.C., Whittaker, D.J., *et al.* (2012). Boldness behavior and stress physiology in a novel urban environment suggest rapid correlated evolutionary adaptation. *Behavioral Ecology*, doi: 10.1093/beheco/ars059.

Atwell, J.W., Cardoso, G.C., Whittaker, D.J., Price, T.D. and Ketterson, E.D. (2014). Hormonal, behavioral, and life-history traits exhibit correlated shifts in relation to population establishment in a novel environment. *The American Naturalist*, 184, E147–160.

Badyaev, A. (2013a). Reconciling innovation and adaptation during recurrent colonization of urban environments: molecular, genetic, and developmental bases. *Avian Urban Ecology: Behavioural and Physiological Adaptations*, 155.

Badyaev, A.V. (2013b). 'Homeostatic hitchhiking': a mechanism for the evolutionary retention of complex adaptations. *Integrative and Comparative Biology*, ict084.

Badyaev, A.V. (2014). Epigenetic resolution of the 'curse of complexity 'in adaptive evolution of complex traits. *The Journal of Physiology*, 592, 2251–2260.

Badyaev, A.V. and Uller, T. (2009). Parental effects in ecology and evolution: mechanisms, processes and implications. *Philosophical Transactions of the Royal Society B: Biological Sciences*, 364, 1169–1177.

Bonier, F. (2012). Hormones in the city: endocrine ecology of urban birds. *Hormones and Behavior*, 61, 763–772.

Bonneaud, C., Balenger, S.L., Russell, A.F., *et al.* (2011). Rapid evolution of disease resistance is accompanied by functional changes in gene expression in a wild bird. *Proceedings of the National Academy of Sciences, USA*, 108(19), 7866–7871.

Bonneaud, C., Balenger, S.L., Hill, G.E. and Russell, A.F. (2012). Experimental evidence for distinct costs of pathogenesis and immunity against a natural pathogen in a wild bird. *Molecular Ecology*, 21(19), 4787–4796.

Bossdorf, O., Richards, C.L. and Pigliucci, M. (2008). Epigenetics for ecologists. *Ecology Letters*, 11, 106–115.

Braby, C.E. and Somero, G.N. (2006). Following the heart: temperature and salinity effects on heart rate in native and invasive species of blue mussels (genus *Mytilus*). *Journal of Experimental Biology*, 209, 2554–2566.

Bradley, C.A. and Altizer, S. (2007). Urbanization and the ecology of wildlife diseases. *Trends in Ecology and Evolution*, 22, 95–102.

Cabezas, S., Carette, M., Tella, J.L., Marchant, T.A. and Bortolotti, G.R. (2013). Differences in acute stress responses between wild-caught and captive-bred birds: a physiological mechanism contributing to current avian invasions? *Biological Invasions*, 15(3), 521–527.

Chapple, D.G., Simmonds, S.M. and Wong, B. (2012). Can behavioral and personality traits influence the success of unintentional species introductions? *Trends in Ecology and Evolution*, 27, 57–64.

Chown, S.L., Slabber, S., McGeoch, M.A., Janion, C. and Leinaas, H.P. (2007). Phenotypic plasticity mediates climate change responses among invasive and indigenous arthropods. *Proceedings of the Royal Society B: Biological Sciences*, 274, 2531–2537.

Cohen, A.A., Martin, L.B., Wingfield, J.C., McWilliams, S.R. and Dunne, J.A. (2012). Physiological regulatory networks: ecological roles and evolutionary constraints. *Trends in Ecology and Evolution*, 27(8), 428–435.

Colautti, R.I. and MacIsaac, H.J. (2004). A neutral terminology to define 'invasive' species. *Diversity and Distributions*, 10, 135–141.

Cote, J., Clobert, J., Meylan, S. and Fitze, P.S. (2006). Experimental enhancement of corticosterone levels positively affects subsequent male survival. 49, 320–327.

Crespi, E.J., Williams, T.D., Jessop, T.S. and Delehanty, B. (2013). Life history and the ecology of stress: how do glucocorticoid hormones influence life-history variation in animals? *Functional Ecology*, 27, 93–106.

Dantzer, B., Newman, A.E., Boonstra, R., *et al.* (2013). Density triggers maternal hormones that increase adaptive offspring growth in a wild mammal. *Science*, 340, 1215–1217.

Dell'Omo, G. and Palmery, M. (2002). Fertility control in vertebrate pest species. *Contraception*, 65, 273–275.

Denardo, D.F. and Sinervo, B. (1994). Effects of steroid–hormone interaction on activity and home-range size of male lizards. *Hormones and Behavior*, 28, 273–287.

Denver, R.J. (2009). Stress hormones mediate environment–genotype interactions during amphibian development. *General and Comparative Endocrinology*, 164, 20–31.

Dickens, M.J., Delehanty, D.J. and Romero, L.M. (2009). Stress and translocation: alterations in the stress physiology of translocated birds. *Proceedings of the Royal Society B: Biological Sciences*, 276, 2051–2056.

Duckworth, R.A. and Sockman, K.W. (2012). Proximate mechanisms of behavioural inflexibility: implications for the evolution of personality traits. *Functional Ecology*, 26, 559–566.

Dufty, A.M., Clobert, J. and Møller, A.P. (2002). Hormones, developmental plasticity and adaptation. *Trends in Ecology and Evolution*, 17, 190–196.

Evans, K.L., Hatchwell, B.J., Parnell, M. and Gaston, K.J. (2010). A conceptual framework for the colonisation of urban areas: the blackbird *Turdus merula* as a case study. *Biological Reviews*, 85, 643–667.

Flatt, T. and Heyland, A. (2011). *Mechanisms of Life History Evolution: The Genetics and Physiology of Life History Traits and Trade-offs*. Oxford, UK: Oxford University Press.

Francis, D., Diorio, J., Liu, D. and Meaney, M.J. (1999). Nongenomic transmission across generations of maternal behavior and stress responses in the rat. *Science*, 286, 1155–1158.

Garland, T. and Adolph, S.C. (1994). Why not to do 2-species comparative-studies: limitations on inferring adaptation. *Physiological Zoology*, 67, 797–828.

Gervasi, S.S., Civitello, D.J., Kilvitis, H.J. and Martin, L.B. (2015). The context of host competence: a role for plasticity in host-parasite dynamics. *Trends in Parasitology*, 31, 419–425.

Ghalambor, C.K., McKay, J.K., Carroll, S.P. and Reznick, D.N. (2007). Adaptive versus nonadaptive phenotypic plasticity and the potential for contemporary adaptation in new environments. *Functional Ecology*, 21, 394–407.

Graham, S.P., Kelehear, C., Brown, G.P. and Shine, R. (2012). Corticosterone–immune interactions during captive stress in invading Australian cane toads (*Rhinella marina*). *Hormones and Behavior*, 62(2), 146–153.

Groothuis, T.G. and Schwabl, H. (2008). Hormone-mediated maternal effects in birds: mechanisms matter but what do we know of them? *Philosophical Transactions of the Royal Society B: Biological Sciences*, 363, 1647–1661.

Hahn, T.P., Wingfield, J.C., Mullen, R. and Deviche, P.J. (1995). Endocrine bases of spatial and temporal opportunism in Arctic-breeding birds. *American Zoologist*, 35, 259–273.

Hau, M. (2007). Regulation of male traits by testosterone: implications for the evolution of vertebrate life histories. *Bioessays*, 29, 133–144.

Hau, M., Wikelski, M., Gwinner, H. and Gwinner, E. (2004). Timing of reproduction in a Darwin's finch: temporal opportunism under spatial constraints. *Oikos*, 106, 489–500.

Hau, M., Gill, S.A. and Goymann, W. (2008). Tropical field endocrinology: ecology and evolution of testosterone concentrations in male birds. *General and Comparative Endocrinology*, 157, 241–248.

Higgins, S.I. and Richardson, D.M. (2014). Invasive plants have broader physiological niches. *Proceedings of the National Academy of Sciences, USA*, 111, 10610–10614.

Holway, D.A., Suarez, A.V. and Case, T.J. (1998). Loss of intraspecific aggression in the success of a widespread invasive social insect. *Science*, 282, 949–952.

Jessop, T.S., Letnic, M., Webb, J.K. and Dempster, T. (2013). Adrenocortical stress responses influence an invasive vertebrate's fitness in an extreme environment. *Proceedings of the Royal Society B: Biological Sciences*, 280(1768), 20131444.

Juliano, S.A., O'Meara, G.F., Morrill, J.R. and Cutwa, M.M. (2002). Desiccation and thermal tolerance of eggs and the coexistence of competing mosquitoes. *Oecologia*, 130, 458–469.

Kearney, M. and Porter, W. (2009). Mechanistic niche modelling: combining physiological and spatial data to predict species' ranges. *Ecology Letters*, 12, 334–350.

Kimball, M.E., Miller, J.M., Whitfield, P.E. and Hare, J.A. (2004). Thermal tolerance and potential distribution of invasive lionfish (*Pterois volitans/miles* complex) on the east coast of the United States. *Marine Ecology Progress Series*, 283, 269–278.

Kolbe, J.J., Kearney, M. and Shine, R. (2010). Modeling the consequences of thermal trait variation for the cane toad invasion of Australia. *Ecological Applications*, 20, 2273–2285.

Krause, S.K., Kelt, D.A., Gionfriddo, J.P. and Van Vuren, D.H. (2014). Efficacy and health effects of a wildlife immunocontraceptive vaccine on fox squirrels. *The Journal of Wildlife Management*, 78, 12–23.

Ledón-Rettig, C.C., Richards, C.L. and Martin, L.B. (2013). Epigenetics for behavioral ecologists. *Behavioral Ecology*, 24, 311–324.

Lee, K.A. and Klasing, K.C. (2004). A role for immunology in invasion biology. *Trends in Ecology and Evolution*, 19, 523–529.

Lee, K.A., Martin, L.B. and Wikelski, M.C. (2005). Responding to inflammatory challenges is less costly for a successful avian invader, the house sparrow (*Passer domesticus*), than its less-invasive congener. *Oecologia*, 145, 244–251.

Lee, K.A., Martin, L.B., Hasselquist, D., Ricklefs, R.E. and Wikelski, M. (2006). Contrasting adaptive immune defenses and blood parasite prevalence in closely related *Passer* species. *Oecologia*, 150, 383–392.

Lema, S.C. and Kitano, J. (2013). Hormones and phenotypic plasticity: Implications for the evolution of integrated adaptive phenotypes. *Current Zoology*, 59, 506–525.

Li, D.M., Wang, G., Wingfield, J.C., *et al.* (2008). Seasonal changes in adrenocortical responses to acute stress in Eurasian tree sparrow (*Passer montanus*) on the Tibetan Plateau: comparison with house sparrow (*P. domesticus*) in North America and with the migratory *P. domesticus* in Qinghai Province. *General and Comparative Endocrinology*, 158(1), 47–53.

Liebl, A.L. and Martin, L.B. (2012). Exploratory behaviour and stressor hyper-responsiveness facilitate range expansion of an introduced songbird. *Proceedings of the Royal Society B: Biological Sciences*, 279, 4375–4381.

Liebl, A.L. and Martin, L.B. (2013). Stress hormone receptors change as range expansion progresses in house sparrows. *Biology Letters*, 9(3).

Liebl, A.L. and Martin, L.B. (2014). Living on the edge: range edge birds consume novel foods sooner than established ones. *Behavioral Ecology*, 25, 1089–1096.

Liebl, A.L., Schrey, A.W., Richards, C.L. and Martin, L.B. (2013). Patterns of DNA methylation throughout a range expansion of an introduced songbird. *Integrative and Comparative Biology*, 53, 351–358.

Love, O.P., McGowan, P.O. and Sheriff, M.J. (2013). Maternal adversity and ecological stressors in natural populations: the role of stress axis programming in individuals, with implications for populations and communities. *Functional Ecology*, 27, 81–92.

Lynn, S.E. (2008). Behavioral insensitivity to testosterone: why and how does testosterone alter paternal and aggressive behavior in some avian species but not others? *General and Comparative Endocrinology*, 157, 233–240.

MacDougall-Shackleton, S.A., Watts, H.E. and Hahn, T.P. (2014). Biological timekeeping: individual variation, performance, and fitness. In *Integrative Organismal Biology*, ed. Martin, L.B., Ghalambor, C.K. and Woods, H.A. Hoboken, NJ: Wiley & Sons, Inc., pp. 235–255.

Martin, L.B. (2009). Stress and immunity in wild vertebrates: timing is everything. *General and Comparative Endocrinology*, 163, 70–76.

Martin, L.B. and Boruta, M. (2014). The impacts of urbanization on avian disease transmission and emergence. In *Avian Urban Ecology*, ed. Gil, D. and Brumm, H. Oxford, UK: Oxford University Press.

Martin, L.B. and Cohen, A. (2014). Physiological regulatory networks: the orchestra of life? In *Integrative Organismal Biology*, ed. Martin, L.B., Ghalambor, C.K. and Woods, H.A. Hoboken, NJ: Wiley & Sons, Inc., pp. 137–152.

Martin, L.B. and Liebl, A.L. (2014). Physiological flexibility in an avian range expansion. *General and Comparative Endocrinology*, 206, 227–234.

Martin, L.B., Alam, J.L., Imboma, T. and Liebl, A.L. (2010a). Variation in inflammation as a correlate of range expansion in Kenyan house sparrows. *Oecologia*, 164, 339–347.

Martin, L.B., Hopkins, W.A., Mydlarz, L. and Rohr, J.R. (2010b). The effects of anthropogenic global changes on immune functions and disease resistance. *Annals of the New York Academy of Sciences*, 1195, 129–148.

Martin, L.B., Liebl, A.L., Trotter, J.H., *et al.* (2011). Integrator networks: illuminating the black box linking genotype and phenotype. *Integrative and Comparative Biology*, 51, 514–527.

Martin, L.B., Coon, C.A.C., Liebl, A.L. and Schrey, A.W. (2014). Surveillance for microbes and range expansion in house sparrows. *Proceedings of the Royal Society B: Biological Sciences*, 281.

Martin, L.B., Liebl, A.L. and Kilvitis, H.J. (2015). Covariation in stress and immune gene expression in a range expanding bird. *General and Comparative Endocrinology*, 211, 14–19.

McGlothlin, J.W. and Ketterson, E.D. (2008). Hormone-mediated suites as adaptations and evolutionary constraints. *Philosophical Transactions of the Royal Society B: Biological Sciences*, 363, 1611–1620.

Medzhitov, R. (2008). Origin and physiological roles of inflammation. *Nature*, 454, 428–435.

Moore, I.T., Greene, M.J., Lerner, D.T., *et al.* (2005). Physiological evidence for reproductive suppression in the introduced population of brown tree snakes (*Boiga irregularis*) on Guam. *Biological Conservation*, 121(1), 91–98.

Noble, D., Jablonka, E., Joyner, M., *et al.* (2014). The integration of evolutionary biology with physiological science. *Journal of Physiology*, 592, 2237–2244.

Partecke, J., Van't Hof, T.J. and Gwinner, E. (2004). Differences in the timing of reproduction between urban and forest European blackbirds (*Turdus merula*): result of phenotypic flexibility or genetic differences? *Proceedings of the Royal Society B*, 271, 1995–2001.

Phillips, B.L., Brown, G.P. and Shine, R. (2010). Life-history evolution in range-shifting populations. *Ecology*, 91, 1617–1627.

Ricklefs, R.E. and Wikelski, M. (2002). The physiology/life-history nexus. *Trends in Ecology and Evolution*, 17, 462–468.

Schrey, A.W., Coon, C.A.C., Grispo, M.T., *et al.* (2012). Epigenetic variation may compensate for decreased genetic variation with introductions: a case study using house sparrows (*Passer domesticus*) on two continents. *Genetics Research International*, 2012, 7.

Schrey, A.W., Liebl, A.L., Richards, C.L. and Martin, L.B. (2014). Range expansion of house sparrows (*Passer domesticus*) in Kenya: evidence of genetic admixture and human-mediated dispersal. *Journal of Heredity*, 105, 60–69.

Sheriff, M.J., Krebs, C.J. and Boonstra, R. (2009). The sensitive hare: sublethal effects of predator stress on reproduction in snowshoe hares. *Journal of Animal Ecology*, 78, 1249–1258.

Sinervo, B. and Calsbeek, R. (2003). Physiological epistasis, ontogenetic conflict and natural selection on physiology and life history. *Integrative and Comparative Biology*, 43, 419–430.

Torchin, M.E., Lafferty, K.D., Dobson, A.P., McKenzie, V.J. and Kuris, A.M. (2003). Introduced species and their missing parasites. *Nature*, 421, 628–630.

Weaver, I.C.G., Cervoni, N., Champagne, F.A., *et al.* (2004). Epigenetic programming by maternal behavior. *Nature Neuroscience*, 7, 847–854.

West-Eberhard, M. (2003). *Developmental Plasticity and Evolution*. Oxford, UK: Oxford University Press.

Woods, H.A. and Wilson, J.K. (2014). An elephant in the fog: unifying concepts of physiological stasis and change. In *Integrative Organismal Biology*, ed. Martin, L.B., Ghalambor, C.K. and Woods, H.A. Hoboken, NJ: Wiley & Sons, Inc., pp. 119–136.

Young, L.J. and Crews, D. (1995). Comparative neuroendocrinology of steroid receptor gene expression and regulation: relationship to physiology and behavior. *Trends in Endocrinology and Metabolism*, 6, 317–323.

5 Life History, Behaviour and Invasion Success

Daniel Sol and Joan Maspons

At the most fundamental level, biological invasions hinge on the fate of individuals surviving and reproducing in novel environments. If individuals are able to reproduce at a higher rate than they die, the invader will increase in numbers and can eventually become established and spread; if the balance is negative, however, the population will decrease over time and end up extinct. Because the rates of birth and death are ultimately determined by how organisms allocate their limited time and energy to reproduction and survival (Stearns, 1992), life history theory has long been deemed essential to understanding the success of invaders (Lewontin, 1965).

Surprisingly, however, life history theory has achieved little success in predicting the outcome of organisms' introduction (Blackburn *et al.*, 2009; Sol *et al.*, 2012b). One reason is the excessive focus on the small population paradigm, implicitly assuming that demographic stochasticity is the main driver of extinction in introduced populations. This has led to the widespread belief that successful invaders are characterized by high fecundity that reduces the exposure to demographic stochasticity by enhancing population growth (Moulton *et al.*, 1986). Nevertheless, for an invader coming from a distant region, unfamiliarity and insufficient adaptation to the new resources, enemies and other hazards are also likely to increase the risk of extinction by negative population growth. Life history theory offers additional mechanisms that may help mitigate these effects, such as bet-hedging (Starrfelt and Kokko, 2012) and the storage effect (Warner and Chesson, 1985; Caceres, 1997). As we argue in this chapter, if we want to fully understand how life history affects animal invasions, we also need to explicitly consider the role of behaviour.

Our argument for the need to better integrate behaviour into life history theory is founded upon three main principles. The first is the fact that behavioural responses are part of the adaptive machinery of animals to cope with uncertainties and evolutionary disequilibria related to novel environments. While the idea is not new (Mayr, 1965), recent theoretical and empirical advances provide a strong foundation for moving forward. The second argument is the growing evidence that behaviour affects and is affected by life history, which implies that both may be part of a same adaptive strategy. Thus, when we examine how life history affects invasion success we are considering not only life history mechanisms but also mechanisms related to behavioural responses to

Biological Invasions and Animal Behaviour, eds J.S. Weis and D. Sol. Published by Cambridge University Press. © Cambridge University Press 2016.

novel environments. The last argument is that behaviour mediates some life history mechanisms of response to novel environments, particularly those related to environmental uncertainty and adaptive mismatch. By clearly delineating these mechanisms, we can better infer when it is necessary to consider behaviour.

Altogether, the above principles create a new way to understand how life history influences population growth in novel environments, potentially contributing to a more predictive theory. Such a theory is necessary not only to better understand the invasion process, but also as a basis to help prevent and mitigate the ecological and economic impact of biological invasions (Kolar and Lodge, 2002; Vall-llosera and Sol, 2009; Leung *et al.*, 2012). Because invaders represent unique opportunities to evaluate how organisms cope with sudden environmental changes, the new theory should also be of great importance in predicting extinction risk associated with human-induced rapid environmental changes like habitat destruction and climate change (Sæther and Bakke, 2000; Sih *et al.*, 2011).

Life History Mechanisms Influencing Invasion Success

Before trying to integrate behaviour into life history theory, we will describe the variety of mechanisms by which life history can directly influence invasion success, highlighting the problems of classic theory and discussing mechanisms that extend the framework to scenarios beyond the small population paradigm.

Limitations of Classic Theory

The debate regarding how suites of life history characteristics affect invasion success has been dominated by two opposed perspectives, both related to the so-called fast–slow continuum of life history variation (Figures 5.1 and 5.2). The most popular is the population growth hypothesis, proposed by Lewontin (1965) more than 45 years ago. This hypothesis predicts that successful invaders should be characterized by a fast-lived strategy in which fecundity is prioritized over survival (Figure 5.1); such a strategy allows rapid population growth, thereby reducing the period at which the founder population remains small and hence more vulnerable to extinction by demographic stochasticity (Moulton *et al.*, 1986; Figure 5.3). Demographic stochasticity arises because the discrete and probabilistic nature of the birth and death of each individual leads to random fluctuations in population size, which in a small population can largely increase the risk of accidental extinction.

The opposite view to the population growth perspective is the hypothesis of the life history buffer (Pimm, 1991; Saether *et al.*, 2004; Forcada *et al.*, 2008). This hypothesis argues that successful invaders should be characterized by slow-lived strategies because their populations are less prone to large stochastic population fluctuations that can accidentally result in extinction. The mechanism is again the discrete and probabilistic nature of the birth and death of individuals. Thus, although a slow-lived species will recover more slowly from a small population size than a fast-lived species, this

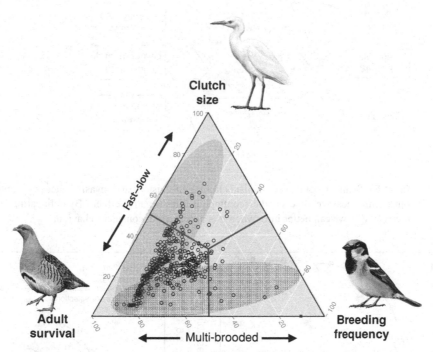

Figure 5.1 Animals exhibit an enormous diversity of life history strategies, which are thought to reflect trade-offs that change the optimal combination of rate of reproduction, age at maturity and longevity with respect to their environments (Stearns, 1992; Roff, 2002). The most well-known trade-off is that between reproduction and survival (Stearns, 1983), which results in a 'fast–slow' continuum ranging from organisms with high reproductive rates and low survival prospects (fast lived), at one end, to those that produce few offspring but survive well (slow lived), at the other. While allocating all reproductive effort in a few attempts allows for rapid population growth under favourable conditions, this strategy is risky when conditions are unfavourable because it reduces the chances of breeding again in the future if there is a reproductive failure. A way to maintain a high productivity with a lower cost of losing a breeding attempt is to have multiple broods per year, which defines a second important axis of life history variation.

advantage may be in part countered by a lower risk of extinction through population fluctuations.

Although the population growth and the life history buffer hypotheses are both based on solid demographic theory (Saether *et al.*, 2004), attempts to pinpoint their relative importance on empirical grounds have failed to draw clear conclusions (Blackburn *et al.*, 2009; Sol *et al.*, 2012b). As already advanced, one limitation of these theories is that they assume that demographic stochasticity is the primary source of extinction of introduced populations (Sol *et al.*, 2012b). Demographic stochasticity is certainly important in the invasion process, as suggested by demographic models and by the fact that the likelihood of establishment in introductions increases with the number of individuals released (Cassey *et al.*, 2004; Lockwood *et al.*, 2005). However, if in the new environment individuals cannot find appropriate resources and/or tolerate the physical (e.g. extreme climate) and biotic (e.g. high pressure from enemies) adversities, then

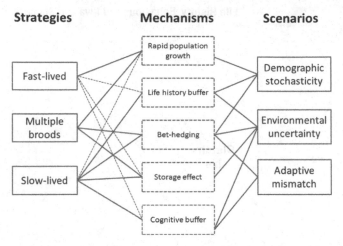

Figure 5.2 Summary of the mechanisms linking life history and invasion success. Continuous lines reflect positive effects and discontinuous lines negative effects. By delineating these mechanisms, we can better infer when it is necessary to consider behaviour.

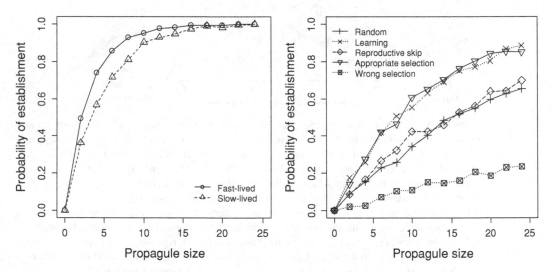

Figure 5.3 Simulations from an agent-based model (Maspons and Sol, unpublished) showing the role of behavioural plasticity in how life history affects establishment success in novel, stochastic environments. The model starts with the introduction of a certain number of adults (propagule size) of a species with a certain life history in a novel environment. The dynamics of each starting population is monitored over time, and the outcome is the proportion of simulations per propagule size that are successful. In the left graph, there is no adaptive mismatch and fast-lived strategies outperform slow-lived strategies when population size is small. In the right graph, the novel region has two habitats, one good and another bad, as reflected by differences in adult mortality. This is used to simulate the existence of an adaptive mismatch, which leads to a decrease in population fitness represented by the higher probability of extinction. Through plastic behavioural adjustments individuals may improve their fitness, although only slow-lived animals generally benefit from it. Population fitness improvements may happen if individuals (i) have an innate preference for one of the habitats (activational plasticity); (ii) breed in a different habitat after a reproductive failure (learning); and (iii) can skip a breeding attempt, thereby enhancing survivorship and increasing the time available for exploration. In contrast, an ecological trap can occur if a lack of plasticity in the use of habitat selection cues leads individuals to settle in the wrong habitat.

the more likely is that they have difficulties in surviving and reproducing and that the population goes extinct by negative population growth irrespective of the degree of demographic stochasticity (Figure 5.3). Because invaders come from distant regions and have had little opportunity to adapt to their new environments, the existence of adaptive mismatching should be pervasive in biological invasions. Indeed, some of the best examples of contemporary evolution involve invasive species (Huey *et al.*, 2000; Reznick and Ghalambor, 2001).

The extent to which an invader is in evolutionary disequilibrium depends on whether there is environmental matching between the regions of origin and introduction. If there is environmental matching, then the fitness decrease should be lower as the species would already possess the adaptations needed to persist there (Williamson, 1996). Yet even under such a scenario, a certain decrease in fitness is generally expected as a result of the uncertainties the invader encounters in the novel environment. Uncertainties are generally associated with environmental stochasticity, but can also arise from a lack of information. For an invader, the ignorance about food sources, enemies and other hazards largely increases the risk of reproductive failure and may affect survival as well. This is evidenced by the low success of species that are translocated or reintroduced within their geographic range (Griffin *et al.*, 2000), a result that cannot be attributed to adaptive mismatches. In African elephants (*Loxodonta africana*), for example, individuals that are translocated to new environments for management or conservation purposes exhibit higher death rates than local elephants, in part due to the lack of familiarity of moved individuals with the new habitat (Pinter-Wollman *et al.*, 2009; Figure 5.4). The existence of uncertainties implies that even when an invader is able to proliferate in the new environment, the rate of population growth should be lower than it would be in a more familiar environment. Such a slowdown in population growth should expose the invaders to demographic stochasticity for a longer period, thereby increasing the probability of extinction by accident.

Extending the Classic Framework

By ignoring the importance of environmental uncertainties and adaptive mismatches, the classical framework dismisses the importance of additional mechanisms through which life history can affect invasion success. These mechanisms include bet-hedging, the storage effect and buffer effects associated with certain adaptations (Figure 5.2).

Bet-hedging. This is a strategy to minimize risks that reduces the temporal variance in fitness at the expense of a lowered arithmetic mean fitness (Stearns and Crandall, 1981; Starrfelt and Kokko, 2012). Perhaps the most widespread form of bet-hedging is to distribute the reproductive effort in many events rather than in a few ones, either through a long reproductive life or through a higher frequency of reproductions per year (Figure 5.1). This reduces the variability of the fitness by the law of large numbers, and hence increases geometric population growth. The advantage of multiple reproductions is exemplified in the invasion of an urbanized environment by dark-eyed juncos, which originated from mountainous regions. In their new environment, reproductive success is

Figure 5.4 In slow-lived species, like African elephants, behaviour is expected to play a central role in assisting individuals in novel environments. When translocated to new reserves for management or conservation, the released elephants exhibit higher death rates than local elephants, in part due to the lack of familiarity of individuals with the new habitat. However, behaviour can improve survival at least in two ways. First, individuals tend to settle in habitats that more resembled their natal ones. Second, over time they also tend to converge in behaviour towards that of the local elephants (presumably through social learning). Such behavioural decisions and adjustments allow reducing adaptive mismatch and environmental uncertainties (photograph: D. Sol).

poor, yet this can be in part compensated by a higher number of broods per season (Yeh and Price, 2004).

Storage effect. In species that give low value to a breeding attempt, the costs of skipping a reproduction are also reduced. This allows individuals to engage in reproductive activities only when conditions are favourable (Williams, 1966), a phenomenon that in community ecology has been referred to as the storage effect (Warner and Chesson, 1985; Caceres, 1997). Delaying or skipping a breeding attempt has been reported in several long-lived species (Lima, 2009), and can be another life history mechanism to deal with novel situations. As an example, Cubaynes *et al.* (2011) reported that red-footed boobies (*Sula sula*) are more likely to skip a breeding event in El-Niño years. This is to be expected if skipping is an adaptive strategy to reduce mortality of adults when the chances of reproducing successfully are low.

Adaptive buffer. Extrinsic mortality differentially affecting juveniles and adults is considered a major driver of life history evolution (Schaffer, 1974; Charlesworth, 1980; Stearns, 1992, 2000; Reznick *et al.*, 2002). A high adult mortality selects for earlier reproduction and higher fecundity so as to reduce the risk of dying without reproducing. Conversely, a low adult mortality selects for a longer reproductive life. This latter strategy can further evolve when animals develop adaptations that reduce variance in adult mortality (Stearns, 2000). During unfavourable periods, these adaptations can be used to survive until conditions improve. The existence of 'buffer adaptations' ensures that survival is high in most years (Morris and Doak, 2004), a requisite for attaining a long reproductive life, and hence may be a mechanism through which life history enhances invasion success. One of the buffer adaptations classically considered in the literature is increased body size, which can provide competitive advantages and protection from enemies. However, evidence that larger species are more successful invaders is contradictory (e.g. Blackburn *et al.*, 2009). Stronger evidence is nonetheless available for behavioural responses, an issue that will be examined in the next sections.

Role of Behaviour in the Response to Novel Environments

Life history theory has largely been developed under the view that organisms are passive subjects of selection. However, behaviour mediates how animals interact with their environment and, by virtue of their plastic nature, can modify the nature of these interactions, shaping the biotic and abiotic pressures that act upon them (Futuyma and Moreno, 1988; Losos *et al.*, 2004). In the next sections, we will discuss the role of behaviour in the response to novel environments, which is essential to better understanding how life history affects invasion success.

In novel environments, animals must decide when and where to breed or what foods are good and what are bad, often with little previous information and lack of specialized adaptations. Behavioural plasticity allows such challenges to be addressed. The idea that behaviour, through cognitive and neural machinery, allows behavioural solutions to unusual or new problems to be devised is known as the cognitive buffer hypothesis (Allman *et al.*, 1993; van Schaik and Deaner, 2003; Sol, 2009a, b).

Mayr (1965) was among the first to propose that plastic behaviours buffer animals when invading novel regions, suggesting that successful invaders should be characterized by a tendency to discover unoccupied habitats and shift habitat preferences. More recent formal developments suggest two forms of behavioural plasticity that are relevant during invasions, namely, activational and developmental plasticity.

Activational Plasticity

Activational plasticity refers to the expression of behaviour and describes the innate response to stimuli that elicit a shift to an alternative behaviour through the activation of a neural network (Snell-Rood, 2013) Because of its immediacy and reversibility, such forms of plasticity allow individuals to efficiently respond to environmental uncertainties by enabling rapid modulation of, or transitions between, behaviours as a function of the individuals' needs (Snell-Rood, 2013; Sol *et al.*, 2013). These include fleeing in the presence of a predator, being attracted to new food opportunities and relaxing mate preferences when population density is low.

Activational behavioural plasticity is particularly important in novel environments for its role in mediating habitat and resource choice, which allows environmental uncertainties and adaptive mismatches to be reduced (Figure 5.3). Where the environment is heterogeneous, for example, animals can enhance their fitness by choosing habitats that suit better to their phenotype. Indeed, newly released animals often reject the habitat near the release site and travel long distances away before settling (Stamps and Swaisgood, 2007). They are also often very selective in habitat choice, despite behaving as ecological generalists in their regions of origin. In the highly invasive monk parakeet (*Myiopsitta monachus*), a generalist parrot introduced to Europe and North America, invading individuals not only exhibit a strong tendency to nest in a particular type of tree, even when other suitable trees are also available, but among them they tend to select the tallest ones, perhaps as a way to reduce the risk of unknown predators (Sol *et al.*, 1997).

Many animals released in new environments exhibit a consistent preference to settle in familiar types of habitats (i.e. those containing stimuli comparable to those in their natal habitat), a phenomenon called natal habitat preference induction (Stamps and Swaisgood, 2007). In Kenya, for example, elephants that were translocated between reserves tended to settle in habitats that were similar to those used in the source site more than did the local population (Pinter-Wollman *et al.*, 2009). The phenomenon of natal habitat preference induction, as a mechanism promoting niche conservatism, is an effective way to reduce adaptive mismatches and environmental uncertainties.

Animals select their habitats based on a variety of environmental cues that indicate qualities related to their niche requirements. In a novel environment, these cues may change yet they can still be perceived as informative by means of a variety of cognitive processes. Categorization, for example, involves classifying cues based upon perceptual or conceptual similarity (Greggor *et al.*, 2014). This allows discrimination between safe versus unsafe categories so as to minimize costly avoidance behaviours or to selectively respond to particularly dangerous predators (Greggor *et al.*, 2014). However, the risk of perceptual errors can lead to ecological traps (Kokko and Sutherland, 2001; Greggor *et al.*, 2014). Mayflies, for example, use polarized lights to decide where to lay the eggs. Yet, some types of asphalt polarize light horizontally in a way that mimics a highly polarized water surface, and as a result mayflies end up laying their eggs on an inappropriate substrate where they are unable to hatch successfully (Kriska *et al.*, 1998). This later example emphasizes that activational plasticity, being the expression of a prewired genetic programme, is often insufficient to deal with novel challenges (Figure 5.3). However, behaviour can also be plastically modified through developmental mechanisms, which allows further improving the response to novel environmental challenges.

Developmental Behavioural Plasticity

Animals can confront novel challenges, like the need to obtain new types of food or avoid unfamiliar predators, by modifying or inventing behaviours, a process known as developmental behavioural plasticity (Snell-Rood, 2013). Developmental behavioural plasticity is not so immediate as activational plasticity, because it involves changes in the nervous system that alter motor responses. However, it has the advantage that it allows animals to construct responses to unfamiliar or novel problems. One of the main mechanisms behind developmental behavioural plasticity is learning, the acquisition of new information influencing performance in behaviour (Dukas, 1998). Instead of consistently expressing the same behaviour to a particular stimulus, learning allows animals to devise innovative behavioural responses or to improve already established behaviours on the basis of experience (Lefebvre *et al.*, 1997; Dukas, 1998; Reader and Laland, 2002; van Schaik and Deaner, 2003; Ricklefs, 2004). Learning is particularly relevant in facilitating access to novel resources. Over the past years, for example, Lefebvre and co-workers have documented hundreds of observations of animals using novel foraging techniques to exploit foods that would otherwise be difficult to exploit (Lefebvre *et al.*, 1997; Overington *et al.*, 2009). Examples include house sparrows (*Passer domesticus*) using automatic sensors to open a bus station door, green jays (*Cynaocorax yncas*) using twigs as probes and levers, and herring gulls (*Larus argentatus*) catching small rabbits and killing them in preparation for eating by dropping them on rocks.

Learning can also mitigate the effect of environmental uncertainties in a number of ways. There is, for example, evidence showing that training captive-bred animals to avoid predators or to use food items they are more likely to encounter once released can improve the success of reintroduction programmes (Griffin *et al.*, 2000; Stamps and Swaisgood, 2007). The cognitive processes by which animals can reduce uncertainties include associative and social learning. Associative learning can, for instance, facilitate the use of cues that indirectly predict the presence of predators, including physical cues or the presence of other species (Griffin *et al.*, 2000). Associative learning also allows the incorporation of new cues to assess the quality of unfamiliar habitats. Seppänen *et al.* (2011), for example, showed that European flycatchers (*Ficedula* spp.) can use associative learning to adopt an arbitrary sign situated in a nest-box as a reliable cue of nest site quality. Social learning, on the other hand, can facilitate the transmission of learned behaviours within and among species. The fitness benefits of novel behaviours can for instance be transmitted to other members of the population by this mechanism (Lefebvre, 2013). Social learning can also be important in shaping niche preferences through imprinting or cultural transmission (Slagsvold and Wiebe, 2011). This can favour natal habitat preference induction (Slagsvold and Wiebe, 2007). By means of a cross-fostering experiment between blue tits (*Cyanistes caeruleus*) and great tits (*Parus major*), Slagsvold and Wiebe (2007) demonstrated that early learning can cause a shift in the feeding niche in the direction of the foster species. Some animals may even adopt new cues to assess habitat quality based on public information inadvertently produced by the presence or breeding performance of individuals of other species with similar environmental needs (Danchin *et al.*, 2004; Parejo *et al.*, 2008).

Evidence for a Role of Behavioural Plasticity in the Invasion Process

Although the hypothesis that plastic behaviours can assist animals in invading novel regions dates back from Mayr (1965), until recently the hypothesis was backed by little empirical evidence. Recent attempts to explore the issue have progressed in three distinct directions. The first is based on broad comparative analyses of historical introductions of animals outside their native ranges. While such historical data have certain limitations, reliable results may still be obtained by using appropriate modelling techniques that control for the effects of introduction effort and other confounding factors (Sol *et al.*, 2007b). The analysis of such quasi-experimental data has revealed that the propensity to invent new behaviours and the underlying neural substrate (see next section) enhance the likelihood of establishment in novel environments in birds, mammals, reptiles and amphibians (Sol *et al.*, 2005, 2008; Amiel *et al.*, 2011). Evidence is, however, absent for fish (Drake, 2007).

The second approach to exploring the role of behavioural plasticity in invasion success has focused on evaluating whether behavioural adjustments are necessary in the situations where invaders attain higher success. For example, many vertebrate invaders attain higher densities in human-altered habitats than in more natural habitats, and there is ample evidence that living in these environments often requires changes in behaviour (Lowry *et al.*, 2012; Sol *et al.*, 2013). Thus, urban animals frequently differ in foraging behaviour (e.g. more readily adopting new foods derived from human activities or developing new foraging techniques), activity patterns (e.g. reducing flight initiation distances or becoming more active during the night) and even in the way they communicate with others (e.g. reducing song frequency in noisy places or singing in the dawn) relative to those from surrounding habitats. Although some of these behavioural differences can reflect evolutionary responses or a sorting process (in which only individuals with proper behaviours are able to colonize the environment), others can unambiguously be attributed to behavioural plasticity (Sol *et al.*, 2013). Laboratory experiments on common mynas (*Sturnus tristis*) introduced to Australia have, for instance, revealed that the species has a remarkable ability to explore and adopt new food types, and that these cognitive abilities are more pronounced in individuals inhabiting the disturbed habitats in which they attain higher success (Sol *et al.*, 2011, 2012a).

The last approach providing insight into the role of behavioural plasticity in the invasion process consists of examining differences in key components of behavioural plasticity between populations of a same species exposed to conditions in which plasticity is expected to be either highly or little relevant. A classic example is the reduction in intraspecific aggression observed in some invasive ants, which has contributed to their invasion success by favouring expansive supercolonies (Holway and Suarez, 1999). Other examples include experiments in birds measuring plasticity in invasive and non-invasive populations. For example, experiments in house sparrows introduced to North America suggest that individuals that are invading a new region are more likely to approach and consume novel foods than those that have been in the region for a longer time (Martin and Fitzgerald, 2005). This latter result can be interpreted in terms of the costs of behavioural plasticity, which predicts that selection should reduce plasticity over time as the invader increasingly becomes locally adapted.

Why do we Need Behaviour to Better Understand how Life History Affects Invasion Success?

Behaviour cannot be ignored as one of the mechanisms by which life history affects invasion success for two main reasons. First, behavioural plasticity is not only an important mechanism of response to novel challenges but its benefits depend on the species' life history. Second, behaviour directly mediates some life history mechanisms, particularly those related to environmental uncertainty and adaptive mismatch like the storage effect and bet-hedging. We will examine both issues in the next sections.

Behavioural Plasticity and the Fast–Slow Continuum

The fast–slow continuum implies a differential need to collect and use information (van Schaik and Deaner, 2003; Sol, 2009a; Sih and Del Giudice, 2012). Animals at the 'slow' extreme of the fast–slow continuum tend to explore more accurately and often exhibit better performance in problem solving and learning. One reason is that they tend to have disproportionately larger brains, which has been shown to enhance the capacity to innovate and learn (Lefebvre *et al.*, 1997; Reader and Laland, 2002; Overington *et al.*, 2009; Reader *et al.*, 2011) and hence improve their survival prospects (Sol *et al.*, 2007a; Kotrschal *et al.*, 2015). Although a slow life history can be developed with no need of a large brain, a large brain can only evolve in animals at the slow extreme of the fast–slow continuum. This is because a large brain takes longer to grow, and hence imposes a developmental constraint in terms of extended growth and maturation (Barton and Capellini, 2011). In addition, the benefits of large brains in providing a 'cognitive buffer' against sources of extrinsic mortality should select for and be selected by a slower life history (Allman, 2000; van Schaik and Deaner, 2003; Sol, 2009a, b). Exploring and learning about the environment is more beneficial for a long-lived than for short-lived species because individuals are more likely to be exposed to environmental changes during their lifetime. In addition, the learned behaviours can provide benefits for longer periods. The costs are also reduced, at least in terms of time constraints. Having more time allows animals to gather more and better information, improve decision-making (Stamps *et al.*, 2005, Mabry and Stamps, 2008) and problem solve.

Nevertheless, animals with high future reproductive prospects are more likely to develop risk-averse behaviours. In a novel environment, this should affect how they resolve the conflict between the need to approach and to explore new resource opportunities and, at the same time, avoid unnecessary risks (Greenberg, 2003; Sol *et al.*, 2011). While this could compromise their chances of succeeding in the new environment, this is not necessarily true for two reasons. First, the balance between approach and avoidance is expected to differ depending on the environmental context. For example, in urbanized environments where exposure to novel feeding opportunities is common-place and risks associated with specialized predators are low, invaders should generally favour exploration over avoidance (Sol *et al.*, 2011). Common mynas introduced to Australia fit well with this expectation, exhibiting lower neophobia and higher neophilia in highly urbanized environments (Sol *et al.*, 2011). Second, both theoretical and empirical evidence suggest that risk-averse animals should explore slowly but more accurately (Marchetti and Drent, 2000). This is because the existence of a speed–accuracy trade-off improves accuracy when exploration is slower (Sih and Del Giudice, 2012). Thus, a risk-averse strategy does not necessarily prevent the animal from adopting novel ecological opportunities and successfully establishing in the new environment.

Behavioural Plasticity and Life History Mechanisms

Plastic behavioural responses can mediate life history mechanisms affecting success in novel or uncertain environments in a variety of ways. This is clear for the storage effect, as the decision whether to breed or not when conditions are uncertain is itself a behavioural decision. Skipping a reproduction also provides more time for exploring the new environment and adjusting behaviour through learning (Figure 5.3).

Our understanding of bet-hedging mechanisms can also benefit from considering behavioural responses. Learning, in particular, can improve breeding performance in species that spread the risk of reproductive failure over several breeding attempts. A change in nest site choice is a well-documented response to nest predation (Lima, 2009). In the Brewer's sparrow (*Spizella breweri*), for example, the cost of a reproductive failure is compensated in part by new nesting attempts (Chalfoun and Martin, 2010). However, previous experience largely modifies subsequent breeding decisions. Thus, pairs move sequential nest sites slightly farther away from their initial site following an episode of nest predation. Moreover, they also change nest patch attributes (e.g. shrub height, shrub density) to a greater extent following nest predation.

Finally, behaviour may affect predictions derived from the small population paradigm. In experimental field introductions of the Argentine ant (*Linepithema humile*), Sagata and Lester (2009) found that propagule pressure was a relatively poor predictor of establishment success owing to the ability of this species to modify its behaviour according to environmental conditions and resource availability. By shaping the extent to which the population is affected by Allee effects, some behaviours can also influence the susceptibility of the population to extinction by demographic stochasticity. Blackburn *et al.* (2009) found that species that can reduce Allee effects through behaviours like dispersal tendency or mate choice are more likely to establish successfully in new environments.

Concluding Remarks

To characterize the life history of successful invaders, classical theory emphasizes the importance of life history mechanisms related to small population dynamics. While we do not deny the importance of these mechanisms, we highlight here the need to also consider alternative mechanisms related to responses to adaptive mismatches and environmental uncertainties. This forces us to take into consideration the role of behaviour.

By its plastic nature, behaviour acts to reduce environmental uncertainties and adaptive mismatches, thereby directly affecting invasion success. However, the importance of behaviour is not the same for all life history strategies. Those that give more value to adults than to offspring presumably benefit more from behavioural plasticity because they are more likely to possess the adaptive machinery underlying plastic responses and have more opportunities to assemble and respond to environmental information. The connection between behaviour and life history suggests thus that both should be considered as part of a same adaptive strategy of response to environmental changes. This idea is further emphasized when considering that behaviour can also mediate some of the mechanisms by which life history influences the response to novel environments, notably the storage effect and bet-hedging.

The existence of a variety of mechanisms by which life history and behavioural plasticity affect population dynamics in novel environments suggests that there is no single strategy to be a successful invader. Thus, the optimal strategy can vary according to the degree of demographic stochasticity, adaptive mismatch and environmental uncertainty (Figure 5.2). This may explain why unravelling the life history of successful invaders has proved difficult. Despite this, if some scenarios are more frequent than others we can still make some generalizations regarding the life history strategies of successful invaders. In vertebrates, for example, many invaders are primarily restricted to human-altered environments (Case, 1996; Sol *et al.*, 2012a; Barnagaud *et al.*, 2013), which facilitate invasions by simultaneously opening new resource opportunities and decreasing the pressure of competition and predation from native species. In these environments, animals are continuously exposed to novel ecological challenges and uncertainties, which should favour a 'slower' strategy of high future reproductive prospects. Indeed, a recent global comparative analysis of avian introductions evidenced that although rapid population growth may be advantageous during invasions when propagule pressure is low, successful invaders are generally characterized by life history strategies in which individuals give priority to future rather than current reproduction, either by means a long reproductive life or a higher frequency of reproductions (Sol *et al.*, 2012b). The more recent finding that urban dwellers also tend to spread reproductive effort across many breeding attempts (Sol *et al.*, 2014) further confirms the importance of this strategy in dealing with environmental uncertainties and adaptive mismatches.

Despite progress, there remain important issues to resolve to better integrate behaviour into a life history framework for biological invasions. A first issue that requires further attention is how cognition is functionally related to variation in life history, and how together these mediate the response of organisms to environmental changes. In this chapter, we have described several such mechanisms, yet others are also possible. For example, animals with enhanced learning abilities tend to be highly social, whether because sociality selects for these abilities (Shultz and Dunbar, 2007) or because both characteristics are a by-product of selection for longer lives (Sol, 2009b). The social environment might further buffer inexperienced individuals in adverse conditions by facilitating behavioural innovations (Liker and Bókony, 2009; Morand-Ferron and Quinn, 2011) and reducing predation risk (Krams *et al.*, 2010). To date, however, the importance of sociality during the invasion process has been difficult to evaluate in part because most introduced animals tend to be social.

The second important issue to resolve is the extent to which life history trade-offs are stable or can be relaxed or even broken. Although the existence of trade-offs is undisputed (Stearns, 1992; Roff, 2002), there are a number of situations that can relax the trade-offs and even change their sign (Stearns, 1989), notably the existence of genotype × environment interactions (Reznick *et al.*, 2000). In addition, natural populations often consist of phenotypically diverse individuals (Araújo *et al.*, 2011), a situation we have largely ignored in this chapter. The existence of individual differences in life history and behaviour (e.g. Réale *et al.*, 2007; Biro and Stamps, 2008; Bolnick *et al.*, 2011; González-Suárez and Revilla, 2013) can generate variation in the demographic parameters of the population, potentially affecting success in novel environments (Bolnick *et al.*, 2011; Phillips and Suarez, 2012). Interestingly, theoretical developments predict that variation in life history and behavioural types should be linked. Biro and Stamps (2008), for example, suggested that variation in exploratory behaviour can be maintained by trade-offs between early and late fecundity. Likewise, Wolf *et al.* (2007) argued that animals varying in future prospects are more likely to develop differences in risk-averse behaviours, which can lead to stable individual variation in aggressiveness and boldness. All these complexities need to be accommodated in a more general theory.

The last aspect for resolution is the extent to which the life history of the invader can evolve towards the optimal strategy for the novel environment. Life history is known to evolve quickly in some invaders following trajectories predicted by life history theory (Huey *et al.*, 2000; Reznick and Ghalambor, 2001; Reznick *et al.*, 2008), yet whether these changes are sufficient to deal with the pressures of novel environments is less clear. Moreover, if individuals respond behaviourally to the new challenges this can hide genetic variation from selection and slow down the adaptation process (Huey *et al.*, 2003). Yet, if responding behaviourally to the most common challenges hides genetic variation from selection, this should enhance the genetic variation available for selection when conditions are more adverse and behavioural responses are less efficient.

These and other potentially fruitful areas of research, if pursued, have the power to reshape the field in the near future, shedding new light into the importance of behaviour in biological invasions and, more generally, contributing to better understand the response of animals to human-induced rapid environmental changes.

Acknowledgements

We are grateful to Julie Lockwood, Louis Lefebvre and an anonymous reviewer for useful comments on the chapter; to Martí Franch for the nice bird drawings of Figure 5.1; and to Louis Lefebvre, Richard Duncan, Tim Blackburn, Phill Cassey, Oriol Lapiedra, Miquel Vall-llosera, Montse Vilà, Sven Bacher and Simon Reader for fruitful discussions over the past years. This work was supported by the Spanish government (Proyecto de Investigación CGL2013-47448-P).

References

Allman, J. (2000). *Evolving Brains*. New York: Scientific American Library.
Allman, J.M., McLaughlin, T. and Hakeem, A. (1993). Brain structures and life-span in primate species. *Proceedings of the National Academy of Sciences, USA*, 90, 3559–3563.
Amiel, J.J., Tingley, R. and Shine, R. (2011). Smart moves: effects of relative brain size on establishment success of invasive amphibians and reptiles. *PLoS ONE*, 6, e18277.

Araújo, M.S., Bolnick, D.I. and Layman, C.A. (2011). The ecological causes of individual specialisation. *Ecology Letters*, 14, 948–958.

Barnagaud, J., Barbaro, L. and Papaïx, J. (2013). Habitat filtering by landscape and local forest composition in native and exotic New Zealand birds. *Ecology*, 95, 78–87.

Barton, R.A. and Capellini, I. (2011). Maternal investment, life histories, and the costs of brain growth in mammals. *Proceedings of the National Academy of Sciences, USA* 108, 6169–6174.

Biro, P.A. and Stamps, J.A. (2008). Are animal personality traits linked to life-history productivity? *Evolution*, 23, 361–368.

Blackburn, T.M., Cassey, P. and Lockwood, J.L. (2009). The role of species traits in the establishment success of exotic birds. *Global Change Biology*, 15, 2852–2860.

Bolnick, D.I., Amarasekare, P., Araújo, M.S., *et al.* (2011). Why intraspecific trait variation matters in community ecology. *Trends in Ecology and Evolution*, 26, 183–192.

Caceres, C.E. (1997). Temporal variation, dormancy and coexistence: A field test of the storage effect. *Proceedings of the National Academy of Science, USA*, 94, 9171–9175.

Case, T.J. (1996). Global patterns in the establishment and distribution of exotic birds. *Biological Conservation*, 78, 69–96.

Cassey, P., Blackburn, T.M., Sol, D., Duncan, R.P. and Lockwood, J.L. (2004). Global patterns of introduction effort and establishment success in birds. *Proceedings of the Royal Society of London, Series B*, 271, S405–S408.

Chalfoun, A.D. and Martin, T.E. (2010). Facultative nest patch shifts in response to nest predation risk in the Brewer's sparrow: a 'win-stay, lose-switch' strategy? *Oecologia*, 163, 885–892.

Charlesworth, B. (1980). *Evolution in Age Structured Populations*. Cambridge, UK: Cambridge University Press.

Cubaynes, S., Doherty, P.F., Schreiber E.A. and Gimenez, O. (2011). To breed or not to breed: a seabird's response to extreme climatic events. *Biology Letters*, 7, 303–306.

Danchin, E., Giraldeau, L.-A., Valone, T.J. and Wagner, R.H. (2004). Public information: from noisy neighbors to cultural evolution. *Science*, 305, 487–491.

Drake, J.M. (2007). Parental investment and fecundity, but not brain size, are associated with establishment success in introduced fishes. *Functional Ecology*, 21, 963–968.

Dukas, R. (1998). Evolutionary ecology of learning. In *Cognitive Ecology: The Evolutionary Ecology of Information Processing and Decision Making*, ed. Dukas, R. Chicago, IL: University of Chicago Press, pp. 129–174.

Forcada, J., Trathan, P.N. and Murphy, E.J. (2008). Life history buffering in Antarctic mammals and birds against changing patterns of climate and environmental variation. *Global Change Biology*, 14, 2473–2488.

Futuyma, D.J. and Moreno, G. (1988). The evolution of ecological specialization. *Annual Review of Ecology and Systematics*, 19, 207–234.

González-Suárez, M. and Revilla, E. (2013). Variability in life-history and ecological traits is a buffer against extinction in mammals. *Ecology Letters*, 16, 242–251.

Greenberg, R. (2003). The role of neophobia and neophilia in the development of innovative behaviour of birds. In *Animal Innovation*, ed. Reader, S.M. and Laland, K.N. Oxford, UK: Oxford University Press, pp. 176–196.

Greggor, A.L., Clayton, N.S., Phalan B. and Thornton, A. (2014). Comparative cognition for conservationists. *Trends in Ecology and Evolution*, 29, 489–495.

Griffin, A.S., Blumstein, D.T. and Evans, C.S. (2000). Training captive-bred or translocated animals to avoid predators. *Conservation Biology*, 14, 1317–1326.

Holway, D. and Suarez, A. (1999). Animal behavior: an essential component of invasion biology. *Trends in Ecology and Evolution*, 5347, 12–14.

Huey, R.B., Gilchrist, G.W., Carlson, M.L., Berrigan, D. and Serra, L. (2000). Rapid evolution of a geographic cline in size in an introduced fly. *Science*, 287, 308–309.

Huey, R.B., Hertz, P.E. and Sinervo, B. (2003). Behavioral drive versus behavioral inertia in evolution: a null model approach. *American Naturalist*, 161, 357–366.

Kokko, H. and Sutherland, W.J. (2001). Ecological traps in changing environments: Ecological and evolutionary consequences of a behaviourally mediated Allee effect. *Evolutionary Ecology*, 3, 537–551.

Kolar, C.S. and Lodge, D.M. (2002). Ecological predictions and risk assessment for alien fishes in North America. *Science*, 298, 1233–1236.

Kotrschal, A., Buechel, S.D., Zala, S.M., Corral, A., Penn, D.J. and Kolm, N. (2015). Brain size affects female but not male survival under predation threat. *Ecology Letters*, 8(7), 646–652.

Krams, I., Bērziņs, A., Krama, T., Wheatcroft, D., Igaune, K. and Rantala, M.J. (2010). The increased risk of predation enhances cooperation. *Proceedings of the Royal Society of London B*, 277, 513–518.

Kriska, G., Horváth, G. and Andrikovics, S. (1998). Why do mayflies lay their eggs en masse on dry asphalt roads? Water-imitating polarized light reflected from asphalt attracts Ephemeroptera. *The Journal of Experimental Biology*, 201, 2273–2286.

Lefebvre, L. (2013). Brains, innovations, tools and cultural transmission in birds, non-human primates, and fossil hominins. *Frontiers in Human Neuroscience*, 7, 245.

Lefebvre, L., Whittle, P., Lascaris, E. and Finkelstein, A. (1997). Feeding innovations and fore-brain size in birds. *Animal Behaviour*, 53, 549–560.

Leung, B., Roura-Pascual, N., Bacher, S., *et al.* (2012). TEASIng apart alien species risk assessments: a framework for best practices. *Ecology Letters*, 15, 1475–1493.

Lewontin, R.C. (1965). Selection for colonizing ability. In *The Genetics of Colonizing Species*, ed. Baker, H. and Stebbins, G. London: Academic Press, pp. 77–94.

Liker, A. and Bókony, V. (2009). Larger groups are more successful in innovative problem solving in house sparrows. *Proceedings of the National Academy of Sciences, USA*, 106, 7893–7898.

Lima, S.L. (2009). Predators and the breeding bird: behavioral and reproductive flexibility under the risk of predation. *Biological reviews of the Cambridge Philosophical Society*, 84, 485–513.

Lockwood, J.L., Cassey, P. and Blackburn, T.M. (2005). The role of propagule pressure in explaining species invasions. *Trends in Ecology and Evolution*, 20, 223–228.

Losos, J.B., Schoener, T.W. and Spiller, D.A. (2004). Predator-induced behaviour shifts and natural selection in field experimental lizard populations. *Nature*, 432, 505–508.

Lowry, H., Lill, A. and Wong, B.B.M. (2012). Behavioural responses of wildlife to urban environments. *Biological Reviews of the Cambridge Philosophical Society*, 88, 537–549.

Mabry, K.E. and Stamps, J.A. (2008). Searching for a new home: decision making by dispersing brush mice. *The American Naturalist*, 172, 625–634.

Marchetti, C. and Drent, P. (2000). Individual differences in the use of social information in foraging by captive great tits. *Animal Behaviour*, 60, 131–140.

Martin, L.B. and Fitzgerald, L. (2005). A taste for novelty in invading house sparrows, *Passer domesticus*. *Behavioral Ecology*, 16, 702–707.

Mayr, E. (1965). The nature of colonising birds. In *The Genetics of Colonizing Species*, ed. Baker, H.G. and Stebbins, G.L. New York: Academic Press, pp. 29–43.

Morand-Ferron, J. and Quinn, J.L. (2011). Larger groups of passerines are more efficient problem solvers in the wild. *Proceedings of the National Academy of Sciences, USA*, 108, 15898–
15903.

Morris, W.F. and Doak, D.F. (2004). Buffering of life histories against environmental stochasticity: accounting for a spurious correlation between the variabilities of vital rates and their
contributions to fitness. *American Naturalist*, 163, 579–590.

Moulton, M.P., Pimm, S.L., Mooney, H.A. and Drake, J.A. (1986). Species introductions to
Hawaii. In *Ecology of Biological Invasions in North America and Hawaii*, ed. Mooney, H.A.
and Drake, J.A. New York: Springer, pp. 231–249.

Overington, S.E., Morand-Ferron, J., Boogert, N.J. and Lefebvre, L. (2009). Technical innovations drive the relationship between innovativeness and residual brain size in birds. *Animal
Behaviour*, 78, 1001–1010.

Parejo, D., Danchin, É., Silva, N., White, J.F., Dreiss, A.N. and Avilés, J.M. (2008). Do great tits
rely on inadvertent social information from blue tits? A habitat selection experiment. *Behavioral Ecology and Sociobiology*, 62, 1569–1579.

Phillips, B. and Suarez, A. (2012). The role of behavioural variation in the invasion of new areas.
In *Behavioural Responses to a Changing World: Mechanisms and Consequences*, ed. Candolin,
U. and Wong, B.B.M. Oxford: Oxford University Press, pp. 190–200.

Pimm, S.L. (1991). *The Balance of Nature? Ecological Issues in the Conservation of Species and
Communities*. Chicago, UK: University of Chicago Press.

Pinter-Wollman, N., Isbell, L.A. and Hart, L.A. (2009). Assessing translocation outcome: comparing behavioral and physiological aspects of translocated and resident African elephants (*Loxodonta africana*). *Biological Conservation*, 142, 1116–1124.

Reader, S.M. and Laland, K.N. (2002). Social intelligence, innovation, and enhanced brain size
in primates. *Proceedings of the National Academy of Science, USA*, 99, 4436–4441.

Reader, S.M., Hager, Y. and Laland, K.N. (2011). The evolution of primate general and cultural intelligence. *Philosophical Transactions of the Royal Society B: Biological Sciences*, 366,
1017–1027.

Réale, D., Reader, S.M., Sol, D. McDougall, P.T. and Dingemanse, N.J. (2007). Integrating animal
temperament within ecology and evolution. *Biological Reviews of the Cambridge Philosophical Society*, 82, 291–318.

Reznick, D. and Ghalambor, C. (2001). The population ecology of contemporary adaptations:
what empirical studies reveal about the conditions that promote adaptive evolution. *Genetica*,
112–113, 183–198.

Reznick, D., Nunney, L. and Tessier, A. (2000). Big houses, big cars, superfleas and the costs of
reproduction. *Trends in Ecology and Evolution*, 15, 421–425.

Reznick, D., Bryant, M.J. and Bashey, F. (2002). r- and K-selection revisited: the role of population regulation in life-history evolution. *Ecology*, 83, 1509–1520.

Reznick, D.N., Ghalambor, C.K. and Crooks, K. (2008). Experimental studies of evolution in guppies: a model for understanding the evolutionary consequences of predator removal in natural
communities. *Molecular Ecology*, 17, 97–107.

Ricklefs, R.E. (2004). The cognitive face of avian life histories. *Wilson Bulletin*, 116, 119–
196.

Roff, D.A. (2002). *Life History Evolution*. Sunderland, MA: Sinauer Associates, Inc.

Sæther, B.-E. and Ø. Bakke. (2000). Avian life history variation and contribution of demographic
traits to the population growth rate. *Ecology*, 81, 642–653.

Saether, B.-E., Engen, S., Pape Møller, A., et al. (2004). Life history variation predicts the effects of demographic stochasticity on avian population dynamics. *The American Naturalist*, 164, 793–802.

Sagata, K. and Lester, P.J. (2009). Behavioural plasticity associated with propagule size, resources, and the invasion success of the Argentine ant *Linepithema humile*. *Journal of Applied Ecology*, 46, 19–27.

Schaffer, W.M. (1974). Selection for optimal life histories: the effects of age structure. *Ecology*, 55, 291–303.

van Schaik, C.P. and Deaner, R.O. (2003). Life history and cognitive evolution in primates. In *Animal Social Complexity*, ed. de Waal, F.B.M. and Tyack, P.L. Cambridge, MA: Harvard University Press, pp. 5–25.

Seppänen, J.-T., Forsman, J.T., Mönkkönen, M., Krams, I. and Salmi, T. (2011). New behavioural trait adopted or rejected by observing heterospecific tutor fitness. *Proceedings of the Royal Society B: Biological Sciences*, 278, 1736–1741.

Shultz, S. and Dunbar, R.I.M. (2007). Evolution in the social brain. *Science*, 317, 1344–1347.

Sih, A. and Del Giudice, M. (2012). Linking behavioural syndromes and cognition: a behavioural ecology perspective. *Philosophical Transactions of the Royal Society of London B: Biological Sciences*, 367, 2762–2772.

Sih, A., Ferrari, M.C.O. and Harris, D.J. (2011). Evolution and behavioural responses to human-induced rapid environmental change. *Evolutionary Applications*, 4, 367–387.

Slagsvold, T. and Wiebe, K.L. (2007). Learning the ecological niche. *Proceedings of the Royal Society of London B: Biological Sciences*, 274, 19–23.

Slagsvold, T. and Wiebe, K.L. (2011). Social learning in birds and its role in shaping a foraging niche. *Philosophical Transactions of the Royal Society of London B: Biological Sciences*, 366, 969–977.

Snell-Rood, E.C. (2013). An overview of the evolutionary causes and consequences of behavioural plasticity. *Animal Behaviour*, 85, 1004–1011.

Sol, D. (2009a). The cognitive-buffer hypothesis for the evolution of large brains. In *Cognitive Ecology*, ed. Dukas, R. and Ratcliffe, R.M. Chicago, IL: Chicago University Press, pp. 111–134.

Sol, D. (2009b). Revisiting the cognitive buffer hypothesis for the evolution of large brains. *Biology Letters*, 5, 130–133.

Sol, D., Santos, D.M.M., Feria, E. and Clavell, J. (1997). Habitat selection by the Monk Parakeet during colonization of a new area in Spain. *Condor*, 99, 39–46.

Sol, D., Duncan, R.P., Blackburn, T. M., Cassey, P. and Lefebvre, L. (2005). Big brains, enhanced cognition, and response of birds to novel environments. *Proceedings of the National Academy of Science, USA*, 102, 5460–5465.

Sol, D., Liker, A., Lefebvre, L. and Székely, T. (2007a). Big-brained birds survive better in nature. *Proceedings of the Royal Society of London B: Biological Sciences*, 274, 763–769.

Sol, D., Vilà, M. and Kühn, I. (2007b). The comparative analysis of historical alien introductions. *Biological Invasions*, 10, 1119–1129.

Sol, D., Bacher, S., Reader, S.M. and Lefebvre, L. (2008). Brain size predicts the success of mammal species introduced into novel environments. *The American Naturalist*, 172 Suppl., S63–71.

Sol, D., Griffin, A.S., Bartomeus, I. and Boyce, H. (2011). Exploring or avoiding novel food resources? The novelty conflict in an invasive bird. *PLoS ONE*, 6, e19535.

Sol, D., Bartomeus, I. and Griffin, A.S. (2012a). The paradox of invasion in birds: competitive superiority or ecological opportunism? *Oecologia*, 169, 553–564.

Sol, D., Maspons, J. and Vall-llosera, M., *et al.* (2012b). Unraveling the life history of successful invaders. *Science*, 337, 580–583.

Sol, D., Lapiedra, O. and González-Lagos, C. (2013). Behavioural adjustments for a life in the city. *Animal Behaviour*, 85, 1101–1112.

Sol, D., González-Lagos, C., Moreira, D., Maspons J. and Lapiedra, O. (2014). Urbanisation tolerance and the loss of avian diversity. *Ecology Letters*, 17, 942–995.

Stamps, J.A. and Swaisgood, R.R. (2007). Someplace like home: experience, habitat selection and conservation biology. *Applied Animal Behaviour Science*, 102, 392–409.

Stamps, J., Krishnan, V. and Reid, M. (2005). Search costs and habitat selection by dispersers. *Ecology*, 86, 510–518.

Starrfelt, J. and Kokko, H. (2012). Bet-hedging – a triple trade-off between means, variances and correlations. *Biological Reviews of the Cambridge Philosophical Society*, 87, 742–755.

Stearns, S. (1983). The influence of size and phylogeny on patterns of covariation among life-history traits in the mammals. *Oikos*, 41, 173–187.

Stearns, S.C. (1989). Trade-offs in life-history evolution. *Functional Ecology*, 3, 259–268.

Stearns, S.C. (1992). *The evolution of life histories*. Oxford, UK: Oxford University Press.

Stearns, S.C. (2000). Life history evolution: successes, limitations and prospects. *Die Naturwissenschaften*, 87, 476–486.

Stearns, S.C. and Crandall, R. (1981). Bet-hedging and persistence as adaptations of colonizers. *Evolution Today*, 371–383.

Vall-llosera, M. and Sol, D. (2009). A global risk assessment for the success of bird introductions. *Journal of Applied Ecology*, 46, 787–795.

Warner, R.R. and Chesson, P.L. (1985). Coexistence mediated by recruitment fluctuations: a field guide to the storage effect. *American Naturalist*, 125(6), 769–787.

Williams, G.C. (1966). *Adaptation and Natural Selection*. Princeton, NJ: Princeton University Press.

Williamson, M.H. (1996). *Biological Invasions*. London: Chapman and Hall.

Wolf, M., van Doorn, G.S.S., Leimar, O. and Weissing, F. J. (2007). Life-history trade-offs favour the evolution of animal personalities. *Nature*, 447, 581–584.

Yeh, P.J. and Price, T.D. (2004). Adaptive phenotypic plasticity and the successful colonization of a novel environment. *The American Naturalist*, 164, 531–542.

6 Behaviour on Invasion Fronts, and the Behaviour of Invasion Fronts

Ben L. Phillips

Introduction

The spatial spread of non-indigenous organisms has been of interest to ecologists at least since the 1950s, with the publication of Elton's landmark book (Elton, 1958) and the first analytical model of spread by Skellam (Skellam, 1951). Since this time, the study of biological invasions has become its own sub-discipline: there has been the accumulation of many more empirical systems (Lockwood *et al.*, 2007) and advances in the theory of biological invasions (Shigesada and Kawasaki, 1997; Hastings *et al.*, 2005). The briefest consideration of the process of invasion in animals leads one to conclude that much of the interesting variation in the success of animal invasions – their probability of establishment, and their rate of spread – is mediated by behaviour. Moreover, there is a growing realization that it is the variation in behaviour – both plastic and genetic – that is the critical element determining the success or otherwise of an animal invasion.

When we admit variation in behaviour, it is clear that the process of biological invasion – beginning from successful transportation and establishment, and ranging through to the rate of spread in the new range – imposes selection on behaviour (Chapple *et al.*, 2012). Some behaviours are more likely to be transported, of these, some are more likely to survive the voyage, of these some are more likely to successfully establish, and so on. At every stage of a biological invasion, selection and trait sorting potentially operates (Blackburn and Duncan, 2001; Simons, 2003). But also, by dint of the small number of individuals involved, invasion involves substantial stochasticity (Williamson, 1996; Phillips *et al.*, 2010b). Thus, individuals that establish invasive populations are a biased subset of those from the native range, but they are different for reasons both deterministic (selection) and stochastic (founder events).

Here, I largely ignore the processes leading up to successful establishment, which is covered by others chapters of this volume, and instead focus on the process of spread within the introduced range. The same rules apply here as well: behavioural variation admits selection and founder events, so the individuals that make up an invasion front show rapid trait divergence from those at the site of introduction. The processes driving these trait divergences on an advancing range edge are relatively simple, but they give

Biological Invasions and Animal Behaviour, eds J.S. Weis and D. Sol. Published by Cambridge University Press. © Cambridge University Press 2016.

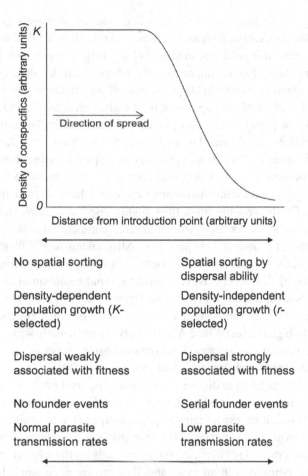

Figure 6.1 The spread of an alien happens as a consequence of dispersal and population growth, and results in a density gradient that moves through space. This density gradient imposes eco-evolutionary forces that are very distinct from those operating in the core of the range.

rise to an astonishing array of biological consequences, some of which are undoubtedly still to be uncovered.

The Evolutionary Dynamics of Spread: Theory

Populations spread as a consequence of two processes: dispersal and population growth (Fisher, 1937). Dispersal moves individuals through space (some of these individuals move into uncolonized territory), and population growth causes the low-density edge populations to increase in density over time (Figure 6.1). Together, dispersal and population growth cause a wave of advance, and the rate at which this wave moves is, of course, governed by rates of dispersal and population growth (Fisher, 1937; Hastings *et al.*, 2005): if dispersal or population growth rates increase, so too does the rate of spread.

The invasion front is, by definition, a region where population density grades from some equilibrium density ('carrying capacity') to zero. And although this density gradient moves through space, it is persistent in time: as long as the population is spreading, this density gradient exists. Several important things happen on this density gradient. To begin with individuals are sorted by dispersal ability: only the most dispersive individuals make it to the edges of the population in any given generation, and so we see assortative mating by dispersal ability. This process – 'spatial sorting' – on its own can cause the runaway evolution of dispersal on an invasion front (Phillips *et al.*, 2008; Shine *et al.*, 2011; Benichou *et al.*, 2012; Bouin *et al.*, 2012). Spatial sorting, however, also sets up a negative correlation between dispersal phenotype and density on the invasion front. That is, highly dispersive individuals consistently find themselves in areas of low conspecific density. On the invasion front, then, highly dispersive individuals may get a consistent absolute fitness advantage through lowered competition with conspecifics (Phillips *et al.*, 2008). Of course, if there are strong Allee effects (e.g. difficulty finding mates at low density), the opposite might also be true. Together, spatial sorting and natural selection (driven by density effects) can cause the rapid evolution of dispersal; an interaction dubbed 'spatial selection' (Travis and Dytham, 2002; Phillips *et al.*, 2008). In a uniformly favourable environment, and in the absence of Allee effects, dispersal will evolve rapidly to higher rates. Where Allee effects are sufficiently strong, however, dispersal rates may remain static, trapped between opposing forces of spatial sorting and natural selection (Travis and Dytham, 2002; Burton *et al.*, 2010).

If we admit heritable variation in dispersal behaviour, dispersal rates will almost certainly evolve on an expanding range edge. But what about the other important process – population growth – would we expect rates of population growth to evolve as well? If we assume a density-regulated population, the answer is yes, and the reason is, again, that persistent density gradient (Phillips, 2009). Individuals on the edge of the expanding range front find themselves in an exponentially growing population. Individuals behind the invasion front find themselves in a density-regulated population. Thus, the invasion front represents a cline between r-selected and non-r-selected environments (MacArthur and Wilson, 1967; Phillips, 2009; Phillips *et al.*, 2010b). Because the invasion front is an exponentially growing population, anything that increases an individual's basic reproductive rate on the invasion front will be favoured. By contrast, behind the invasion front, reproduction is more strongly governed by density effects such as competition with conspecifics. As the invasion front propagates forward in space, there will always be a subset of genes sitting in this r-selected environment on the invasion front. In this low-density region, continuous r-selection will be strong for much longer than would be possible in a contained space: r-selection will proceed as long as the invasion progresses (Burton *et al.*, 2010; Phillips *et al.*, 2010b). Thus, on the invasion front, we would expect reproductive rate to increase over time, and this change will potentially manifest in many aspects of the organism, including its behaviour.

The evolution of population growth rate has consequences, not just for behaviours relating to growth rate, but also for dispersal. There is a positive feedback between an evolving growth rate and the spatial selection that drives dispersal evolution (Perkins

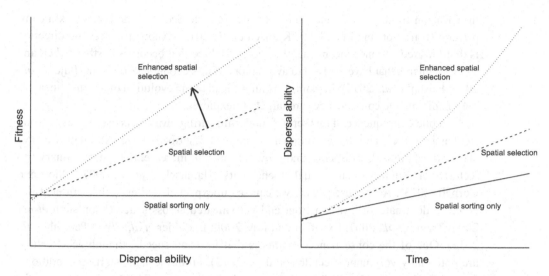

Figure 6.2 Spatial processes operating on the evolution of dispersal on an expanding range edge. The first panel shows the correlation between dispersal phenotype and relative fitness, and the second panel shows the expected trajectory of dispersal phenotypes on the invasion front over time. Spatial sorting happens on invasion fronts and can lead to increased dispersal even in the absence of fitness differentials. Nonetheless, there may often also be a fitness advantage to dispersing (through density release), and so spatial sorting and natural selection may interact ('spatial selection') to increase the rate of dispersal evolution. Finally, if reproductive rates also increase as a consequence of *r*-selection on the invasion front, the fitness advantage from density release gets bigger over time ('enhanced spatial selection').

et al., 2013). If we recall that dispersal evolves upwards partly because of the fitness advantage associated with density release, we can see that this fitness advantage becomes larger as the intrinsic growth rate of the population increases. Thus, the increasing population growth rates leads to increasing selection on dispersal; an effect that has been dubbed 'enhanced spatial selection' (Perkins *et al.*, 2013; Figure 6.2).

As well as these direct processes of selection and sorting, the evolutionary dynamics on expanding range edges are also strongly influenced by drift and founder events. Invasion fronts are made up of relatively few individuals, and dispersal effectively samples these few individuals and/or their offspring to make up the invasion front in the next generation. Thus, the invasion front can be seen as a series of founder events. This serial foundering has many interesting consequences. First, and most obviously, it will lead to a reduction in genetic diversity on the invasion front over time. This effect has been recognized for a long time and is regularly used to infer range expansion from population genetic data (Excoffier *et al.*, 2009). Recently, however, researchers have begun to wonder at the consequences of this lowered variation on invasion fronts, and there are several interesting new directions emerging. One such observation is that relatedness increases on invasion fronts, and so we might expect the emergence of cooperative behaviour to be more likely following a period of range expansion (Van Dyken *et al.*, 2013). High relatedness can also directly contribute to selection for higher dispersal on the invasion front, as individuals move further so as to decrease the levels of kin competition (Hamilton and May, 1977; Kubisch *et al.*, 2013). Another interesting direction is that lowered variance means that the costs of inbreeding becomes vastly weaker, and so mechanisms that have evolved to avoid inbreeding become less relevant (Pujol *et al.*, 2009; Peischl *et al.*, 2015): in plants this might lead to the evolution of asexual lineages, but in animals the consequences are largely unexplored.

A second consequence of this serial foundering during invasion is that we see a strong stochastic process atop the evolutionary dynamics described above. Although we have clear expectations that dispersal and growth rates will increase, these evolutionary trajectories are subject to substantial stochasticity (Hallatschek *et al.*, 2007; Elliott and Cornell, 2013). As a consequence, we can see unexpected, and even deleterious, traits come to dominate an invasion front and get smeared across space (Klopfstein *et al.*, 2006; Travis *et al.*, 2007; Excoffier and Ray, 2008; Excoffier *et al.*, 2009; Peischl *et al.*, 2015). One of the consequences of this is that invasion speeds, though accelerating, are potentially very unpredictable (Phillips, 2015). From the same starting conditions, very different trait values and rates of range shift can evolve. This has implications for management (how do we deal with the uncertainty around rates of spread), but also for basic research (genetic or behavioural variation across invasion fronts may have no clear adaptive explanation).

A third consequence of serial foundering during invasion is that the invasion front may lose its parasites and pathogens. It is well recognized that the founder event associated with initial introduction often leaves introduced populations free of parasites (Mitchell and Power, 2003; Torchin *et al.*, 2003). Less appreciated, however, is that these founder events continue as the population spreads in its new range (Phillips *et al.*, 2010c). So even those parasites that are introduced with their host may still be lost on the invasion front. Why does this happen? By chance – particularly for pathogens and parasites that are naturally at low prevalence – individuals arriving on the edge of the expanding range edge may all be free of a particular parasite. When this happens, particularly if that parasite also has fitness consequences, the parasite may never again regain the invasion front, but will lag behind the invasion of its host (Phillips *et al.*, 2010c). This lag is further compounded if the parasite has density-dependent transmission. Thus, individuals on invasion fronts may find themselves free of parasites and pathogens. Given the strong role of parasites and pathogens in shaping the behaviour and demographics of their host (e.g. Hudson *et al.*, 1998; Barber and Dingemanse, 2010), such a situation clearly has manifold ramifications. Again, we might expect to see faster reproductive rates, changes in the mating system, or in levels of sociality, but again, these possibilities are yet to be explored empirically.

So at this stage, we have an expectation that, on an expanding range edge, dispersal rates will evolve upwards, and population growth rate will similarly increase (both subject to stochasticity and Allee effects). As well as this, average relatedness increases, and rates of parasitism drop, which may respectively lead to further increases in dispersal and reproductive rates. Increases in dispersal and population growth, of course, lead to faster spread. So the evolutionary dynamics on an invasion front will often cause that invasion front to accelerate over time (Perkins, 2012; Perkins *et al.*, 2013).

The Evolutionary Dynamics of Spread: Empirical Results

So much for the theory. Does this theory actually play out in the real world? And how are abstractions such as dispersal and population growth rates manifest in animal behaviour? Over the last 10 years, there has been a steady accumulation of studies suggesting rapid trait evolution on invasion fronts. By far the most well developed of these systems is the invasion of the cane toad across northern Australia, but empirical work now spans animal taxa ranging from ants to crustaceans, fish and birds. In this section, the author will focus largely on the cane toad work, but will point to how results from the toad system are paralleled elsewhere.

Cane toads (*Rhinella marina*, formerly *Bufo marinus*) were introduced along the north-eastern coastline of Australia in 1937 in a bid to provide biological control of beetles impacting the sugar industry (Lever, 2001). The founders of this population were 101 animals brought from Hawaii (itself an introduced population, originally from Guyana via Puerto Rico). These 101 founders rapidly bred (toads produce tens of thousands of eggs per clutch) and toads quickly established themselves and began to spread. In the first few decades, the toad invasion spread at around 10 km y^{-1} (Phillips *et al.*, 2006). In northern Australia, toads found a massive region of suitable habitat, and the invasion has been progressing steadily across northern Australia ever since. At the time of writing, the invasion front is now just over 2000 km from where it began in 1937. Examination of the historical spread rate shows that the invasion steadily accelerated: while it progressed at 10 km y^{-1} 75 years ago, it now moves forward around five times faster at 45–60 km y^{-1} (Phillips *et al.*, 2006; Urban *et al.*, 2008).

The Evolution of Dispersal

The toad invasion is a biological invasion on an enormous scale. More than 75 generations have elapsed in which the evolutionary pressures described above have been in play. We can observe that the invasion has accelerated, but has this been driven by shifts in dispersal and population growth? The short answer is yes (Perkins *et al.*, 2013). First, it is clear that dispersal rates have evolved. By collecting animals from across northern Australia and observing their rates of dispersal in the field we know that toads from the invasion front disperse more rapidly than their counterparts in longer-established populations (Phillips *et al.*, 2008; Alford *et al.*, 2009). Moreover, these patterns were also present in their captive-reared offspring, suggesting that dispersal is heritable (Phillips *et al.*, 2010a). At the level of behaviour, greater dispersal rates are achieved by invasion-front animals moving more often and following much straighter paths than their conspecifics in older populations (Brown *et al.*, 2014). While there are clearly shifts in path straightness, there is nonetheless no inherent directionality evolving: toads from the invasion front are as likely to disperse in an easterly direction (back towards the bulk of the population as in a westerly direction (into uncolonized territory; Brown *et al.*, 2015). This latter result appears to be driven by the fact that directionality is not heritable.

While increased dispersal during invasion has been very clearly demonstrated in toads, there are many other systems in which shifts in dispersal-relevant traits have been observed during range shift. On expanding range edges: crickets have a higher proportion of winged individuals (Simmons and Thomas, 2004); butterflies have larger wings and thoraxes (Hughes *et al.*, 2007); ants have a higher proportion of dispersal castes (Léotard *et al.*, 2009); bluebirds have higher dispersal tendency (Duckworth and Badyaev, 2007); common mynahs have larger wings (Berthouly-Salazar *et al.*, 2012); ladybirds have greater flight speed (Lombaert *et al.*, 2014). The list of systems in which we see increased dispersal on expanding ranges is large and growing. Many of these systems examine dispersal-relevant morphological traits, so the assumption is that shifts in these traits will be paralleled by shifts in dispersal behaviour. Certainly, behaviour tends to account for much greater variation in dispersal than morphology alone (big wings are useful, but if you choose not to use them . . .), but dispersal behaviour can be difficult to measure. Because of the inherent difficulty in measuring dispersal, for many species we will have to content ourselves with shifts in dispersal-relevant traits. Nonetheless as technology for tracking animals gets better and smaller (e.g. Bridge *et al.*, 2011), we can expect more studies to demonstrate shifts in dispersal behaviour rather than in dispersal-associated traits.

The Evolution of Growth Rates

The hypothesis that growth rate should increase during invasion is less widely supported by empirical evidence than the prediction about dispersal rates. The hypothesis also rests on slightly shakier assumptions in that it requires a population to be strongly density regulated. While many populations will be (Brook and Bradshaw, 2006), it is also likely that many populations are governed in a density-independent manner. Also, measuring population growth rate is difficult, and so we will again typically look for proxies such as rates of maturation and fecundity. Despite these difficulties, there is also growing empirical support for increased growth rates during invasion.

Cane toads are one such example. There is little doubt that toad populations are strongly density regulated, and this regulation happens primarily in the juvenile stages. Tadpoles compete with each other (Alford, 1994), but also actively target and consume conspecific eggs (Crossland and Shine, 2011a) and secrete chemicals that interfere with the development of any eggs that they cannot physically destroy (Crossland and Shine, 2011b). On land, juvenile toads are strongly cannibalistic, even to the point of having an evolved behaviour – vibrating their toes to lure smaller conspecifics – that facilitates this cannibalistic behaviour (Pizzatto and Shine, 2008). On the invasion front, where conspecific density is low (and competition for food is also lower; Brown *et al.*, 2013), much of this conspecific regulation is absent. So instead of investing in strategies to minimize the impact of conspecifics, toads can simply invest in rapidly producing offspring. Life history theory tells us that the most effective way of increasing reproductive rate is to mature quickly (e.g. Lewontin, 1965; Reznick *et al.*, 2002) and there is very clear evidence that toads on the invasion front do exactly this. When grown individually, but under common conditions, tadpole and juvenile toads from the invasion front grew around 30% faster than their conspecifics from long-established populations (Phillips, 2009). This is a large difference in growth rate and, all else being equal, would result in substantially greater rates of reproduction in invasion-front toads. Although behavioural aspects of this increased growth rate are unknown, we would expect them to be there.

Changes in population growth rate (or associated traits) have been observed in several other invasive systems besides toads. Typically these differences are observed from field-collected specimens, and so it is difficult to separate environmental effects from evolutionary shifts. One exception to this is the higher individual growth rates in a common environment observed in invasion-front populations of a damselfly (Therry *et al.*, 2014a, b). When we expand our scope to admit observations of field-collected populations, however, the list rapidly grows. Fish are particularly well-represented, with evidence for greater growth or reproductive rates on the invasion front for invasive jewelfish (Lopez *et al.*, 2012); Vendace (Bøhn *et al.*, 2004, Amundsen *et al.*, 2012), white perch (Feiner *et al.*, 2012) and round gobies (Gutowsky and Fox, 2012).

While evolutionary shifts in growth/reproduction might be achieved by re-allocation of physiological investment, they can equally be influenced by behaviours. Increased foraging effort, increased sociality and increased 'boldness', for example, might be the behavioural drivers of shifts towards higher growth and reproduction rates, and there are hints that these behavioural traits may well shift during invasions (Duckworth and Badyaev, 2007; Rodriguez *et al.*, 2010).

Empirical Opportunities

The decreased genetic variation of invasion-front animals, and their loss of parasites, potentially have manifold ramifications for the evolution of behaviour, but there is, to the author's knowledge, no empirical work on these possibilities to date. In particular, it seems likely that mating systems may be subject to change during invasion. Certainly, evidence is mounting that mating systems in plants are often modified during invasion, and this is thought to be driven by the decreased variance (and so lowered costs of inbreeding) on the invasion front (Pujol *et al.*, 2009). The same effect happens in animals too (e.g. Facon *et al.*, 2011), but the consequences of this for the evolution of the mating system in animals remains to be investigated. Animal mating systems are also shaped to some extent by parasite transmission (it is costly to have many mates if there is a good chance of catching something with each mating; Agrawal and Lively, 2001). So, examination of mating systems on invasion fronts remains an intriguing avenue for future research.

Behavioural Syndromes

From the foregoing, it should be abundantly clear that the process of invasion has an astonishing number of consequences for the evolution of the vanguard population. Given sufficient time to expand, the invasion-front population will be radically different from the population at the point of introduction. This is certainly the case with cane toads in Australia. As well as differences in dispersal and growth across the invasion history, research has uncovered differences in leg-length, toxicity, the immune system, parasite and disease prevalence, physiology and competitive ability. The sheer number of traits that vary with invasion history is astonishing, and undoubtedly there are more to be uncovered. In this system and others, many of these differences will have adaptive explanations, but some may not.

There is much current interest in the behavioural and life history literature with 'syndromes', otherwise known as correlations between traits (e.g. Sih *et al.*, 2004b; Stevens *et al.*, 2013). In behaviour, there has been a surge of interest around personality, the idea that individuals have unique and relatively consistent behaviours that are distributed along a few axes of variation in personalities (Dingemanse and Réale, 2005). 'Behavioural syndromes' describe the situation of a correlation between behavioural traits, and the major axis of variation is often between 'bold' and 'shy' individuals. That is, across contexts, some individuals tend to be bolder than others, showing faster exploratory behaviour, taking more risks and so on (Sih *et al.*, 2004a).

While the existence of behavioural syndromes is becoming increasingly clear, their genesis remains murky (Biro and Stamps, 2008). Why does there seem to be a major axis in behavioural variation, consistent across many taxa, between bold and shy individuals? Against the backdrop of evolution on range edges, it is very tempting to speculate that behavioural syndromes (and indeed broader life history syndromes) have their genesis in periods of range shift (Phillips and Suarez, 2012). Certainly, evolutionary pressures on an invasion front should select for bolder individuals: individuals that are more active, take greater risks and are neophilic. Thus, range shift should magnify variation along the fast–slow axis of life history variation, and this axis has already been noted as a possible driver of variation in personality (Réale *et al.*, 2010). Given that most species shift their range at some time or another (Vermeij, 2005), or alternatively exist in metapopulations (in which, to some extent, the evolutionary dynamics of range shift are internal to the system), behavioural syndromes may often have a genesis in the evolutionary dynamics of range shift. If we accept that range shift can create substantial variation along the bold–shy axis, the question is not one of how variation is generated so much as how it is maintained.

Management Implications

Another outcome of the astonishing divergence between vanguard and long-established populations is that management strategies – aimed at controlling and eradicating invasive species – may be more or less effective, depending on where they are executed. A strategy that works well in long-established populations may have little impact on the expanding range edge. As an example, a strategy of culling individuals might achieve reasonably long-lasting reductions in local abundance in a long-established population, but will have only a very temporary success in a vanguard population because the treated area will be rapidly recolonized.

Similarly, efforts at biological control may be more or less effective between vanguard and long-established populations. If we imagine the introduction of a disease agent, then there is little doubt that differences in transmission (driven by host movement, mating system and sociality) and pathogenicity (driven by changes in the host immune system) would emerge between old and new populations. Moreover, on the expanding range edge, it is very likely that our agent of biological control would simply be left behind (Phillips *et al.*, 2010c).

As well as creating complications, an appreciation of the evolutionary dynamics of invasion potentially creates new opportunities. One such opportunity is to use gene flow to create barriers to spread. As invasions progress, the vanguard population typically becomes more dispersive. As a consequence, landscape features that may once have acted as a barrier to dispersal (e.g. rivers, mountains, valleys) will be less effective. Because of this, the invasion may spread into regions it would not otherwise have achieved. If such barriers can be identified, then a management option to renew the effectiveness of these barriers would be to translocate individuals from long-established populations to the soon-to-be-invaded side of the potential barrier. The barrier remains effective against the slow-dispersing phenotypes from the range core, and these phenotypes spread back towards the oncoming invasion. If the phenotypes from the range core are more competitive (and we would expect them to be), then we set up a situation of a 'genetic backburn': the highly dispersive phenotypes never reach the barrier (Phillips *et al.*, 2016).

Interestingly, exactly this situation is about to play out in the cane toads' invasion of northern Australia. The toads will soon reach an arid barrier that has the potential to stop them reaching a further 268 000 km^2 of the Australian mainland (Florance *et al.*, 2011). Whether the barrier will be effective depends on careful management of artificial waterbodies along a thin strip of coastline (Tingley *et al.*, 2013). Given that invasion-front toads move more than five times further each wet season than their long-established counterparts, one way to make this barrier substantially more effective would be to execute a genetic backburn. That is, we could use rapid behavioural evolution to ensure the barrier's effectiveness.

Conclusion

Behavioural variation and selection on that variation not only lead to altered phenotypes on invasion fronts, but also change the behaviour of the invasion front itself. If invasion-front populations develop greater dispersal and reproductive rates (as often seems to occur), the invasion will accelerate over time. As well as this, control strategies that might work in the core of the range may be more or less effective on the expanding range edge. Finally, the extremely stochastic nature of the eco-evolutionary dynamics on the invasion front means that predicting the time course of spread or trait evolution is fraught with uncertainty. Together, these outcomes mean that the evolution of behaviour during invasion is at once a fascinating and challenging study for the theoretically inclined, while also being of immense practical relevance.

References

Agrawal, A.F. and Lively, C.M. (2001). Parasites and the evolution of self-fertilization. *Evolution*, 55, 869–879.

Alford, R.A. (1994). Interference and exploitation competition in larval *Bufo marinus*. *Advances in Ecology and Environmental Sciences*, 297–306.

Alford, R.A., Brown, G.P., Schwarzkopf, L., Phillips, B.L. and Shine, R. (2009). Comparisons through time and space suggest rapid evolution of dispersal behaviour in an invasive species. *Wildlife Research*, 36, 23–28.

Amundsen, P.-A., Salonen, E., Niva, T., *et al.* (2012). Invader population speeds up life history during colonization. *Biological Invasions*, 14, 1501–1513.

Barber, I. and Dingemanse, N.J. (2010). Parasitism and the evolutionary ecology of animal personality. *Philosophical Transactions of the Royal Society B: Biological Sciences*, 365, 4077–4088.

Benichou, O., Calvez, V., Meunier, N. and Voituriez, R. (2012). Front acceleration by dynamic selection in Fisher population waves. *Physical Review E*, 86, 041908.

Berthouly-Salazar, C., van Rensburg, B.J., Le Roux, J.J., van Vuuren, B.J. and Hui, C. (2012). Spatial sorting drives morphological variation in the invasive bird, *Acridotheris tristis*. *PLoS ONE*, 7, e38145.

Biro, P.A. and Stamps, J.A. (2008). Are animal personality traits linked to life-history productivity? *Trends in Ecology and Evolution*, 23, 361–368.

Blackburn, T.M. and Duncan, R.P. (2001). Establishment patterns of exotic birds are constrained by non-random patterns in introduction. *Journal of Biogeography*, 28, 927–939.

Bøhn, T., Terje Sandlund, O., Amundsen, P.-A. and Primicerio, R. (2004). Rapidly changing life history during invasion. *Oikos*, 106, 138–150.

Bouin, E., Calvez, V., Meunier, N., *et al.* (2012). Invasion fronts with variable motility: phenotype selection, spatial sorting and wave acceleration. *Comptes Rendus Mathematique*, 350, 761–766.

Bridge, E.S., Thorup, K., Bowlin, M.S., *et al.* (2011). Technology on the move: recent and forthcoming innovations for tracking migratory birds. *BioScience*, 61, 689–698.

Brook, B.W. and Bradshaw, C.J.A. (2006). Strength of evidence for density dependence in abundance time series of 1198 species. *Ecology*, 87, 1445–1451.

Brown, G.P., Kelehear, C. and Shine, R. (2013). The early toad gets the worm: cane toads at an invasion front benefit from higher prey availability. *Journal of Animal Ecology*, 82, 854–862.

Brown, G.P., Phillips, B.L. and Shine, R. (2014). The straight and narrow path: the evolution of straight-line dispersal at a cane toad invasion front. *Proceedings of the Royal Society B: Biological Sciences*, 281, 20141385.

Brown, G.P., Phillips, B.L. and Shine, R. (2015). Directional dispersal has not evolved during the cane toad invasion. *Functional Ecology*, 29, 830–838.

Burton, O.J., Travis, J.M.J. and Phillips, B.L. (2010). Trade-offs and the evolution of life-histories during range expansion. *Ecology Letters*, 13, 1210–1220.

Chapple, D.G., Simmonds, S.M. and Wong, B. (2012). Can behavioral and personality traits influence the success of unintentional species introductions? *Trends in Ecology and Evolution*, 27, 57–64.

Crossland, M.R. and Shine, R. (2011a). Cues for cannibalism: cane toad tadpoles use chemical signals to locate and consume conspecific eggs. *Oikos*, 120, 327–332.

Crossland, M.R. and Shine, R. (2011b). Embryonic exposure to conspecific chemicals suppresses cane toad growth and survival. *Biology Letters*, rsbl20110794.

Dingemanse, N.J. and Réale, D. (2005). Natural selection and animal personality. *Behaviour*, 142, 1159–1184.

Duckworth, R.A. and Badyaev, A.V. (2007). Coupling of dispersal and aggression facilitates the rapid range expansion of a passerine bird. *Proceedings of the National Academy of Sciences, USA*, 104, 15017–15022.

Elliott, E.C. and Cornell, S.J. (2013). Are anomalous invasion speeds robust to demographic stochasticity? *PLoS ONE*, 8, e67871.

Elton, C.S. (1958). *The Ecology of Invasions by Animals and Plants*. London: Methuen.

Excoffier, L. and Ray, N. (2008). Surfing during population expansions promotes genetic revolutions and structuration. *Trends in Ecology and Evolution*, 23, 347–351.

Excoffier, L., Foll, M. and Petit, R.J. (2009). Genetic consequences of range expansions. *Annual Review of Ecology and Systematics*, 2009, 40.

Facon, B., Hufbauer, R.A., Tayeh, A., *et al.* (2011). Inbreeding depression is purged in the invasive insect *Harmonia axyridis*. *Current Biology*, 21, 424–427.

Feiner, Z., Aday, D.D. and Rice, J. (2012). Phenotypic shifts in white perch life history strategy across stages of invasion. *Biological Invasions*, 14, 2315–2329.

Fisher, R.A. (1937). The wave advance of advantageous genes. *Annals of Eugenics*, 7, 355–369.

Florance, D., Webb, J.K., Dempster, T., *et al.* (2011). Excluding access to invasion hubs can contain the spread of an invasive vertebrate. *Proceedings of the Royal Society B: Biological Sciences*, 278, 2900–2908.

Gutowsky, L.F.G. and Fox, M.G. (2012). Intra-population variability of life-history traits and growth during range expansion of the invasive round goby, *Neogobius melanostomus*. *Fisheries Management and Ecology*, 19, 78–88.

Hallatschek, O., Hersen, P., Ramanathan, S. and Nelson, D.R. (2007). Genetic drift at expanding frontiers promotes gene segregation. *Proceedings of the National Academy of Sciences*, 104, 19926–19930.

Hamilton, W.D. and May, R.M. (1977). Dispersal in stable habitats. *Nature*, 269, 578–581.

Hastings, A., Cuddington, K., Davies, K.F., *et al.* (2005). The spatial spread of invasions: new developments in theory and evidence. *Ecology Letters*, 8, 91–101.

Hudson, P.J., Dobson, A.P. and Newborn, D. (1998). Prevention of population cycles by parasite removal. *Science*, 282, 2256–2258.

Hughes, C.L., Dytham, C. and Hill, J.K. (2007). Modelling and analysing evolution of dispersal in populations at expanding range boundaries. *Ecological Entomology*, 32, 437–445.

Klopfstein, S., Currat, M. and Excoffier, L. (2006). The fate of mutations surfing on the wave of range expansion. *Molecular Biology and Evolution*, 23, 482–490.

Kubisch, A., Fronhofer, E.A., Poethke, H.J. and Hovestadt, T. (2013). Kin competition as a major driving force for invasions. *The American Naturalist*, 181, 700–706.

Léotard, G., Debout, G., Dalecky, A., *et al.* (2009). Range expansion drives dispersal evolution in an equatorial three-species symbiosis. *PLoS ONE* 4:e5377.

Lever, C. (2001). *The Cane Toad. The History and Ecology of a Successful Colonist*. Yorkshire, UK: Westbury Academic and Scientific Publishing.

Lewontin, R.C. (1965). Selection for colonizing ability. In *The Genetics of Colonizing Species*, ed. Baker, H. and Stebbins, G. London: Academic Press, pp. 77–94.

Lockwood, J.L., Hoopes, M.F. and Marchetti, M.P. (2007). *Invasion Ecology*. Maiden, MA: Blackwell.

Lombaert, E., Estoup, A. Facon, B., *et al.* (2014). Rapid increase in dispersal during range expansion in the invasive ladybird *Harmonia axyridis*. *Journal of Evolutionary Biology*, 27, 508–517.

Lopez, D.P., Jungman, A.A. and Rehage, J.S. (2012). Nonnative African jewelfish are more fit but not bolder at the invasion front: a trait comparison across an Everglades range expansion. *Biological Invasions*, 14, 2159–2174.

MacArthur, R.H. and Wilson, E.O. (1967). *The Theory of Island Biogeography*. Princeton, NJ: Princeton University Press.

Mitchell, C.E. and Power, A.G. (2003). Release of invasive plants from fungal and viral pathogens. *Nature*, 421, 625–627.

Peischl, S., Kirkpatrick, M. and Excoffier, L. (2015). Expansion load and the evolutionary dynamics of a species range. *American Naturalist*, 185(4), E81–93.

Perkins, T.A. (2012). Evolutionarily labile species interactions and spatial spread of invasive species. *The American Naturalist*, 179, E37–E54.

Perkins, T.A., Phillips, B.L., Baskett, M.L. and Hastings, A. (2013). Evolution of dispersal and life-history interact to drive accelerating spread of an invasive species. *Ecology Letters*, 16, 1079–1087.

Phillips, B.L. (2009). The evolution of growth rates on an expanding range edge. *Biology Letters*, 5, 802–804.

Phillips, B.L. (2015). Evolutionary processes make invasion speed difficult to predict. *Biological Invasions*, 17, 1949–1960.

Phillips, B.L. and Suarez, A.V. (2012). The role of behavioural variation in the invasion of new areas. In *Behavioural Responses to a Changing World: Mechanisms and Consequences*, ed. Candolin, U. and Wong, B.B.M. Oxford: Oxford University Press, pp. 190–200.

Phillips, B.L., Brown, G.P., Webb, J.K. and Shine, R. (2006). Invasion and the evolution of speed in toads. *Nature*, 439, 803.

Phillips, B.L., Brown, G.P., Travis, J.M.J. and Shine, R. (2008). Reid's paradox revisited: the evolution of dispersal in range-shifting populations. *The American Naturalist*, 172, S34–S48.

Phillips, B.L., Brown, G.P. and Shine, R. (2010a). Evolutionarily accelerated invasions: the rate of dispersal evolves upwards during range advance of cane toads. *Journal of Evolutionary Biology*, 23, 2595–2601.

Phillips, B.L., Brown, G.P. and Shine, R. (2010b). Life-history evolution in range-shifting populations. *Ecology*, 91, 1617–1627.

Phillips, B.L., Kelehear, C., Pizzatto, L., *et al.* (2010c). Parasites and pathogens lag behind their host during periods of host range-advance. *Ecology*, 91, 872–881.

Phillips, B.L., Tingley, R. and Shine, R. (2016). Genetic backburning to halt invasions. *Proceedings of the Royal Society B: Biological Sciences*, 283, 20153037.

Pizzatto, L. and Shine, R. (2008). The behavioral ecology of cannibalism in cane toads (*Bufo marinus*). *Behavioral Ecology and Sociobiology*, 63, 123–133.

Pujol, B., Zhou, S.-R., Vilas, J.S. and Pannell, J.R. (2009). Reduced inbreeding depression after species range expansion. *Proceedings of the National Academy of Sciences, USA*, 106, 15379–15383.

Réale, D., Garant, D., Humphries, M.M., *et al.* (2010). Personality and the emergence of the pace-of-life syndrome concept at the population level. *Philosophical Transactions of the Royal Society B: Biological Sciences*, 365, 4051–4063.

Reznick, D., Bryant, M.J. and Bashey, F. (2002). r- and K-selection revisited: the role of population regulation in life-history evolution. *Ecology*, 83, 1509–1520.

Rodriguez, A., Hausberger, M. and Clergeau, P. (2010). Flexibility in European starlings' use of social information: experiments with decoys in different populations. *Animal Behaviour*, 80, 965–973.

Shigesada, N. and Kawasaki, K. (1997). *Biological Invasions: Theory and Practice*. Oxford, UK: Oxford University Press.

Shine, R., Brown, G.P. and Phillips, B.L. (2011). An evolutionary process that assembles phenotypes through space rather than through time. *Proceedings of the National Academy of Sciences, USA*, 108, 5708–5711.

Sih, A., Bell, A. and Johnson, J.C. (2004a). Behavioral syndromes: an ecological and evolutionary overview. *Trends in Ecology and Evolution*, 19, 372–378.

Sih, A., Bell, A.M., Johnson, J.C. and Ziemba, R.E. (2004b). Behavioral syndromes: an integrative overview. *The Quarterly Review of Biology*, 79, 241–277.

Simmons, A.D. and Thomas, C.D. (2004). Changes in dispersal during species' range expansions. *American Naturalist*, 164, 378–395.

Simons, A.M. (2003). Invasive aliens and sampling bias. *Ecology Letters*, 6, 278–280.

Skellam, J.G. (1951). Random dispersal in theoretical populations. *Biometrika*, 38, 196–218.

Stevens, V.M., Trochet, A., Blanchet, S., Moulherat, S., Clobert, J. and Baguette, M. (2013). Dispersal syndromes and the use of life-histories to predict dispersal. *Evolutionary Applications*, 6, 630–642.

Therry, L., Lefevre, E., Bonte, D. and Stoks, R. (2014a). Increased activity and growth rate in the non-dispersive aquatic larval stage of a damselfly at an expanding range edge. *Freshwater Biology*, 59, 1266–1277.

Therry, L., Nilsson-Örtman, V., Bonte, D. and Stoks, R. (2014b). Rapid evolution of larval life history, adult immune function and flight muscles in a poleward-moving damselfly. *Journal of Evolutionary Biology*, 27, 141–152.

Tingley, R., Phillips, B.L., Letnic, M., *et al.* (2013). Identifying optimal barriers to halt the invasion of cane toads *Rhinella marina* in arid Australia. *Journal of Applied Ecology*, 50, 129–137.

Torchin, M.E., Lafferty, K.D., Dobson, A.P., McKenzie, V.J. and Kuris, A.M. (2003). Introduced species and their missing parasites. *Nature*, 421, 628–630.

Travis, J.M.J. and Dytham, C. (2002). Dispersal evolution during invasions. *Evolutionary Ecology Research*, 4, 1119–1129.

Travis, J.M.J., Münkemüller, T., Burton, O.J., Best, A., Dytham, C. and Johst, K. (2007). Deleterious mutations can surf to high densities on the wave front of an expanding population. *Molecular Biology and Evolution*, 24, 2334–2343.

Urban, M.C., Phillips, B.L., Skelly, D.K. and Shine, R. (2008). A toad more travelled: the heterogeneous invasion dynamics of cane toads in Australia. *The American Naturalist*, 171, E134–E148.

Van Dyken, J.D., Müller, M.J., Mack, K.M. and Desai, M.M. (2013). Spatial population expansion promotes the evolution of cooperation in an experimental prisoner's dilemma. *Current Biology*, 23, 919–923.

Vermeij, G. (2005). Invasion as expectation: a historical fact of life. In *Species Invasions: Insights into Ecology, Evolution, and Biogeography*, ed. Sax, D.F., Stachowicz, J.J. and Gaines, S.D. Sunderland, MA: Sinauer Associates, pp. 315–339.

Williamson, M. (1996). *Biological Invasions*. London: Chapman and Hall.

7 The Role of Dispersal Behaviour and Personality in Post-establishment Spread

Jennifer S. Rehage, Julien Cote and Andrew Sih

Introduction

Invasions occur as a sequence of steps that start with the transport of a non-native species outside its geographic range, followed by introduction into a novel environment, establishment (i.e. survival and reproduction) and then spread (Williamson and Fitter, 1996; Blackburn *et al.*, 2011; Chapple and Wong, 2015). Once they are able to spread beyond the point of introduction, non-native species are typically considered 'invasive', and may become a conservation concern, as a subset of them will go on to have a negative impact on native communities. The extent of spread or range occupied is an essential part of how we quantify an invader's impact on a recipient community (Parker *et al.*, 1999; Ricciardi *et al.*, 2013), emphasizing the importance of understanding the mechanisms underlying invasive spread.

Spread entails that the introduced individuals are dispersing, surviving and reproducing at multiple locations across the invaded range. A spreading non-native must overcome both environmental and dispersal barriers in order to be successful (Blackburn *et al.*, 2011). As it spreads, it will likely encounter novel abiotic and biotic conditions that it must cope with in order to become established at each new location (e.g. Sih *et al.*, 2011). The invader must be a good disperser, dispersing and colonizing new habitat patches in an unfamiliar landscape. Dispersal is fundamentally a behavioural process that allows organisms to move from one reproductive patch to another (Clobert *et al.*, 2009), and in an invasion context, it allows an invader to find suitable conditions in new habitats, driving spread. Indeed, invasive species have been found to be better dispersers than their non-invasive relatives (Rehage and Sih, 2004; Bubb *et al.*, 2006; Table 7.1). In this chapter, we examine the behaviours associated with dispersal in the spread phase of invasions and discuss the role of intraspecific behavioural variation and animal personalities in dispersal and invasive spread. We overview the nature of dispersal and implications for invasions, outline dispersal mechanisms and behaviours responsible for spread in the handful of systems where it has been explored, and highlight insights from invasive mosquitofish (*Gambusia* spp.).

Biological Invasions and Animal Behaviour, eds J.S. Weis and D. Sol. Published by Cambridge University Press. © Cambridge University Press 2016.

Table 7.1 Variation in dispersal and underlying behaviours in invasion studies. Studies compare dispersing versus resident individuals, invasion front versus core populations, introduced versus native populations, and native/expanding versus non-native/displaced species

Behavioural traits	Population/Individual comparison	Directionality of trait variation	Taxa (Source)
Dispersal behaviour/ tendency	Native vs. non-native species	Non-native species moves greater distances	Signal vs. white-clawed crayfish (Bubb et al., 2006)
	Native vs. non-native species	Non-native species has a greater dispersal tendency	Four mosquitofish species (Rehage and Sih, 2004)
	Invasion front vs. core populations	Individuals from invasion front have higher flight speeds but similar flight endurance and motivation to fly off	Ladybird (Lombaert et al., 2014)
	Invasion front vs. core populations	Individuals from invasion front have similar dispersal tendencies	African jewelfish (Lopez et al., 2012)
	Invasion front vs. core populations	Individuals from invasion front follow straighter paths, move more often and further, and have higher endurance	Cane toads (Alford et al., 2009; Llewelyn et al., 2010)
Aggressiveness (intraspecific)	Invasion front vs. core populations	Individuals from invasion front are more aggressive	Western bluebird (Duckworth and Badyaev, 2007), round goby (Groen et al., 2012)
	Non-native vs. native populations	Individuals from non-native populations are more aggressive	Signal crayfish (Pintor et al., 2008)
	Non-native vs. native populations	Individuals from non-native populations are less aggressive	Argentine ant (Holway et al., 1998; Suarez et al., 1999)
	Invasion front vs. core populations	Individuals from invasion front are less aggressive	Signal crayfish (Hudina et al., 2014)
Aggressiveness (interspecific)	Expanding vs. displaced species	Expanding species is more aggressive	Western bluebird vs. mountain bluebird (Duckworth and Badyaev, 2007)
Boldness	Dispersing vs. resident individuals	Individuals from bolder populations disperse more	Western mosquitofish (Cote et al., 2011)
	Introduced vs. native populations	Individuals from introduced populations are bolder	Signal crayfish (Pintor et al., 2008)
	Invasion front vs. core populations	Individuals from invasion front are equally bold	African jewelfish (Lopez et al., 2012), round goby (Groen et al., 2012)
	Native vs. non-native species	Non-native species are bolder	Four mosquitofish species (Rehage and Sih, 2004)

(cont.)

Table 7.1 (*cont.*)

Behavioural traits	Population/Individual comparison	Directionality of trait variation	Taxa (Source)
Behavioural flexibility/ neophobia	Spreading population vs. well-established populations	Individuals from an actively spreading population respond more slowly to novel foods but not novel objects	House sparrows (Martin and Fitzgerald, 2005)
Activity/ exploratory behaviour	Dispersing vs. resident individuals	Dispersers from more active/exploratory populations disperse more	Western mosquitofish (Cote *et al.*, 2011)
	Native vs. non-native species	Non-native species explore more	Four mosquitofish species (Rehage and Sih, 2004)
Sociability	Dispersing vs. resident individuals	Dispersers are less social	Western mosquitofish (Cote *et al.*, 2011)

The Nature of Dispersal and Implications for Invasive Spread

Dispersal typically involves three stages: departure from a living area, moving between areas (transience) and settlement in a new area (Bowler and Benton, 2005; Clobert *et al.*, 2009). Dispersal is different from daily movements since it involves a clear departure decision from an established living area to integrate into an existing population or, in the case of an invasion, to establish a new population. Dispersal is a behavioural response resulting from a balance between the costs and benefits of movement. Many biotic and abiotic factors will affect these costs and benefits along the three dispersal stages (Bowler and Benton, 2005; Ronce, 2007; Clobert *et al.*, 2009). Dispersers' benefits can derive from leaving habitats with unsuitable conditions; e.g. those with high intraspecific or interspecific competition, predation risk or inbreeding risk. On the other hand, dispersers are also faced with costs and risks (Bonte *et al.*, 2012). Dispersal is a costly behaviour, as it may require a significant amount of energy and time, and a risky behaviour because, in many cases, dispersers face both high risks of mortality while dispersing and uncertainty about finding suitable habitats and/or populations to settle in.

The benefits and costs of dispersal are likely to be context dependent and can act similarly or differently on the three stages of dispersal. First, dispersal benefits and costs may vary with the context of dispersal. A well-known example is landscape fragmentation. Dispersal costs increase with population isolation and matrix unsuitability (Bonte *et al.*, 2012). Dispersers are more visible to predators and the probability of finding a suitable habitat patch is lower (Bowne and Bowers, 2004). Second, benefits and costs at each stage can be interrelated, resulting in carryover effects of departure decisions on decisions and success in subsequent dispersal stages (reviewed in Clobert *et al.*, 2009). Because of the sequential nature of dispersal, departure decisions should impact behaviour during transience (Clobert *et al.*, 2009). For example, common dwarf spiders (*Erigoneatra*) display variation in transience behaviour (Bonte *et al.*, 2009). Some individuals disperse over long distances, by ballooning, while others disperse over short distance, by rappelling. Interestingly, these decisions are also context dependent, as thermal conditions in the initial habitat explain the frequency of long- and short-distance dispersal, with riskier long-distance ballooning being more frequent under cooler, spring-like conditions when suitable habitats are homogeneously distributed (Bonte *et al.*, 2008).

In an adaptive framework, organisms should disperse when leaving yields higher expected fitness than staying. Thus, ecological contexts inducing departure should be associated with a high probability of dispersal success. For example, in the large white butterfly (*Pieris brassicae*), male-biased sex ratios increase both the probability of emigration and the probability of surviving transience (Trochet *et al.*, 2013). Similar predictions can be made on settlement decisions and success. Natal habitats are likely to shape habitat preferences during settlement. On the one hand, individuals may choose a habitat with cues similar to the ones encountered in their natal area (i.e. natal habitat preference induction), either because they may be better prepared for them or may locate them more quickly (Davis and Stamps, 2004). In flying squirrels (*Pteromysvolans L.*), settlement choices by juveniles can be explained by natal patch size and nest localization (Selonen *et al.*, 2007). On the other hand, in some situations adaptive dispersal predicts that individuals should disperse and choose a novel habitat in order to maximize their fitness (Edelaar *et al.*, 2008). If individuals leave a habitat patch to escape unfavourable local conditions, they should avoid habitat patches with similar ecological conditions when settling. For instance, in the common lizard (*Zootoca vivipara*), dispersers leaving low-density populations are more likely to settle in high-density populations (Cote and Clobert, 2007a), where they, in fact, have higher fitness (Cote *et al.*, 2008).

In the context of invasions, non-native species face the difficult task of making good decisions about settlement despite a lack of experience with local habitats and the ecological cues required to make settlement decisions (Davis and Stamps, 2004). In addition, because, by definition, invasions involve movements into areas that lack conspecifics, early invaders cannot rely on social cues to detect good habitats (Doligez *et al.*, 2003; Cote and Clobert, 2007b). Furthermore, after settling in novel habitats, non-native individuals may encounter novel species (i.e. predators, parasites, competitors) that they are not well adapted to (e.g. Sih *et al.*, 2011). And, this could induce high dispersal rates from newly colonized sites.

The above dispersal costs can be offset somewhat by the benefits that non-native species encounter when introduced at a low density and thus experience low intraspecific competition. Conversely, low initial densities can also reduce fitness due to Allee effects and increased kin competition. When a small number of individuals are introduced, high relatedness and kin competition are expected at the site of introduction, which might have negative effects on fitness and potentially result in faster invasion spread to avoid inbreeding and competition with kin (Cote *et al.*, 2007; Kubisch *et al.*,

2013). Additionally, natal habitat preference induction in newly invaded sites could make the offspring of colonizers more likely to prefer empty habitats and facilitate range expansion (Davis and Stamps, 2004).

The Importance of Individual Variation in Dispersal

A common idea is that, for a given environmental context, a species exhibits set dispersal behaviours, and dispersers are a random subset of individuals from the source population. In recent years, however, we have shifted focus to recognize that individual variation in dispersal decisions and behaviour matters. Faced with the same environmental conditions, different individuals will make different dispersal decisions (Clobert *et al.*, 2009). In fact, dispersers often differ from non-dispersers in phenotype (Bowler and Benton, 2005; Clobert *et al.*, 2009). Morphological, physiological and/or behavioural traits relate to how individuals deal with abiotic and biotic factors, and thus affect the benefits and costs of dispersal at the individual level. This creates a suite of phenotypic traits that differs between dispersers and non-dispersers known as a 'dispersal syndrome' (Bowler and Benton, 2005; Clobert *et al.*, 2009). Dispersal syndromes often include life history traits, such as higher fecundity and survival, which might reflect life history strategies (Clobert *et al.*, 2009; Stevens *et al.*, 2014). These dispersal syndromes vary with the ecological drivers and contexts of dispersal (Cote *et al.*, 2010a). For example, in common lizards, individuals dispersing due to high kin competition/relatedness are longer and heavier than residents, but this pattern disappears when kin competition/relatedness within the population is low (Cote *et al.*, 2007).

Although dispersal syndromes are often studied at the departure stage, phenotypic specializations may also influence patterns of transience and settlement. For example, transience duration, distance travelled and path tortuosity can vary consistently among individuals, and can be linked to both genetic and phenotypic variation (Stamps *et al.*, 2007; Delgado *et al.*, 2010). In eagle owls (*Bubo bubo*), dispersers in poorer condition take straighter paths and travel shorter distances (Delgado *et al.*, 2010). A disperser's phenotype can also explain settlement success, as shown in bank voles (*Myodes glareolus*), where post-settlement reproductive success increases with immigrants' aggressiveness and decreases with their sociability (Rémy *et al.*, 2014). Phenotypic specialization observed at departure can be maintained through the entire dispersal process and impact the success of decisions in the transience and settlement stages, linking the three stages together in an adaptive manner (Clobert *et al.*, 2009). For example, disperser phenotypes may increase competitive ability in the settlement patch when dealing with both conspecifics and heterospecifics (Bonte *et al.*, 2014; Duckworth and Badyaev, 2007), or improve colonizing ability in an empty habitat patch (Cote *et al.*, 2007). For instance, high relatedness of common lizards induces the departure of larger individuals, which consequently increases the colonization success of empty habitats (Cote *et al.*, 2007).

Because colonization is at the root of invasions, dispersal syndromes may be highly important to the invasion process. Indeed, while many studies focus on the number of propagules, invasion success can depend as much on the phenotype of individuals

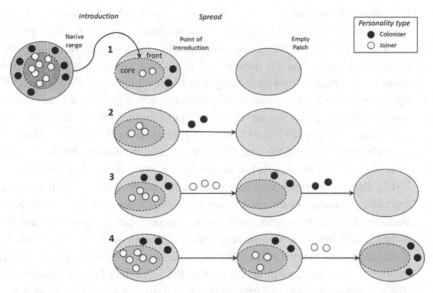

Figure 7.1 Conceptual diagram of spread from the point of introduction in an invasion to uncolonized habitat patches with two personality types, a colonizer and a joiner, and shown across time steps (1–4). Colonizers may be the bolder, more aggressive, active/exploratory and/or asocial individuals relative to joiners. Colonizers also occupy the front of the range, while joiners remain in core populations (modified from Clobert *et al.*, 2009; Cote *et al.*, 2010).

colonizing as on their number (e.g. Burgess and Marshall, 2011). Invading empty habitats may require phenotypic specialization to move over longer distance, to forage in novel places and on novel prey, to avoid new predators, and to more easily acquire individual (as opposed to social) information. The existence of such phenotypic attributes at the invasion front should increase the speed and success of invasions (Figure 7.1). These phenotypic specializations and dispersal propensity may have therefore coevolved in ways that promote the invasion process (Phillips *et al.*, 2006; Duckworth, 2008; Cote *et al.*, 2010b). A well-known example is the cane toad (*Rhinella marina*) invasion in Australia. Toads at the invasion front have longer legs than toads in older populations, a morphological trait conferring higher locomotion ability and longer dispersal distances (Phillips *et al.*, 2006). As dispersal ability is partly genetically driven, the range expansion is accelerated by spatial assortment of dispersal phenotypes across the range (Philips *et al.*, 2010). Using this dispersal syndrome framework, invasiveness may be correlated with an *r*-strategy (e.g. high fecundity, early maturity, short life expectancy) allowing invasive pests to grow rapidly to high density and have high impacts on invaded communities. Studies using this basic framework, however, often neglect the behavioural dimension of a dispersal syndrome.

Personality-dependent Dispersal and its Effect on Invasive Spread

Personality refers to suites of behaviours that are consistent across contexts and time (Sih *et al.*, 2004; Réale *et al.*, 2010). These correlations can be positive or negative, exist both within and between individuals, and result in individuals having distinct behavioural or personality types. Some individuals may have a bolder personality than other individuals when faced with predators, but will also be bolder when encountering food, mates and caring for their young (e.g. Johnson and Sih, 2007). While an individual's personality is not necessarily fixed over a lifetime, a consequence of personality types can be limited behavioural plasticity, which can result in important tradeoffs across situations and affect an individual's performance (Sih *et al.*, 2012).

Personality types are linked to foraging, predator avoidance, reproduction, competitive abilities (Biro and Stamps, 2008; Smith and Blumstein, 2008) and dispersal tendencies, and thus are relevant in an invasion context (Chapple *et al.*, 2012; Sih *et al.*, 2012). Personality types will affect how well individuals fair in all stages of dispersing and colonizing: the decision to depart or stay at the point of introduction, how well invaders do in the transience phase, the decision of when and where to settle in an empty habitat patch, and how well invaders do in the patch selected. Due to the behavioural correlations that make a personality or a 'behavioural syndrome', individuals that are the better dispersers and most likely to be the colonists of new patches will have traits that can promote successful establishment and greater overall invasion success (Figure 7.1), as well as increase the probability of impact on native species (e.g. more aggressive individuals, Hudina *et al.*, 2014).

In individuals that exhibit personality-dependent dispersal, their dispersal tendency is associated with boldness, aggressiveness, activity/exploratory behaviour, and/or sociability (Cote *et al.*, 2010a; Table 7.1). Previous studies show that when dispersers are compared to residents, dispersers are more active (Ducatez *et al.*, 2012), more or less aggressive (O'Riain *et al.*, 1996; Schoepf and Schradin, 2012), more or less bold/exploratory (Fraser *et al.*, 2001; Rasmussen and Belk, 2012), and more or less social (Cote and Clobert, 2007a; Schoepf and Schradin, 2012). For example, Fraser *et al.* (2001) showed that the dispersal distance of Trinidad killifish (*Rivulus hartii*) in streams, is positively correlated with their boldness and exploratory behaviour in a novel habitat.

Only a small number of studies have examined dispersal behaviour and personality-dependent dispersal in invasive species (Table 7.1). Successful colonization of empty habitat patches during an invasion is expected to depend on the presence of multiple personality types in the invading population, and on individuals with certain personality traits, such as those that are more aggressive, bolder/more exploratory or asocial at the invasion front (Figure 7.1; Cote *et al.*, 2010a). These behavioural traits should reduce the costs of dispersal (Clobert *et al.*, 2009), promote the colonization of new habitats and speed up the spread of invasions (Chapple *et al.*, 2012; Chapple and Wong, 2015).

Greater boldness and activity/exploratory behaviour has been detected among individuals, populations and species that differ in invasiveness (Table 7.1). These traits should be adaptive during the spread phase, when organisms are colonizing environments that are likely to be novel and risky, where familiarity is low, and cues that are typically indicative of good habitat quality may be absent (e.g. presence of conspecifics; Stamps, 2001). Increased activity and exploratory behaviour can counteract potential Allee effects by allowing invaders to find mates at the low densities found at the edge of an invasion. Similarly, boldness may allow individuals to better cope with the novel conditions (habitats, prey, competitors and predators) encountered as they spread into new areas. Here, individuals are likely to experience conditions that they themselves have never experienced (i.e. ecologically novel) and/or that they have no evolutionary history with (i.e. evolutionary novel). Martin and Fitzgerald (2005) show that house sparrows (*Passer domesticus*) from an actively expanding population are less fearful of novel foods than individuals from a population established for 150 years.

Aggressiveness often confers an advantage in competition for limited resources (Huntingford, 2011). Aggressiveness increases the probability of resource acquisition, benefiting growth, survival and reproduction (Hudina *et al.*, 2014). During spread, aggressiveness should increase the probability of successful settlement. For example, Pintor *et al.* (2008) showed that invasive signal crayfish (*Pacifastacus leniusculus*) were more aggressive in introduced habitats where prey availability was low. Intraspecific aggression was also highly correlated with foraging activity and boldness across nonnative populations. In western bluebirds (*Sialia mexicana*), dispersers are larger and highly aggressive, which allows them to outcompete and displace less aggressive mountain bluebirds (*Sialia currucoides*; Duckworth and Badyaev, 2007). The aggression levels of western bluebirds are heritable, repeatable and increase from inner populations to the leading edge of the range. A similar pattern of increasing aggression at the invasion front and associated dominance is present in round gobies (*Neogobius melanostomus*) invading the Great Lakes (Groen *et al.*, 2012).

Yet, dispersers are not always the more aggressive and dominant individuals (Bekoff, 1977). Dispersers may also be less aggressive, subordinate individuals that are forced out by dominant individuals. In this case, less aggressive individuals are excluded from high density, established populations, and accumulate at the periphery of the range or emigrate. Non-native signal crayfish from the invasion front are consistently less aggressive than individuals from established populations (Hudina *et al.*, 2014). Reduced aggression could be beneficial to population growth (e.g. better parental care; Duckworth and Badyaev, 2007) and to achieving the high population densities that ultimately displace native taxa (e.g. Argentine ants; Holway and Suarez, 1999).

Sociability is a relatively underexplored component of personality-dependent dispersal and successful spread. Work on common lizards shows that dispersal tendency depends on the interaction between local density and personality (Cote and Clobert, 2007a; Cote *et al.*, 2008). Asocial individuals are sensitive to crowding (disperse when density is high) and preferentially settle in low-density patches, whereas social individuals seek out conspecifics, have higher fitness at high densities and disperse when densities are low (Figure 7.2). Among invasive taxa, the role of sociability has only been explored in mosquitofish (Table 7.1 and section below).

Importance of Variation in Personality Type in Invasions

Variation in personality type among individuals within an invading population can also play a major role in enhancing invasion success (both the likelihood of establishment and the speed of invasion) by expanding the range of conditions where the species can

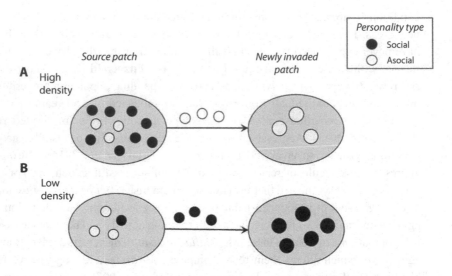

Figure 7.2 Conceptual diagram depicting context-dependency in personality-dependent dispersal. In A, the source patch has a high density of social individuals, which promotes asocials to disperse and colonize empty patches, where they thrive at low density (depicted by large-size dots). In B, the source patch is low density, and social individuals disperse away from it in search of a high-density patch. Social individuals achieve higher fitness in a newly colonized high-density patch.

be successful (Sih *et al.*, 2012). This is analogous to the effects of genetic variation in allowing species to cope with fluctuating environments. Inter-individual variation in personality type can counterbalance the potential negative effects of the across-situation personality trade-offs that may limit the invasion success of any one personality type (Sih *et al.*, 2012). In addition, variation in personality type can affect the impact of invaders since it can increase the number and magnitude of interactions with native taxa (Juette *et al.*, 2014).

Within-species variation in personalities can be particularly important because successful spread typically requires repeated cycles of: (i) dispersal, followed by (ii) colonization of empty patches and (iii) population growth to achieve high densities. Success across these three stages might often require a mix of personality types (Sih *et al.*, 2012; Figure 7.1). Fogarty *et al.* (2011) explored this idea by simulating invasions with varying mixes of social and asocial individuals. In their model, asocial individuals disperse from established, high-density patches to colonize empty patches. Their presence in newly occupied patches facilitated social individuals to join and increase density, and the high density then induces asocial individuals to depart and colonize additional empty patches. Elliott and Cornell (2012), in a similar modelling effort, showed that invasions can occur twice as fast with dispersal polymorphisms than with a single personality type. In a well-known empirical example, more aggressive western bluebirds are the dispersers at the invasion front, but they are poor parents, and successful spread results from selection favouring lower aggression in individuals behind the invasion front (Duckworth, 2008).

Context Dependency of Personality-dependent Dispersal

Although we have emphasized that personality-dependent dispersal can facilitate invasion success, this need not always be true. To understand how and when personality-dependent dispersal might enhance invasion success, it is important to emphasize that both personality-dependent dispersal and personality-dependent spread are likely to be context dependent. Both whether bolder versus shyer individuals disperse more readily from an established 'source' site, and whether bolder versus shyer individuals have higher fitness in a newly invaded site, will typically depend on the ecological or social conditions in both the source site and the newly invaded site (e.g. predation risk). A key factor then is the similarity of conditions in source versus newly invaded sites (Sih *et al.*, 2011; Figure 7.2). Assuming that animals tend to disperse when local conditions are unsuitable for their personality, the personalities that leave a source site (versus those that stay) should be the ones that are not well adapted to the conditions in the source site. If the invaded site has similar unsuitable conditions, then the disperser will still be poorly adapted to those conditions and thus unlikely to establish well. In contrast, if the invaded site has better conditions relative to the abandoned site, the disperser may be well adapted to the new conditions and thus more likely to establish well. The latter situation clearly facilitates a more successful invasion.

When might we expect source and invaded sites to be similar versus different? In an optimal world, organisms should only disperse if they will likely do better in invaded sites. While this might be true if organisms have a long, stable evolutionary history guiding adaptive dispersal tendencies, in the new world of human-induced environmental change and invasions into new habitats, organisms might still be well adapted to leave when current conditions are poor, but might not have good information about the quality of conditions in novel places where they might settle.

One situation where personality-dependent dispersal might be particularly likely to enhance invasion success involves the interaction between population density and dispersal of asocial versus social personalities. Source and newly invaded sites almost always differ in density; source sites having higher, often much higher densities than newly invaded sites (Figure 7.2). As noted above, assuming that social individuals thrive at high densities and avoid low densities, whereas asocial individuals avoid (disperse away from) crowded, high-density conditions and thrive at low densities, then an expected repeated pattern would be one where asocial individuals leave established high-density source sites, and are well adapted to colonize and thrive in invaded sites that by definition tend to have very low initial densities of that species (e.g. Fogarty *et al.*, 2011; Figure 7.2). These patches will eventually attract social individuals, thereby increasing density and promoting the dispersal of asocial individuals to colonize new empty patches.

Patterns involving other ecological drivers could either enhance or reduce invasion success. For example, if predation risk is a key threat in a source site, we expect dispersers to be individuals with personalities that are particularly poor at coping with predators. They might then only do well in those newly invaded sites that have low predation pressure; i.e. providing a mechanism for 'enemy release' promoting invasion success (Colautti *et al.*, 2004). If invaded sites still have numerous predators or, worse yet, novel predators, the combination of predation-risk-dependent and personality-dependent dispersal might reduce invasion success. Note that if a key factor explaining establishment success is the match between the disperser's personality and conditions in the invaded site, if potential invaded sites vary substantially in ecological conditions, then adaptive personality-dependent settlement choice should enhance invasion success, regardless of the drivers that induced personality-dependent dispersal.

Finally, just as newly invaded sites almost always have a low density of conspecifics, they also often have novel conditions – new biotic or abiotic challenges that are distributed in space and time in patterns that are initially poorly known by the invader. Thus invasion success can require personalities that are flexible, innovative (Sol *et al.*, 2002), and responsive. An important point that has rarely been examined is which conditions in the source site would tend to drive these types of personalities to disperse.

Lessons from Mosquitofish

In this section, we illustrate how personality types and behavioural syndromes can affect invasion, dispersal and spread using *Gambusia* as a model system. *Gambusia* includes approximately 45 species of small, livebearing fishes (Poeciliidae). From this genus, the eastern and western mosquitofish (*G. holbrooki* and *G. affinis*) have been introduced for mosquito control worldwide and have spread successfully to over 40 countries (Welcomme, 1992). Mosquitofish are efficient foragers, feeding on a wide variety of prey (Goodsell and Kats, 1999; Rehage *et al.*, 2005a), with strong negative impacts on native communities (Courtenay and Meffe, 1989; Webb and Joss, 1997; Goodsell and Kats, 1999). Moreover, mosquitofish are efficient dispersers (Brown, 1985; Congdon, 1994), able to disperse in both deep and shallow waters (3 mm; Alemadi and Jenkins, 2008), and can tolerate or adapt to a large diversity of environmental conditions (Pyke, 2005). Mosquitofish are thus listed among the 100 worst invasive species worldwide (Lowe *et al.*, 2000), and understanding *Gambusia* invasions is of great importance. Thankfully, a large and diverse literature (reviewed in Pyke, 2005) provides considerable information on life history traits (e.g. Brown, 1985; Alemadi and Jenkins, 2008; Chen *et al.*, 2010), morphology (e.g. Langerhans *et al.*, 2004), physiology (e.g. Knapp *et al.*, 2011; Seebacher *et al.*, 2014), behaviour (e.g. Rehage *et al.*, 2005b; Hoysak and Godin, 2007; Burns *et al.*, 2012; Cote *et al.*, 2012; Ward, 2012; Herbert-Read *et al.*, 2013; Biro and Andriaenssens, 2013), and correlations among relevant traits (Horth, 2003; Rehage and Sih, 2004; Cote *et al.*, 2010b; 2011; Wilson *et al.*, 2010; Seebacher *et al.*, 2013; Sinclair *et al.*, 2014).

A growing literature examines mosquitofish personality types and behavioural syndromes (e.g. Rehage and Sih, 2004; Cote *et al.*, 2010b; 2012; Wilson *et al.*, 2010; Herbert-Read *et al.*, 2013; Biro and Adriaenssens, 2013; Seebacher *et al.*, 2013; Sinclair *et al.*, 2014). A first set of studies comparing invasive and non-invasive *Gambusia* species (Rehage and Sih, 2004; Rehage *et al.*, 2005a, b) showed that *Gambusia* species differ in their boldness, exploration, foraging rates and response to novel competitors and predators. Importantly, these studies found behavioural differences between

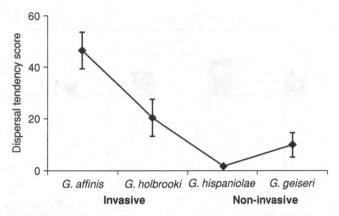

Figure 7.3 Variation in dispersal tendency among invasive mosquitofish species (*Gambusia affinis* and *G. holbrooki*) and non-invasive sister species (*G. geiseri* and *G. hispaniola*) (modified from Rehage and Sih, 2004).

invasive and non-invasive species. Although invasive and non-invasive species did not differ in boldness (i.e. shelter use), invasive species displayed a higher exploratory tendency (Rehage and Sih, 2004; Figure 7.3) and foraging efficiency (Rehage *et al.*, 2005a, b) than their non-invasive counterparts.

Subsequent studies have investigated behavioural differences among individuals in mosquitofish species using similar behavioural assays. These studies measured latency to emerge from a refuge, exploration rate in a novel environment and time spent swimming (Cote *et al.*, 2010b; 2011; 2013; Wilson *et al.*, 2010; Ward, 2012; Sinclair *et al.*, 2013; Biro and Adriaenssens, 2013), proxies of boldness, exploratory tendencies and activity, respectively. Other behavioural types were also measured, including shoaling tendency (Wilson *et al.*, 2010; Cote *et al.*, 2010b; 2012), sociability and/or boldness, and aggressiveness (Sinclair *et al.*, 2014; Seebacher *et al.*, 2013). These behavioural types were consistent over time (Cote *et al.*, 2010b; 2011; Sinclair *et al.*, 2014; Biro and Adriaenssens 2013) and often correlated into a behavioural syndrome (Cote *et al.*, 2010b; 2011; but see Wilson *et al.*, 2010). Despite significant consistency, individual behavioural scores also varied with social context (Ward, 2012), individual level of exercise (Sinclair *et al.*, 2014), reproductive status (Seebacher *et al.*, 2013) and time (Biro and Adriaenssens, 2013). A recent study shows that some individuals consistently vary more than others, suggesting that intra-individual variation can be a repeatable characteristic of individuals (Biro and Adriaenssens, 2013).

Between- and within-species comparisons have shown correlations among behavioural phenotypes and life history traits. First, invasive *Gambusia* species have higher exploratory tendencies (Figure 7.3) and foraging rate than non-invasive *Gambusia* species (Rehage and Sih, 2004; Rehage *et al.*, 2005a). This association highlights an 'invaders syndrome', facilitating species spread and impacts on invaded communities. Second, within-species studies found relationships between boldness, aggressiveness and reproductive traits (fecundity and stage of pregnancy). These relationships could arise from the effect of reproductive status on behavioural expression, from stable life

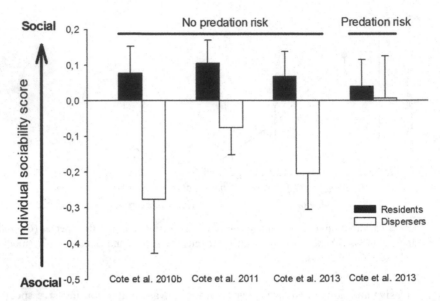

Figure 7.4 Summary of sociability values for *Gambusia affinis* in relation to the dispersal status of individuals (residents vs. dispersers), across predation and no predation contexts and studies.

history strategies that favour associated personality types (Reale *et al.*, 2010) or from reciprocal feedbacks linking life history and behavioural syndromes (Sih *et al.*, 2015). For example, late-stage pregnant eastern mosquitofish invested more in risky aggressive behaviours than early-stage pregnant females (Seebacher *et al.*, 2013), which shows an effect of reproductive stage on behavioural expression. However, in another study, boldness was not correlated with pregnancy stage but was negatively correlated with fecundity (Wilson *et al.*, 2010), which can result from life history strategies or the effect of reproductive status on behavioural expression. Third, in western mosquitofish, we found that survival in the presence of predators was negatively related to social tendencies, with asocial individuals surviving better (Brodin, unpublished data). Finally, individual sociability levels have also been shown to govern dispersal decisions in western mosquitofish (Cote *et al.*, 2010b; Cote *et al.*, 2011; Figure 7.4), and are therefore likely involved in invasive spread.

More recently, we ran a suite of studies examining sociability in *G. affinis*. These studies clearly show that dispersal behaviour varies among individuals in a repeatable manner (Cote *et al.*, 2010b) and is linked to between-individual differences in sociability (Figure 7.4). Individuals vary in their shoaling tendency and this individual tendency is partly stable over several months, a substantial portion of this small fish's lifetime (Cote *et al.*, 2011). Asocial individuals tend to shoal less and exhibit weaker choosiness for shoal characteristics (Cote *et al.*, 2010b; Cote *et al.*, 2012). We do not have information on shoaling behaviour in natural populations, but we expect social individuals to form large shoals and asocial fish to swim alone, joining shoals only occasionally. Using artificial streams, we further showed that asocial individuals were more likely to leave their initial population (Cote *et al.*, 2011, 2013) and disperse farther (Cote *et al.*, 2010b). Furthermore, we showed that group composition in terms of sociability types also matters for individual dispersal decisions (Cote *et al.*, 2011). While asocial individuals are more likely to disperse in general, individuals in groups with a greater number of asocial individuals or of bold individuals are more likely to disperse regardless of their own personality type. These results agree with our earlier point that individuals at the front of an invasion are not a random subset of the source population, but display specific behavioural characteristics. This could result in higher invasion success, as already shown in other species (see above), and can hasten the spread of an invasion (Fogarty *et al.*, 2011). As asocial individuals are also more likely to escape novel predators at low density (Cote, unpublished data), a low shoaling tendency in dispersers might therefore increase resistance against novel predators and thus enhance settlement success at the invasion front.

Finally, our work on mosquitofish also illustrates the context dependency of the relationship between personality type and dispersal. First, the link between the individual sociability levels of mosquitofish and their dispersal decisions depends on predation risk experienced in the initial habitat (Cote *et al.*, 2013; Figure 7.4). When mosquitofish were released into a pool containing a caged predator, sociability was no longer related to dispersal tendency. This result shows that personality-dependent spread is likely to vary with environmental conditions encountered during invasion. Second, the expression of personality types is only partly stable over time, and instead can vary with ecological contexts (Ward, 2012) and individual life histories (Sinclair *et al.*, 2013). In eastern mosquitofish, swimming exercise increases exploration, aggressiveness and locomotion capacity (Sinclair *et al.*, 2013). Sinclair *et al.* (2013) therefore predicted that dispersal tendency is self-reinforcing and, at the front of invasion, greater aggressiveness could facilitate invasion success. Overall, we predict that asocial individuals with enhanced aggressiveness will successfully colonize empty habitats and establish populations. Over time, the relaxation of aggressiveness in colonizers can facilitate a second dispersal wave by social individuals. Such successive dispersal waves should create spatiotemporal variation in the distribution of personality types in the landscape and may increase the speed of invasion spread (Fogarty *et al.*, 2011; Figure 7.2).

Conclusions and Future Directions

Throughout this chapter, we provide evidence for how the dispersal behaviour of invaders, their personality types, and inter-individual variation associated with both, may affect the spread of invasions. Invasions pose a unique challenge to dispersers since novelty, uncertainty and risk may be relatively high in the introduced range, while information and experience with local habitats may be low, and cues related to habitat quality can be absent. Dispersers may find it difficult then to assess the costs and benefits of dispersal, which will influence decisions relating to when and where to depart, transit and settle during an invasion.

In regard to personality type, we argue that personality-dependent dispersal can promote successful and faster spread since dispersers will have a suite of traits selected for while spreading (and beyond, i.e. impact), and which can counteract the high costs of dispersal during an invasion (e.g. boldness when information/familiarity is low). The authors' work on mosquitofish shows that individual variation in aggressiveness, boldness/exploratory behaviour and sociality can promote dispersal, and should be an important factor contributing to the overall invasion success of this notorious invader. Invasions can proceed faster through repeated cycles of (i) dispersal, (ii) colonization and (iii) population growth when there are mixes of personality types, as this can expand the range of conditions under which the invading species can succeed.

We identify several important research gaps and fruitful directions for future research. Variation in personality type among individuals and populations in the invaded range may arise from selection in the novel habitat, biased introductions or from true variation in personality-dependent dispersal, and future studies should aim to distinguish among these mechanisms. The degree of variation in personality types may vary among species, and may be an important determinant of which species become good invaders (i.e. 'winners'), and which ones fail to invade or cope with rapid anthropogenic change ('losers'). Comparative studies examining the role of behavioural polymorphism may be informative in the confluence of behaviour and conservation. Last, the context dependency of personality-dependent dispersal remains poorly understood. The degree to which variation among patches drives how different personalities sort themselves in spaces needs further investigation. For instance, how personality-dependent dynamics interact with invader densities via propagule pressure, known to be a key determinant of invasion success, may be a productive venue of future research.

References

Alemadi, S.D. and Jenkins, D.G. (2008). Behavioural constraints for the spread of the eastern mosquitofish, *Gambusia holbrooki (Poeciliidae)*. *Biological Invasions*, 10, 59–66.

Alford, R.A., Brown, G.P., Schwarzkopf, L., Phillips, B.L. and Shine, R. (2009). Comparisons through time and space suggest rapid evolution of dispersal behaviour in an invasive species. *Wildlife Research*, 36, 23–28.

Bekoff, M. (1977). Mammalian dispersal and the ontogeny of individual behavioural phenotypes. *American Naturalist*, 111, 715–732.

Biro, P.A. and Adriaenssens, B. (2013). Predictability as a personality trait: consistent differences in intraindividual behavioural variation. *American Naturalist*, 182, 621–629.

Biro, P.A. and Stamps, J.A. (2008). Are animal personality traits linked to life-history productivity? *Trends in Ecology and Evolution*, 23, 361–368.

Blackburn, T.M., Pyšek, P., Bacher, S., *et al.* (2011). A proposed unified framework for biological invasions. *Trends in Ecology and Evolution*, 26, 333–339.

Bonte, D., Travis, J.M.J., De Clercq, N., Zwertvaegher, I. and Lens, L. (2008). Thermal conditions during juvenile development affect adult dispersal in a spider. *Proceedings of the National Academy of Sciences*, 105, 17000–17005.

Bonte, D., Clercq, N.D., Zwertvaegher, I. and Lens, L. (2009). Repeatability of dispersal behaviour in a common dwarf spider: evidence for different mechanisms behind short- and long-distance dispersal. *Ecological Entomology*, 34, 271–276.

Bonte, D., Van Dyck, H., Bullock, J.M., *et al.* (2012). Costs of dispersal. *Biological Reviews*, 87, 290–312.

Bonte, D., De Roissart, A., Wybouw, N. and Van Leeuwen, T. (2014). Fitness maximization by dispersal: evidence from an invasion experiment. *Ecology*, 95, 3104–3111.

Bowler, D.E. and Benton, T.G. (2005). Causes and consequences of animal dispersal strategies: relating individual behaviour to spatial dynamics. *Biological Reviews*, 80, 205–225.

Bowne, D.R. and Bowers, M.A. (2004). Interpatch movements in spatially structured populations: a literature review. *Landscape Ecology*, 19, 1–20.

Brown, K.L. (1985). Demographic and genetic characteristics of dispersal in the mosquitofish, *Gambusia affinis (Pisces: Poeciliidae). Copeia*, 1985, 597–612.

Bubb, D.H., Thom, T.J. and Lucas, M.C. (2006). Movement, dispersal and refuge use of co-occurring introduced and native crayfish. *Freshwater Biology*, 51, 1359–1368.

Burgess, S.C. and Marshall, D.J. (2011). Are numbers enough? Colonizer phenotype and abundance interact to affect population dynamics. *Journal of Animal Ecology*, 80, 681–687.

Burns, A.L.J., Herbert-Read, J., Morrell, L.J. and Ward, A.J.W. (2012). Consistency of leadership in shoals of mosquitofish (*Gambusia holbrooki*) in novel and in familiar environments. *PloS ONE*, 7, e36567.

Chapple, D.G. and Wong, B.B.M. (2015). The role of behavioural variation and behavioural syndromes across different stages of the introduction process. In *Behavioural Responses to a Changing World: Mechanisms and Consequences*, ed. Candolin, U. and Wong, B.B.M. Oxford, UK: Oxford University Press, pp. 190–200.

Chapple, D.G., Simmonds, S.M. and Wong, B.B.M. (2012). Can behavioural and personality traits influence the success of unintentional species introductions? *Trends in Ecology and Evolution*, 27, 57–64.

Chen, T., Beekman, M. and Ward, A.J.W. (2010). The role of female dominance hierarchies in the mating behaviour of mosquitofish. *Biology Letters*, 7, 343–345.

Clobert, J., Le Galliard J.F., Cote J., Meylan S. and Massot, M. (2009). Informed dispersal, heterogeneity in animal dispersal syndromes and the dynamics of spatially structured populations. *Ecology Letters*, 12, 197–209.

Colatti, R.I, Ricciardi, A., Grigorovich, I.A., and MacIsaac, H.J. (2004). Is invasion success explained by the enemy release hypothesis. *Ecology Letters*, 7, 721–733.

Congdon, B.C. (1994). Characteristics of dispersal in the eastern mosquitofish *Gambusia holbrooki. Journal of Fish Biology*, 45, 943–952.

Cote, J. and Clobert, J. (2007a). Social personalities influence natal dispersal in a lizard. *Proceedings of the Royal Society B: Biological Sciences*, 274, 383–390.

Cote, J. and Clobert, J. (2007b). Social information and emigration: lessons from immigrants. *Ecology Letters*, 10, 411–417.

Cote, J., Clobert, J. and Fitze, P.S. (2007). Mother-offspring competition promotes colonization success. *Proceedings of the National Academy of Sciences, USA*, 104, 703–708.

Cote, J., Dreiss, A. and Clobert, J. (2008). Social personality trait and fitness. *Proceedings of the Royal Society B: Biological Sciences*, 275, 2851–2858.

Cote, J., Clobert, J., Brodin, T., Fogarty, S. and Sih, A. (2010a). Personality dependent dispersal: characterization, ontogeny and consequences for spatially structured populations. *Philosophical Transactions of the Royal Society B: Biological Sciences*, 365, 4065–4076.

Cote, J., Fogarty, S., Weinersmith, K., Brodin, T. and Sih, A. (2010b). Personality traits and dispersal tendency in the invasive mosquitofish *Gambusia affinis*. *Proceedings of the Royal Society B: Biological Sciences*, 277, 1571–1579.

Cote, J., Fogarty, S., Brodin, T. and Sih, A. (2011). Personality-dependent dispersal in invasive mosquitofish: group composition matters. *Proceedings of the Royal Society B: Biological Sciences*, 278, 1670–1678.

Cote, J., Fogarty, S. and Sih, A. (2012). Individual sociability and choosiness between shoal types. *Animal Behaviour*, 83, 1469–1476.

Cote, J., Fogarty, S., Tymen, B., Sih, A. and Brodin, T. (2013). Personality dependent dispersal cancelled under predation risk. *Proceedings of the Royal Society B: Biological Sciences*, 280, 2013–2349.

Courtenay, W.R., Jr. and Meffe, G.K. (1989). Small fishes in strange places: a review of introduced poeciliids. In *Ecology and Evolution of Livebearing Fishes (Poeciliidae)*, ed. Meffe, G.K. and Snelson, Jr., F.F. Englewood Cliffs, NJ: Prentice-Hall, pp. 319–331.

Davis, J.M. and Stamps, J.A. (2004). The effect of natal experience on habitat preferences. *Trends in Ecology and Evolution*, 19, 411–416.

Delgado, M.D., Penteriani, V., Revilla, E. and Nams, V.O. (2010). The effect of phenotypic traits and external cues on natal dispersal movements. *Journal of Animal Ecology*, 79, 620–632.

Doligez, B., Danchin Cadet, C., Danchin, T. and Boulinier, E. (2003). When to use public information for breeding habitat selection? The role of environmental predictability and density dependence. *Animal Behaviour*, 66, 973–988.

Ducatez, S., Legrand, D., Chaput-Bardy, A., *et al.* (2012). Inter-individual variation in movement: is there a mobility syndrome in the large white butterfly (*Pieris brassicae*)? *Ecological Entomology*, 37, 377–385.

Duckworth, R.A. (2008). Adaptive dispersal strategies and the dynamics of a range expansion. *American Naturalist*, 172, S4–S17.

Duckworth, R.A. and Badyaev, A.V. (2007). Coupling of dispersal and aggression facilitates the rapid range expansion of a passerine bird. *Proceedings of the National Academy of Sciences, USA*, 104, 15017–15022.

Edelaar, P., Siepielski, A.M. and Clobert, J. (2008). Matching habitat choice causes directed gene flow: a neglected dimension in evolution and ecology. *Evolution*, 62, 2462–2472.

Elliott, E.C. and Cornell, S.J. (2012). Dispersal polymorphism and the speed of biological invasions. *PLoS ONE*, 7, e40496.

Fogarty, S., Cote, J. and Sih, A. (2011). Social personality polymorphism and the spread of invasive species: a model. *American Naturalist*, 177, 273–287.

Fraser, D.F., Gilliam, J.F., Daley, M.J., Le, A.N. and Skalski, G.T. (2001). Explaining leptokurtic movement distributions: intrapopulation variation in boldness and exploration. *American Naturalist*, 158, 124–135.

Goodsell, J.A. and Kats, L.B. (1999). Effect of introduced mosquitofish on pacific treefrogs and the role of alternative prey. *Conservation Biology*, 13, 921–924.

Groen, M., Sopinka, N.M., Marentette, J.R., *et al.* (2012). Is there a role for aggression in round goby invasion fronts? *Behaviour*, 149, 685–703.

Herbert-Read, J.E., Krause, S., Morrell, L.J., *et al.* (2013). The role of individuality in collective group movement. *Proceedings of the Royal Society B: Biological Sciences*, 280, 20122564.

Holway, D.A. and Suarez, A.V. (1999). Animal behaviour: an essential component of invasion biology. *Trends in Ecology and Evolution*, 14, 328–330.

Holway, D.A., Suarez, A.V. and Case, T.J. (1998). Loss of intraspecific aggression in the success of a widespread invasive social insect. *Science*, 282, 949–952.

Horth, L. (2003). Melanic body colour and aggressive mating behaviour are correlated traits in male mosquitofish (*Gambusia holbrooki*). *Proceedings of the Royal Society: Biological Sciences*, 270, 1033–1040.

Hoysak, D.J. and Godin, J.G. (2007). Repeatability of male mate choice in the mosquitofish, *Gambusia holbrooki*. *Ethology*, 113, 1007–1018.

Hudina, S., Hock, K. and Zganec, K. (2014). The role of aggression in range expansion and biological invasions. *Current Zoology*, 60, 401–409.

Huntingford, F.A. (2011). *Animal Conflict*. London: Chapman and Hall.

Johnson, J.C. and Sih, A. (2007). Fear, food, sex and parental care: a syndrome of boldness in the fishing spider, *Dolomedes triton*. *Animal Behaviour*, 74, 1131–1138.

Juette, T., Cucherousset, J. and Cote, J. (2014). Animal personality and the ecological impacts of freshwater non-native species. *Current Zoology*, 60, 417–427.

Knapp, R., Marsh-Matthews, E., Vo, L. and Rosencrans, S. (2011). Stress hormone masculinizes female morphology and behaviour. *Biology Letters*, 64, 598–606.

Kubisch, A., Fronhofer, E.A., Poethke, H.J. and Hovestadt, T. (2013). Kin competition as a major driving force for invasions. *American Naturalist*, 181, 700–706.

Langerhans, R.B., Layman, C.A., Shokrollahi, A.M. and DeWitt, T.J. (2004). Predator-driven phenotypic diversification in *Gambusia affinis*. *Evolution*, 58, 2305–2318.

Llewelyn J., Philips B.L., Alford R.A., Schwartzkopf, L. and Shine, R. (2010). Locomotor performance in an invasive species: cane toads from the invasion front have greater endurance, but not speed, compared to conspecifics from long-colonised area. *Oecologia*, 162, 343–348.

Lombaert, E., Estoup, A., Facon, B., *et al.* (2014). Rapid increase in dispersal during range expansion in the invasive ladybird *Harmonia axyridis*. *Journal of Evolutionary Biology*, 27, 508–517.

Lopez, D.P., Jungman A.A. and Rehage, J.S. (2012). Nonnative African jewelfish are more fit but not bolder at the invasion front: a trait comparison across an Everglades range expansion. *Biological Invasions*, 14, 2159–2174.

Lowe, S., Browne, M., Boudjelas, S. and De Poorter, M. (2000). *100 of the World's Worst Invasive Alien Species: A Selection from the Global Invasive Species Database*. The Invasive Species Specialist Group (ISSG) , a specialist group of the Species Survival Commission (SSC) of the World Conservation Union (IUCN), Auckland.

Martin, L.B. and Fitzgerald, L. (2005). A taste for novelty in invading house sparrows *Passer domesticus*. *Behavioural Ecology*, 16, 702–707.

O'Riain, M.J., Jarvis, J.U.M. and Faulkes, C.G. (1996). A dispersive morph in the naked mole-rat. *Nature*, 380, 619–621.

Parker, I.M., Simberloff, D., Lonsdale, W.M., *et al.* (1999). Impact: toward a framework for understanding the ecological effects of invaders. *Biological Invasions*, 1, 3–19.

Phillips, B.L., Brown, G.P., Webb, J.K. and Shine, R. (2006). Invasion and the evolution of speed in toads. *Nature*, 43:803.

Phillips, B.L., Brown, G.P. and Shine, R. (2010). Evolutionarily accelerated invasions: the rate of dispersal evolves upwards during the range advance of cane toads. *Journal of Evolutionary Biology*, 23, 2595–2601.

Pintor, L.M., Sih, A. and Bauer, M. (2008). Differences in aggression, activity and boldness between native introduced populations of an invasive crayfish. *Oikos*, 117, 1629–1636.

Pyke, G.H. (2005). A review of the biology of *Gambusia affinis* and *G. holbrooki*. *Fish Biology and Fisheries*, 15, 339–365.

Rasmussen, J.E. and Belk, M.C. (2012). Dispersal behaviour correlates with personality of a North American fish. *Current Zoology*, 58, 260–270.

Réale, D., Dingemanse, N.J., Kazem, A.J.N. and Wright, J. (2010). Evolutionary and ecological approaches to the study of personality. *Philosophical Transactions of the Royal Society B: Biological Sciences*, 365, 3937–3946.

Rehage, J.S. and Sih, A. (2004). Dispersal behaviour, boldness, and the link to invasiveness: a comparison of four *Gambusia* species. *Biological Invasions*, 6, 379–391.

Rehage, J.S., Barnett, B.K. and Sih, A. (2005a). Foraging behaviour and invasiveness: do invasive *Gambusia* exhibit higher feeding rates and broader diets than their non-invasive relatives? *Ecology of Freshwater Fish*, 14, 352–360.

Rehage, J.S., Barnett, B.K. and Sih, A. (2005b). Behavioural responses to a novel predator and competitor of invasive mosquitofish and their non-invasive relatives (*Gambusia* sp.). *Behavioural Ecology and Sociobiology*, 57, 256–266.

Rémy, A., Le Galliard, J.F., Odden, M. and Andreassen, H.P. (2014). Concurrent effects of age class and food distribution on immigration success and population dynamics in a small mammal. *Journal of Animal Ecology*, 83, 813–822.

Ricciardi A., Hoopes, M. F., Marchetti, M. P. and Lockwood, J. L. (2013). Progress toward understanding the ecological impacts of nonnative species. *Ecological Monographs*, 83, 263–282.

Ronce, O. (2007). How does it feel to be like a rolling stone? Ten questions about dispersal evolution. *Annual Review of Ecology, Evolution, and Systematics*, 38, 251–253.

Schoepf, I. and Schradin, C. (2012). Better off alone! Reproductive competition and ecological constraints determine sociality in the African striped mouse (*Rhabdomys pumilio*). *Journal of Animal Ecology*, 81, 649–656.

Seebacher, F., Ward, A.J.W. and Wilson, R.S. (2013). Increased aggression during pregnancy comes at a higher metabolic cost. *Journal of Experimental Biology*, 216, 771–776.

Seebacher, S., Beaman, J. and Little, A.G. (2014). Regulation of thermal acclimation varies between generations of the short-lived mosquitofish that developed in different environmental conditions. *Functional Ecology*, 28, 137–148.

Selonen, V., Hanski, I.K. and Desrochers, A. (2007). Natal habitat-biased dispersal in the Siberian flying squirrel. *Proceedings of the Royal Society B: Biological Sciences*, 274, 2063–2068.

Sih, A., Bell, A.M., Johnson, J.C. and Ziemba, R.E. (2004). Behavioural syndromes: an integrative overview. *The Quarterly Review of Biology*, 79, 241–277.

Sih, A., Ferrari, M.C.O. and Harris, D.J. (2011). Evolution and behavioural responses to human induced rapid environmental change. *Evolutionary Applications*, 4, 367–387.

Sih, A., Cote, J., Evans, M., Fogarty, S. and Pruitt J. (2012). Ecological implications of behavioural syndromes. *Ecology Letters*, 15, 278–289.

Sih, A., Mathot, K.J., Moiron, M. Montiglio, P.-O. *et al.* (2015). Animal personality and state-behaviour feedbacks: a review and guide for empiricists. *Trends in Ecology and Evolution*, 30, 50–60.

Sinclair, E., Noronha de Souza, C., Ward, A. and Seebacher, F. (2014). Exercise changes behaviour. *Functional Ecology*, 28, 652–659.

Smith, B.R. and Blumstein, D.T. (2008). Fitness consequences of personality: a meta-analysis. *Behavioural Ecology*, 19, 448–455.

Sol, D., Timmermans, S. and Lefebvre, L. (2002). Behavioural flexibility and invasion success in birds. *Animal Behaviour*, 63, 495–502.

Stamps, J.A. (2001). Habitat selection by dispersers: integrating proximate and ultimate approaches. In *Dispersal*, ed. Clobert, J., Danchin, E., Dhondt, A. and Nichols, J. Oxford, UK: Oxford University Press, pp. 230–242.

Stamps, J.A., Davis, J.M., Blozis, S.A. and Boundy-Mills, K.L. (2007). Genotypic variation in refractory periods and habitat selection by natal dispersers. *Animal Behaviour*, 74, 599–610.

Stevens, V.M., Whitmee, S., Le Galliard, J.F., *et al.* (2014). A comparative analysis of dispersal syndromes in terrestrial and semi-terrestrial animals. *Ecology Letters*, 17, 1039–1052.

Suarez, A.V., Tsutsui, N.D., Holway, D.A. and Case, T.J. (1999). Behavioural and genetic differentiation between native and introduced populations of the Argentine ant. *Biological Invasions*, 1, 43–53.

Trochet, A., Legrand, D., Larranaga, N., *et al.* (2013). Population sex ratio and dispersal in experimental two-patch metapopulations of butterflies. *Journal of Animal Ecology*, 82, 946–955.

Ward, A. (2012). Social facilitation of exploration in mosquitofish (*Gambusia holbrooki*). *Behavioural Ecology and Sociobiology*, 66, 223–230.

Webb, C.E. and Joss, J. (1997). Does predation by the fish *Gambusia holbrooki* (Atheriniformes: Poecilidae) contribute to declining frog populations? *Australian Zoologist*, 30, 316–324.

Welcomme, R.L. (1992). A history of international introductions of inland aquatic species. *Marine Science Symposium*, 194, 3–14.

Williamson, M. and Fitter, A. (1996). The varying success of invaders. *Ecology*, 77, 1661–1666.

Wilson, A.D.M., Godin, J.G.J. and Ward, A.J.W. (2010). Boldness and reproductive fitness correlates in the eastern mosquitofish. *Gambusia holbrooki. Ethology*, 116, 96–104.

Part II

Behavioural Interactions Between Invaders and Native Species

8 Invasive Plants as Novel Food Resources, the Pollinators' Perspective

Ignasi Bartomeus, Jochen Fründ and Neal M. Williams

Biological invasions are one of the main drivers of global change and have negatively impacted all biomes and trophic levels (Hobbs, 2000; Vilà *et al.*, 2011). While most introduced species fail to establish, or establish small naturalized populations (hereafter exotic species), a few become invasive and rapidly increase in abundance and/or range (hereafter invasives or invaders; Pyšek *et al.*, 2004). It is these invader species that are most often linked to negative impacts on native/endemic communities. Although most interactions between invasive and native species at the same trophic level result in negative direct impacts (e.g. plant–plant competitive interactions; Vilà *et al.*, 2011), when the invasive plant species can be used as a resource those interactions can also be positive for consumers such as native herbivores, predators or mutualists, at least for some species (Heleno *et al.*, 2009; Bezemer *et al.*, 2014). Entomophilous exotic plant species, for example, not only compete directly for space and light with other plants, but also offer resource opportunities for the native pollinator community (Stouffer *et al.*, 2014). Most research on this topic to date has taken the plant perspective, focusing on how successful plant invaders integrate into the native plant–pollinator interaction networks (Vilà *et al.*, 2009), and how this integration in turn impacts the native plant community (Morales and Traveset, 2009). However, species-specific responses of pollinators to the addition of exotic plants are rarely taken into account. This represents an important gap in our knowledge as pollinator foraging choices determine the structure of interactions within communities, which in turn have important implications for the community stability (Thébault and Fontain, 2010) and functioning (Thomson *et al.*, 2012). How different pollinators respond to the changed composition of floral species within the community that results from exotic plant invaders deserves more attention.

From the pollinators' point of view, exotic plants are novel food resources, and as such their relative abundance, attractiveness, rewards (i.e. nutritional value) and distinctiveness partly determine their use by various pollinators' (Carvalheiro *et al.*, 2014). Although exotics as a group are not preferred or avoided within their new communities (Williams *et al.*, 2011), it might be that particular exotic species are preferred by some pollinators while avoided by others. The intrinsic preferences for different plant hosts is an important factor determining host use. Hence, the direct benefits or costs of a novel resource use will differ among pollinator species. Moreover, in a community

Biological Invasions and Animal Behaviour, eds J.S. Weis and D. Sol. Published by Cambridge University Press. © Cambridge University Press 2016.

context, the preferences of each pollinator affect the other pollinators' choices, potentially leading to indirect effects. For example, some pollinator species may obtain indirect benefits if the invasive plants distract other pollinators from natives, reducing competition. Alternatively, pollinators may pay indirect costs if competition is increased or if invasive plants reduce the availability of a preferred native plant.

We review the evidence on direct and indirect benefits and costs of invasive plants on pollinators and re-analyse Williams *et al.* (2011) dataset on pollinator specific preferences so as to identify species that prefer some exotic plants over native plants and vice versa. This information is crucial to understanding the consequences for the pollinator community because if only some pollinators take advantage of exotic plants this can favour populations of some pollinator species (winners) over others (losers). By using an approach that takes into account both pollinator behavioural responses and interaction network structure, we can better understand the invasion process, with important implications for conservation actions.

Effects of Plant Invasions on Pollinator Populations and Community Structure

The impact of invasive plants interacting with native pollinators has received considerable attention for its potential to disrupt native mutualisms (Traveset and Richardson, 2006). However, most work to date has focused solely on how invasive plant interactions with native pollinators changes the pollination success of native plants. Interestingly, existing data show that invasive plants can have positive, neutral or negative effects on pollination of native plants (Bjerknes *et al.*, 2007). The contrasting results may reflect invasive plant density (Muñoz and Cavieres, 2008; Dietzsch *et al.*, 2011), spatial aggregation (Cariveau and Norton, 2009) or flower morphology and attractiveness (Morales and Traveset, 2009; Carvalheiro *et al.*, 2014).

In contrast, the effects of exotic or invasive species on the pollinator populations and communities have received far less attention (Stout and Morales, 2009). Because most plant–pollinator systems are generalized (Waser *et al.*, 1996), invasive plants are usually well integrated in the plant–pollinator network of interactions (Vilà *et al.*, 2009). Hence, we might expect overall effects on pollinators to mirror the changes (positive or negative) in floral resources offered by the newly invaded community. If the entomophilous exotic or invasive plants increase the resources present in the community, this should also allow the increase of most pollinator populations (Tepedino *et al.*, 2008). Stout and Morales (2009) cite indirect evidence that some social pollinators (e.g. bumblebees) can be favoured by non-native mass flowering crops (Westphal *et al.*, 2003; Herrmann *et al.*, 2007), which may be analogous to the effect of abundant invasive species. However, the same authors note examples where exotic plants are not used by native pollinators due to flower morphology or chemistry. Despite Stout and Morales' (2009) call for more research on this topic, few additional studies have been published since then.

Current evidence suggests that food resource availability may indeed regulate pollinator's populations, at least those of bees (reviewed in Williams and Kremen, 2007;

Roulston and Goodell, 2011; Crone, 2013; but see Steffan-Dewenter and Schiele, 2008, for potential regulation by nesting resources). However, studies of the effect of exotic plants on pollinators' population dynamics are extremely rare, particularly for non-invasive exotic plants that establish small populations. Palladini and Maron (2014) provide one of the few examples of effects of exotic plants (mainly *Euphorbia esula*) on the reproduction of a solitary bee species (*Osmia lignaria*). For this species, the number of nests established and offspring production per female was positively related to native plant abundance and negatively related to exotic plant species. This may be because although *Osmia lignaria* foraged on exotics for nectar, the species did not use *Euphorbia esula* exotic pollen to provision nests. Thus, the specific ability to use the invader resources emerges as a key factor affecting the potential impacts of the invader on pollinators. The only other evidence to date for direct effects of invasive plants through resource augmentation is for bumblebees, whose annual social life history allows demographic responses to be measured within a single season. Within-season abundance can increase almost four times in sites invaded by the plant *Lupinus polyphyllus*, compared to in non-invaded sites (Jakobsson and Padrón, 2014). In addition, the foraging season of *Bombus terrestris* in the UK can be extended into winter through its use of exotic plants that fill a late season phenological niche (Stelzer *et al.*, 2010). Such within-season demographic responses are likely to have longer-term population effects, although no study has quantified such effect to date.

A second group of studies provides indirect evidence that exotic plants affect pollinators by using community approaches to compare invaded with non-invaded sites. These studies show a variety of pollinator responses, including increased abundance and species richness in invaded sites (Lopezaraiza-Mikel *et al.*, 2007), lower abundance and diversity of pollinators in invaded areas (Moroń *et al.*, 2009), or no difference in abundance between invaded and non-invaded sites (Nienhuis *et al.*, 2009). It is therefore not surprising that a recent meta-analysis reported no changes in overall pollinator abundance in invaded sites (Montero-Castaño and Vilà, 2012). However, the studies included in the meta-analysis were not designed to infer population changes, and the result should be interpreted with caution. Moreover, most of the examples concern abundant invasive plants. Plant abundance can strongly influence pollinators decision to incorporate a new plant into its diet (Valdovinos *et al.*, 2010), and thus the results may differ when examining non-invasive exotic plants. Likewise, given the wide array of pollinators ranging from birds and bats to bees and hoverflies, it is unlikely to find a consistent overall response.

The Importance of Behaviour

While exotic plants can represent new resource opportunities for native animals, evidence suggests that only a minority of pollinator species can take advantage of these new opportunities. For example, generalist bees more commonly forage on invasive exotic plants than specialists (Lopezaraiza-Mikel *et al.*, 2007; Tepedino *et al.*, 2008; Padrón *et al.* 2009), or social bumblebees are more common in invaded sites than solitary bees (Nienhuis *et al.* 2009). Bartomeus *et al.* (2008), for example, report that bees were more often recorded on native species than on the invader *Carpobrotus* aff. *acinaciformis*, except for the social bumblebee *Bombus terrestris* and for most beetles, which visited *Carpobrotus* flowers almost exclusively. Hence, pollinators can discriminate between native and exotic plants, and the decision of exploiting one or another can vary across species. Analysing the invasion process as a novel resource availability for pollinators may give us a framework to predict which pollinators can benefit from the invasion process.

Incorporating any novel resource into the diet requires a series of conditions to be met. First, the pollinator must recognize the novel resource as a host. Second, the visitor must be able to use this new resource and, third, the resource must be profitable (i.e. a net benefit) for the visitor to exploit it. Hence, exotic plant use depends on intrinsic traits of the plant and pollinator. We cannot assume that for native pollinators exotic flowering species are always fundamentally different from the native flowering species; nevertheless, plants presenting new colours, shapes or chemical compounds may not be easily used by all pollinators in the community (Stout and Morales, 2009). Furthermore many pollinators have innate preferences for certain colours and/or shapes (Gumbert, 2000; Riffell *et al.*, 2008), which may make novel flowers less attractive than the natives which have coevolved in the community. Neophobic responses also may make some pollinators unlikely to approach and explore novel food opportunities (Forrest and Thomson, 2009). However, some pollinators may learn how to exploit new resources if their behaviour is flexible enough (Chittka and Niven, 2009; Forrest and Thomson, 2009). In particular, bees, which are the main pollinator group both in numbers and in effectiveness (reviewed in Winfree *et al.*, 2011), have a powerful neuronal system able to learn new tasks (Chittka and Niven, 2009) and their behaviour flexibility has been suggested to be linked to their ability to persist in disturbed environments (Kevan and Menzel, 2012). However, the role of learning abilities in incorporating new foraging plants is little explored (Dukas and Real, 1993) and most information comes from a handful of species (mostly social).

Even if the exotic plant is recognized as a potential host, pollinators might not be able to exploit the new plant species if they are unable to handle its flowers (Parker, 1997; Corbet *et al.*, 2001), or to digest its nectar (Adler and Irwin, 2005) or pollen (Sedivy *et al.*, 2011; Palladini and Maron, 2014). Morphological matching between flowers and pollinators may thus be important in determining pollinator visitation patterns (Gibson *et al.*, 2012; Bartomeus, 2013; Stang *et al.*, 2009, see also Pearse *et al.*, 2013, for antagonist insect–plant interactions). Second, even in the cases where pollinators recognize and can use the novel resource, their decision to include it in their diet will depend on its quality and abundance relative to others in the community. The thresholds for switching to a resource based on its quality or abundance have been shown to be variable among different species in birds (Carnicer *et al.*, 2008). Insect pollinators switch between foraging plants depending on the resource availability (Inouye, 1978; Chittka *et al.*, 1997). Like for birds, thresholds for switching are likely to be different for different species; however, the experimental evidence that insect pollinators can discriminate between different resources and learn to forage on the preferred one is limited to a handful of bee species, and the switching strategies among species are mostly unknown.

The Importance of the Community Context

Pollinator–plant interactions do not occur in isolation, but form part of a complex network of interactions (Bascompte and Jordano, 2007). For example, competitive exclusion between pollinator species can drive foraging behaviour patterns (Johnson and Hubell, 1975). Invasive plant species can modify the native network structure not only by creating new interactions with native pollinators but also by modifying the existing plant–pollinator interactions among the native species (Bartomeus *et al.*, 2008; Albrecht *et al.*, 2014). Such changes are especially likely when invaders are abundant and provide accessible resources (i.e. nectar and pollen), thus potentially interacting with a large proportion of the resident pollinator species (Albrecht *et al.*, 2014). In any case, modifications to pairwise interactions can have cascading effects throughout the community. For example, dynamic pollination network models have been used to show that removal of well-established invasive plants negatively affected the persistence of pollinator interactions through the network (Valdovinos *et al.*, 2009; Carvalheiro *et al.*, 2008). Detailed empirical studies also show that co-flowering neighbouring invasive species affect pollinator choices (Cariveau and Norton, 2009; Waters *et al.*, 2014).

The behavioural switching (also called interaction rewiring) between resources has been recently studied at the community level using network theory. There is increased evidence that pollinators can rewire their community-wide interactions depending on the context (Kaiser-Bunbury *et al.*, 2010), such as the addition of a plant invasion. Jackobson and Padrón (2014) speculate that the attraction of bumblebees to the invasive plant *Lupinus polyphyllus* reduced competition for the native plants, allowing an increase of solitary bee visits to natives. Similarly, Montero-Castaño (2014) showed that monopolization of the invasive species *Hedysarum coronarium* by honeybees allowed other bee species to establish interactions with natives that are not realized when the honeybee is present. Context-dependent rewiring is supported by findings that despite consistent species-specific preferences for certain flowers across communities, there is also important variation and flexibility in preferences among different contexts (Fründ *et al.*, 2010). Hence, both direct effects and indirect effects on pollinators are expected after a plant invasion, and those can only be understood in a community-wide framework.

In Figure 8.1, we illustrate a simplified plant–pollinator network with two distinctive scenarios. In the first one (Figure 8.1b), we add an entomophilous exotic plant that does not reduce the abundances of other plants. Pollinator species able to use this plant (identified as p1 and p2) will establish new links with this exotic plant (grey links). Species p1 will have more food resources and can potentially increase its population over time (bigger circle denotes population size increase), whereas species p2 will experience a neutral effect because it changes from foraging on natives to foraging on

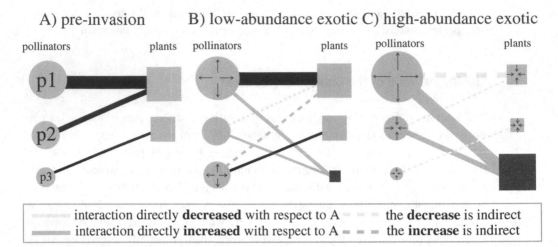

Figure 8.1 Simplified plant–pollinator networks before the invasion process (a) in two distinctive scenarios: (b) the exotic plant (dark grey square) adds a novel resource without affecting the rest of the community, and (c) a superabundant invasive plant (dark grey square) adds an abundant novel resource, while reducing native plants resource availability. In the schemes we can see the interactions established in each case (lines connecting pollinators and plants). The colour of links depicts whether their frequency is increased (grey) or decreased (light grey) with respect to A. The changes in size of the plants or pollinators depict the winners and losers of the invasion process. When the effect is indirect is noted with dashed lines. See text for further explanation.

the exotic. These are the direct (often neutral or positive) effects of the exotic plant on pollinators. Other pollinators may not be able or may not choose to visit the new invasive plant (p3), but as the competition for their preferred resources is changed, they may receive indirect benefits (p3). Experiments removing dominant pollinators have shown that a relaxed competition for resources may lead to diet expansion of some species (Brosi and Briggs, 2013), supporting our example with species p3.

In the second scenario (Fig 8.1c), the exotic plant is an abundant invader that also reduces native plant abundance by direct competition (e.g. for space). In this scenario, only a few pollinators (p1) may benefit, while all others will experience increased competition for resources (p2, p3). This is an oversimplification, and of course the net benefit for pollinators will depend not only on the number of visits, but on the quality of those visits (e.g. reward uptake, nutritional content of the exotic species, etc.). Moreover, some species will require a variety of pollen sources to complete larvae development (Roulston and Cane, 2000) highlighting the importance of maintaining plant diversity. The magnitude of the indirect and direct effects will depend also on the time-scale at which it is evaluated, with functional responses and local switching occurring faster than numerical responses (i.e. population growth). Moreover, the relative phenological timing of plants and bees can modify their mutual influence. All in all, the net costs and benefits are likely to depend on many factors, but this framework supports the scarce information presented above, where some social generalized species tend to increase their abundance after invasion by highly attractive species, but other pollinators have mixed responses.

Case Study: Bee Preferences in California

Can we predict which pollinators will be winners or losers of the invasion process? Measuring population responses or fecundity is a daunting task, especially at the community level; however, we can gain indirect evidence by looking at pollinator preferences. Within a plant community, pollinators do not prefer exotic plants as a group (Williams *et al.*, 2011) or even prefer natives (Chrobock *et al.*, 2013), but individual pollinator preferences have not been explored yet. Preference is defined as using a resource more than expected given its abundance. Conversely, avoidance occurs when a resource is underused relative to its abundance. The null model of no preferences is the case when pollinators visit flowers in proportion to their abundance in the community. Deviations from this null model can help us identify pollinator species that prefer exotic species (hence receiving a potential direct benefit) and species that avoid the exotics (hence, receiving negative, neutral or positive indirect effects in some cases). We recognize that we cannot infer direct fitness consequences, or predict indirect effects from a static network. Ours, nonetheless, is the first attempt to identify direct effects and serve as a proxy for identifying pollinator winners after the invasion process. Most importantly, in this way we can emphasize that pollinators differ in their behaviour, acknowledging that the effects on specific pollinators cannot be generalized. Furthermore, in the future, we can explore what determines pollinator preferences. Are they driven by plant traits, such as abundance or morphology, by pollinator traits, or a combination of both?

To explore this preference-based proxy, we used the same dataset used in Williams *et al.* (2011). For simplicity we show here only seven sites from semi-natural habitats in California. This system is especially suitable to test our questions, because it contains several exotic plants, ranging from abundant invaders to naturalized exotic plants, as well as a variety of pollinator species. We calculated preferences pooling all sites, but we separate our analysis in three sampling periods (early, mid and late season). We treated periods separately because plant turnover was substantial over the season and otherwise might have masked the preference relationships.

First we re-evaluated that pollinators do not prefer exotic plants as a group within a quantitative framework, where expected (E) visitation values are calculated based on plant mean abundance across sites, and observed (O) visitation values are the sums of pollinator visitation to each plant across sites. Chi statistics were used for each of the three tests (i.e. one per season) to assess if there is an overall preference. The Pearson residuals of the χ^2 tests ($O - E/\sqrt{E}$) estimate the magnitude of preference or avoidance for a given plant based on deviation from expected values and its significance was assessed by building Bonferroni confidence intervals (see Neu *et al.*, 1974, and Byers *et al.*, 1984). In order to test for differences between exotics and natives we compared the *electivity* values of exotic and native plants using linear models. *Electivity* values ($E' - O'/E' + O'$, where E' and O' are the proportional expected and observed values;

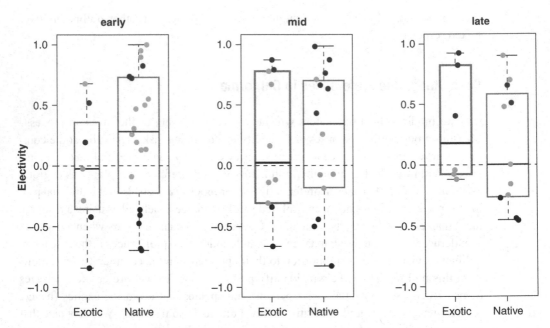

Figure 8.2 Boxplots of the *electivity* indices per plant species separated by plant origin (exotic, native) and season (early, mid, late). Each plant value is plotted in the background in grey when preference is not significant and in black when significant (significance based on χ^2 Pearson residuals test). Positive values indicate plants that are preferred and negative values those that are avoided.

Ivlev, 1964) bound between -1 and $+1$, are easier to interpret and highly correlated with Pearson residuals. The R code to calculate these indexes can be found at https://gist.github.com/ibartomeus/cdddca21d5dbff26a25e.

We show that when pooling visits for all pollinator species, some plants are preferred over others (χ^2 test p-values for early, mid and late seasons < 0.001; *electivity* values range from -0.84 to 0.82 indicating that we find both over-preferred plants and under-preferred plants; Figure 8.2). In accordance with the results reported in Williams *et al.* (2011), pooled pollinators showed no preferences for exotic plants in any season (Figure 8.2; early season: $F_{1,25} = 1.29$, p-value: 0.27; mid-season: $F_{1,22} = 0.06$, p-value: 0.81; late season: $F_{1,15} = 0.46$, p-value: 0.51), but this general trend does not contradict the fact that specific native or exotic plants are indeed preferred (in black in Figure 8.2).

Second, we shift the focus of our analysis to analyse pollinator species-specific preferences. We excluded pollinator species with less than 20 visits recorded per season in order to prevent confounding rarity with specialization (Blüthgen, 2010). We end up with 16 pollinator species, some of them present in several seasons, making 22 pollinator–season combinations. Again, each of the 22 pollinator–season combinations was evaluated using χ^2 tests and *electivity* values were compared between exotic and native plants for each pollinator species using mixed models with season as a random effect. All pollinators showed significant preferences for certain plant species (χ^2 p-values < 0.05). When analysed individually, most pollinators do not show a consistent

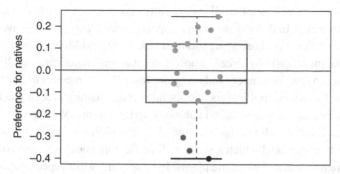

Figure 8.3 Boxplot of the effect sizes (i.e. model estimates) indicating the difference between electivity to exotics and to natives for each pollinator species. Positive values indicate overall preferences for natives, and negative values to exotics. Data points for each pollinator species are indicated in light grey when not significant, in dark grey when $p < 0.1$ and in black when $p < 0.05$.

preference for exotic or native plants, with the exception of three species (Figure 8.3; all species p-values > 0.1, except *Bombus melanopygus* $= 0.05$, *Dialictus incompletum* $= 0.07$ and *Halictus ligatus* $= 0.03$), all of them preferring exotics over natives. However, some pollinators of the 16 analysed do prefer only one or more native plants (four species: *Evylaeus* sp., *Osmia nemoris*, *Ceratina arizonensis*, *Calliopsis fracta*), others prefer only one or more exotic plants (seven species: *Synhalonia actuosa*, *Synhalonia frater*, *Bombus vosnesenskii*, *Halictus ligatus*, *Halictus tripartitus*, *Megachile apicalis*, *Bombus californicus*) and some prefer a mix of exotic and natives (four species: *Bombus melanopygus*, *Ashmeadiella aridula*, *Dialictus incompletum*, *Dialictus tegulariforme*). The differential preferences regarding the exotic status create the basis for expecting winners and losers after an invasion process (see Figure 8.1).

In conclusion, although the overall pattern is no preference for exotic plants, some particular exotic (and native) plants are preferred overall. Similarly, most pollinators do not have overall preferences for exotics, but a few species do favour them. These are social species, usually common and sometimes even considered species typical of disturbed areas (e.g. *Halictus ligatus*, *Dialictus incompletum*). Interestingly, even within the species with no overall preference for exotics, we identify pollinators that prefer particular exotic plants. These pollinators are more likely to be positively affected by the invasion process, the others negatively affected, as their preferred resources will potentially diminish through displacement by invasive plant species.

Relevance for Conservation

We highlighted that pollinator species vary in response to plant invasions, including pollinators use, preference and, in some cases, population dynamic consequences. Assessing the winners and losers in front of the rapid rate of invasive species introductions is crucial for understanding the responses of species groups performing important ecosystem functions, like pollination (Ollerton *et al.*, 2011; Klein *et al.*, 2007). There has been a recent awareness of pollinator declines globally (Potts *et al.*, 2010; González-Varo *et al.*, 2013). However, biological invasions, especially by plant species, have received little attention as a threat (but see Stout and Morales, 2009, and Morales *et al.*, 2013, for effects of animal invasions). We already know that not all pollinators are equally affected by global change, a few are winners and many are losers (Bartomeus *et al.*, 2013). Interestingly, among the winner pollinators, we found species that are able to use flowering crops and tolerate new human-modified habitats (Bartomeus and Winfree, 2013). Gaining information about which species are able also to exploit new exotic plants will be a way forward to understand which species will be flexible enough to survive in novel ecosystems, often dominated by exotic plants. In a changing world, species able to adapt their foraging strategies to use new resources may be the better suited to survive. For example, bumblebees that use the widespread plant invader *Impatiens glandulifera* in the EU are thriving, while endangered bumblebee species do not use it (Kleijn and Raemakers, 2008).

If we are going to manage emerging novel ecosystems, we need to incorporate pollinator-specific responses to different global change drivers, including plant invasions. Some bumblebees and other trophic generalist bees can benefit from exotic plant invasions, as shown by the fact that those can use and even prefer to forage on new exotic plants. This behavioural flexibility may be the key to persisting in a changing world, and maintaining an important ecosystem function. More research is needed on the degree that plant invasions negatively affect those species in comparison with other disturbances that are occurring simultaneously. We need to implement better population monitoring programmes at the community level (so indirect responses can be accounted for), but overall, understanding better which role the pollinator behaviour flexibility and cognitive capabilities play in the process of adapting to novel environments is a promising line of research.

References

Adler, L.S. and Irwin, R.E. (2005). Ecological costs and benefits of defenses in nectar. *Ecology*, 86(11), 2968–2978.

Albrecht, M., Padrón, B., Bartomeus, I. and Traveset, A. (2014). Consequences of plant invasions on compartmentalization and species' roles in plant–pollinator networks. *Proceedings of the Royal Society B: Biological Sciences*, 281(1788), 20140773.

Bartomeus, I. (2013). Understanding linkage rules in plant–pollinator networks by using hierarchical models that incorporate pollinator detectability and plant traits. *PLoS ONE*, 8(7), e69200.

Bartomeus, I. and Winfree, R. (2013). Pollinator declines: reconciling scales and implications for ecosystem services. *F1000Research*, 2.

Bartomeus, I., Vilà, M. and Santamaría, L. (2008). Contrasting effects of invasive plants in plant–pollinator networks. *Oecologia*, 155(4), 761–770.

Bartomeus, I., Ascher, J.S., Gibbs, J., *et al.* (2013). Historical changes in north-eastern US bee pollinators related to shared ecological traits. *Proceedings of the National Academy of Sciences, USA*, 110(12), 4656–4660.

Bascompte, J. and Jordano, P. (2007). Plant-animal mutualistic networks: the architecture of bio-diversity. *Annual Review of Ecology, Evolution, and Systematics*, 567–593.

Bezemer, T.M., Harvey, J.A. and Cronin, J.T. (2014). Response of native insect communities to invasive plants. *Annual Review of Entomology*, 59, 119–141.

Bjerknes, A.L., Totland, Ø., Hegland, S.J. and Nielsen, A. (2007). Do alien plant invasions really affect pollination success in native plant species? *Biological Conservation*, 138(1), 1–12.

Blüthgen, N. (2010). Why network analysis is often disconnected from community ecology: a critique and an ecologist's guide. *Basic and Applied Ecology*, 11(3), 185–195.

Brosi, B.J. and Briggs, H.M. (2013). Single pollinator species losses reduce floral fidelity and plant reproductive function. *Proceedings of the National Academy of Sciences*, 110(32), 13044–13048.

Byers, C.R., Steinhorst, R.K. and Krausman, P.R. (1984). Clarification of a technique for analysis of utilization-availability data. *The Journal of Wildlife Management*, 1050–1053.

Cariveau, D.P. and Norton, A.P. (2009). Spatially contingent interactions between an exotic and native plant mediated through flower visitors. *Oikos*, 118(1), 107–114.

Carnicer, J., Abrams, P.A. and Jordano, P. (2008). Switching behavior, coexistence and diversification: comparing empirical community-wide evidence with theoretical predictions. *Ecology Letters*, 11, 802–808.

Carvalheiro, L.G., Barbosa, E.R.M. and Memmott, J. (2008). Pollinator networks, alien species and the conservation of rare plants: *Trinia glauca* as a case study. *Journal of Applied Ecology*, 45(5), 1419–1427.

Carvalheiro, L.G., Biesmeijer, J.C., Benadi, G., *et al.* (2014). The potential for indirect effects between co-flowering plants via shared pollinators depends on resource abundance, accessibility and relatedness. *Ecology Letters*, 17(11), 1389–1399.

Chittka, L. and Niven, J. (2009). Are bigger brains better? *Current Biology*, 19(21), R995–R1008.

Chittka, L. Gumbert, A. and Kunze, J. (1997). Foraging dynamics of bumblebees: correlates of movements within and between plant species. *Behavioral Ecology*, 8(3), 239–249.

Chrobock, T., Winiger, P., Fischer, M. and van Kleunen, M. (2013). The cobblers stick to their lasts: pollinators prefer native over alien plant species in a multi-species experiment. *Biological Invasions*, 15(11), 2577–2588.

Corbet, S.A., Bee, J., Dasmahapatra, K., *et al.* (2001). Native or exotic? Double or single? Evaluating plants for pollinator-friendly gardens. *Annals of Botany*, 87(2), 219–232.

Crone, E.E. (2013). Responses of social and solitary bees to pulsed floral resources. *The American Naturalist*, 182(4), 465–473.

Dietzsch, A.C., Stanley, D.A. and Stout, J.C. (2011). Relative abundance of an invasive alien plant affects native pollination processes. *Oecologia*, 167(2), 469–479.

Dukas, R. and Real, A.L. (1993). Learning constraints and floral choice behaviour in bumblebees. *Animal Behaviour*, 46, 637–644.

Forrest, J. and Thomson, J.D. (2009). Pollinator experience, neophobia and the evolution of flowering time. *Proceedings of the Royal Society B: Biological Sciences*, 276(1658), 935–943.

Fründ, J., Linsenmair, K.E. and Blüthgen, N. (2010). Pollinator diversity and specialization in relation to flower diversity. *Oikos*, 119(10), 1581–1590.

Gibson, M.R., Richardson, D.M. and Pauw, A. (2012). Can floral traits predict an invasive plant's impact on native plant–pollinator communities? *Journal of Ecology*, 100(5), 1216–1223.

González-Varo, J.P., Biesmeijer, J.C., Bommarco, R., *et al.* (2013). Combined effects of global change pressures on animal-mediated pollination. *Trends in Ecology and Evolution*, 28(9), 524–530.

Gumbert, A. (2000). Color choices by bumblebees (*Bombus terrestris*): innate preferences and generalization after learning. *Behavioral Ecology and Sociobiology*, 48(1), 36–43.

Heleno, R.H., Ceia, R.S., Ramos, J.A. and Memmott, J. (2009). Effects of alien plants on insect abundance and biomass: a food-web approach. *Conservation Biology*, 23(2), 410–419.

Herrmann, F., Westphal, C., Moritz, R.F. and Steffan-Dewenter, I. (2007). Genetic diversity and mass resources promote colony size and forager densities of a social bee (*Bombus pascuorum*) in agricultural landscapes. *Molecular Ecology*, 16(6), 1167–1178.

Hobbs, R.J. (ed.). (2000). *Invasive Species in a Changing World*. Washington DC: Island Press.

Inouye, D.W. (1978). Resource partitioning in bumblebees: experimental studies of foraging behavior. *Ecology*, 59(4), 672–678.

Ivlev, V.S. (1964). *Experimental Ecology of the Feeding of Fishes*. New Haven, CT: Yale University Press.

Jakobsson, A. and Padrón, B. (2014). Does the invasive *Lupinus polyphyllus* increase pollinator visitation to a native herb through effects on pollinator population sizes? *Oecologia*, 174(1), 217–226.

Johnson, L.K. and Hubell, P. (1975). Contrasting foraging strategies and coexistence of two bee species on a single resource. *Ecology*, 56(6), 1398–1406.

Kaiser-Bunbury, C.N., Muff, S., Memmott, J., Müller, C.B. and Caflisch, A. (2010). The robustness of pollination networks to the loss of species and interactions: a quantitative approach incorporating pollinator behaviour. *Ecology Letters*, 13(4), 442–452.

Kevan, P.G. and Menzel, R. (2012). The plight of pollination and the interface of neurobiology, ecology and food security. *The Environmentalist*, 32(3), 300–310.

Kleijn, D. and Raemakers, I. (2008). A retrospective analysis of pollen host plant use by stable and declining bumblebee species. *Ecology*, 89(7), 1811–1823.

Klein, A.M., Vaissiere, B.E., Cane, J.H., *et al.* (2007). Importance of pollinators in changing landscapes for world crops. *Proceedings of the Royal Society B: Biological Sciences*, 274(1608), 303–313.

Lopezaraiza-Mikel, M.E., Hayes, R.B., Whalley, M.R. and Memmott, J. (2007). The impact of an alien plant on a native plant–pollinator network: an experimental approach. *Ecology Letters*, 10(7), 539–550.

Montero-Castaño, A. (2014). Interacciones entre polinizadores y la planta exótica *Hedysarum coronarium* a distintas escalas espaciales. PhD thesis. Madrid, Spain: Universidad Complutense de Madrid.

Montero-Castaño, A. and Vilà, M. (2012). Impact of landscape alteration and invasions on pollinators: a meta-analysis. *Journal of Ecology*, 100(4), 884–893.

Morales, C.L. and Traveset, A. (2009). A meta-analysis of impacts of alien vs. native plants on pollinator visitation and reproductive success of co-flowering native plants. *Ecology Letters*, 12(7), 716–728.

Morales, C.L., Arbetman, M.P., Cameron, S.A. and Aizen, M.A. (2013). Rapid ecological replacement of a native bumblebee by invasive species. *Frontiers in Ecology and the Environment*, 11(10), 529–534.

Moroń, D., Lenda, M., Skórka, P., *et al.* (2009). Wild pollinator communities are negatively affected by invasion of alien goldenrods in grassland landscapes. *Biological Conservation*, 142(7), 1322–1332.

Muñoz, A.A. and Cavieres, L.A. (2008). The presence of a showy invasive plant disrupts pollinator service and reproductive output in native alpine species only at high densities. *Journal of Ecology*, 96(3), 459–467.

Neu, C.W., Byers, C.R. and Peek, J.M. (1974). A technique for analysis of utilization-availability data. *The Journal of Wildlife Management*, 38, 541–545.

Nienhuis, C.M., Dietzsch, A.C. and Stout, J.C. (2009). The impacts of an invasive alien plant and its removal on native bees. *Apidologie*, 40(4), 450–463.

Ollerton, J., Winfree, R. and Tarrant, S. (2011). How many flowering plants are pollinated by animals? *Oikos*, 120(3), 321–326.

Padrón, B., Traveset, A., Biedenweg, T., *et al.* (2009). Impact of alien plant invaders on pollination networks in two archipelagos. *PLoS ONE*, 4(7), e6275.

Palladini, J.D. and Maron, J.L. (2014). Reproduction and survival of a solitary bee along native and exotic floral resource gradients. *Oecologia*, 176(3), 789–798.

Parker, I.M. (1997). Pollinator limitation of *Cytisus scoparius* (Scotch broom), an invasive exotic shrub. *Ecology*, 78(5), 1457–1470.

Pearse, I.S., Harris, D.J., Karban, R. and Sih, A. (2013). Predicting novel herbivore–plant interactions. *Oikos*, 122(11), 1554–1564.

Potts, S.G., Biesmeijer, J.C., Kremen, C., *et al.* (2010). Global pollinator declines: trends, impacts and drivers. *Trends in Ecology and Evolution*, 25(6), 345–353.

Pyšek, P., Richardson, D.M., Rejmánek, M., *et al.* (2004). Alien plants in checklists and floras: towards better communication between taxonomists and ecologists. *Taxon*, 53(1), 131–143.

Riffell, J.A., Alarcón, R., Abrell, L., *et al.* (2008). Behavioral consequences of innate preferences and olfactory learning in hawkmoth–flower interactions. *Proceedings of the National Academy of Sciences, USA*, 105(9), 3404–3409.

Roulston, T.H. and Cane, J.H. (2000). Pollen nutritional content and digestibility for animals. *Plant Systematics and Evolution*, 222(1–4), 187–209.

Roulston, T.A.H. and Goodell, K. (2011). The role of resources and risks in regulating wild bee populations. *Annual Review of Entomology*, 56, 293–312.

Sedivy, C., Müller, A. and Dorn, S. (2011). Closely related pollen generalist bees differ in their ability to develop on the same pollen diet: evidence for physiological adaptations to digest pollen. *Functional Ecology*, 25(3), 718–725.

Stang, M., Klinkhamer, P.G., Waser, N.M., Stang, I. and van der Meijden, E. (2009). Size-specific interaction patterns and size matching in a plant–pollinator interaction web. *Annals of Botany*, 103(9), 1459–1469.

Steffan-Dewenter, I. and Schiele, S. (2008). Do resources or natural enemies drive bee population dynamics in fragmented habitats. *Ecology*, 89(5), 1375–1387.

Stelzer, R.J., Chittka, L., Carlton, M. and Ings, T.C. (2010). Winter active bumblebees (*Bombus terrestris*) achieve high foraging rates in urban Britain. *PLoS ONE*, 5(3), e9559.

Stouffer, D.B., Cirtwill, A.R. and Bascompte, J. (2014). How exotic plants integrate into pollination networks. *Journal of Ecology*, 102(6), 1442–1450.

Stout, J.C. and Morales, C.L. (2009). Ecological impacts of invasive alien species on bees. *Apidologie*, 40(3), 388–409.

Tepedino, V.J., Bradley, B.A. and Griswold, T.L. (2008). Might flowers of invasive plants increase native bee carrying capacity? Intimations from Capitol Reef National Park, Utah. *Natural Areas Journal*, 28(1), 44–50.

Thébault, E. and Fontaine, C. (2010). Stability of ecological communities and the architecture of mutualistic and trophic networks. *Science*, 329(5993), 853–856.

Thompson, R.M., Brose, U., Dunne, J.A., *et al.* (2012). Food webs: reconciling the structure and function of biodiversity. *Trends in Ecology and Evolution*, 27(12), 689–697.

Traveset, A. and Richardson, D.M. (2006). Biological invasions as disruptors of plant reproductive mutualisms. *Trends in Ecology and Evolution*, 21(4), 208–216.

Valdovinos, F.S., Ramos-Jiliberto, R., Flores, J.D., Espinoza, C. and López, G. (2009). Structure and dynamics of pollination networks: the role of alien plants. *Oikos*, 118(8), 1190–1200.

Valdovinos, F.S., Ramos-Jiliberto, R., Garay-Narváez, L., Urbani, P. and Dunne, J.A. (2010). Consequences of adaptive behaviour for the structure and dynamics of food webs. *Ecology Letters*, 13(12), 1546–1559.

Vilà, M., Bartomeus, I., Dietzsch, A.C., *et al.* (2009). Invasive plant integration into native plant–pollinator networks across Europe. *Proceedings of the Royal Society B: Biological Sciences*, 276(1674), 3887–3893.

Vilà, M., Espinar, J.L., Hejda, M., *et al.* (2011). Ecological impacts of invasive alien plants: a meta-analysis of their effects on species, communities and ecosystems. *Ecology Letters*, 14(7), 702–708.

Waser, N.M., Chittka, L., Price, M.V., Williams, N.M. and Ollerton, J. (1996). Generalization in pollination systems and why it matters. *Ecology*, 77(4), 1043–1060.

Waters, S.M., Fisher, S.E. and Hille Ris Lambers, J. (2014). Neighborhood-contingent indirect interactions between native and exotic plants: multiple shared pollinators mediate reproductive success during invasions. *Oikos*, 123(4), 433–440.

Westphal, C., Steffan-Dewenter, I. and Tscharntke, T. (2003). Mass flowering crops enhance pollinator densities at a landscape scale. *Ecology Letters*, 6(11), 961–965.

Williams, N.M. and Kremen, C. (2007). Resource distributions among habitats determine solitary bee offspring production in a mosaic landscape. *Ecological Applications*, 17(3), 910–921.

Williams, N.M., Cariveau, D., Winfree, R. and Kremen, C. (2011). Bees in disturbed habitats use, but do not prefer, alien plants. *Basic and Applied Ecology*, 12(4), 332–341.

Winfree, R., Bartomeus, I. and Cariveau, D.P. (2011). Native pollinators in anthropogenic habitats. *Annual Review of Ecology, Evolution and Systematics*, 42(1), 1.

9 In the Light of Introduction: Importance of Introduced Populations for the Study of Brood Parasite–Host Coevolution

Tomáš Grim and Bård G. Stokke

Introduction

Avian brood parasitism represents an extreme and mindboggling reproductive strategy: parasites avoid various costs of parental care and delegate those to other birds, either conspecific or heterospecific. Causes and consequences of brood parasitism have attracted the attention of evolutionary biologists for decades: few natural antagonistic relationships in nature are so amenable to experimental manipulations and quantification of their pros and cons (Rothstein, 1990).

Although representing a mere single per cent of avian phylogenetic diversity, *c.* 100 species, avian obligate brood parasites receive disproportionately high attention from both amateur ornithologists and scientists in evolutionary ecology (Feeney *et al.*, 2014). Such 'full-time' parasites, like *Cuculus* cuckoos or *Molothrus* cowbirds, never build their own nests and consign their progeny to the care of heterospecific hosts. Naturally, this form of parasitism is always interspecific. The other major form of avian brood parasitism is conspecific parasitism, which has been found in *c.* 250 species. Such intraspecific parasites lay eggs into their own nests, but deposit additional eggs into the nests of conspecific neighbours. These parasites, prevalent in, for instance, waterfowl and colonially breeding birds, are always facultative. Rarely, some species, for example some ducks or *Coccyzus* cuckoos, show a mixed strategy of victimizing both conspecifics and heterospecifics apart from laying eggs into their own nests. In recent decades, such natural-born cheats, either conspecific or heterospecific, became popular models to study coevolution, because they allow the investigation of selective pressures on evolution of coevolved traits, the importance of genes, environment and learning, and the resulting trait expressions at various stages of the breeding cycle (Feeney *et al.*, 2014).

Here we highlight the heuristic value of shifting population ranges and introduced populations for studies on brood parasite–host coevolution. First, we outline basic coevolutionary assumptions in this field of study and point to a fundamental problem

Biological Invasions and Animal Behaviour, eds J.S. Weis and D. Sol. Published by Cambridge University Press. © Cambridge University Press 2016.

of studies concerning evolution of coevolved adaptations: the length of parasite–host contact is, in almost all cases, unknown, being estimated only indirectly. Second, we consider advantages of studies related to expanding, declining and introduced host and parasite populations, and review the few existing studies of populations with known recent length of parasite–host contact. Finally, we discuss prospects for future studies and suggest suitable candidate model systems.

Brood Parasite–Host Studies: Opportunities, Limitations and Solutions

Opportunities

The basic assumption/scenario behind all coevolutionary models is intuitive and simple (Davies and Brooke, 1989). A naïve host, lacking any specific anti-parasite defences, starts to become victimized by a parasite. Parasitism is costly for hosts. Thus, host individuals showing any cognitive and behavioural traits that lower the impact of parasites (e.g. breeding in 'safe' sites, increased nest guarding and aggression, removal of odd eggs or chicks) enjoy higher fitness. Higher host fitness means lower parasite fitness and this selects for parasite counter-adaptations (e.g. furtive laying, mimetic eggs, competitive chicks). If there is enough genetic variation for both host and parasite 'battle' traits, each host adaptation and parasite counter-adaptation would improve (i.e. become more effective) over time. Additionally, new offensive and defensive traits could arise. These principles hold for all developmental stages of the parasite–host battle, i.e. front-line defences against parasite adults (i.e. before they deposit their eggs), then against their eggs, and chicks and finally fledglings (Feeney et al., 2014).

The crucial parameters that affect all aspects of parasite–host coevolution are time (above) and space. Within a specific host, not all populations are necessarily parasitized at the same time (and at the same rate), creating mosaics of coevolutionary hot and cold spots (Thompson, 2005). Knowing when parasites and their hosts came into contact is critical to achieving a better understanding of the adaptations they both may express.

As parasite and host traits coevolve in time and space, the sophistication of host and parasite armoury may provide an indirect measure of the length (time) and intensity (parasitism costs) of coevolution. For example, unsuitable or novel 'hosts' should show no anti-parasite defences, recent hosts in evolutionary time should show poor defences, and hosts used for a long time should show fine-tuned defences (Davies and Brooke, 1989; Moksnes et al., 1991). Such hosts may have even extirpated their parasites, as documented by the almost total rejection of simulated experimental parasitism, combined with historical but not current evidence of natural parasitism (Honza et al., 2004). Importantly, the level of defence among species may very well vary with developmental stage, conditional cues, learning and individual personality, resulting in diverging evolutionary trajectories within and between host–parasite systems (Feeney et al., 2014).

There are many examples of both spatial and temporal variation in adaptations within single host species depending on parasitism pressure, which are likely to reflect genetic differences, phenotypic plasticity or both (Briskie et al., 1992; Nakamura et al., 1998;

Soler *et al.*, 1999; Moskát *et al.*, 2002; Avilés and Møller, 2003; Stokke *et al.*, 2008; Spottiswoode and Stevens, 2012; Kuehn *et al.*, 2014). In turn, this variation may reflect evolutionary lag (Igic *et al.*, 2012) or spatiotemporal changes in parasite adaptations (Thorogood and Davies, 2013). Several studies, however, suggest that egg rejection behaviour may be retained for long time periods even in the absence of parasitism by brown-headed cowbirds, *Molothrus ater* (Rothstein, 2001; Peer *et al.*, 2011). Similarly, several potential hosts of common cuckoos, *Cuculus canorus*, show sophisticated anti-parasite adaptations, even though they are not currently regularly parasitized (Davies and Brooke, 1989; Moksnes *et al.*, 1991; Martín-Vivaldi *et al.*, 2013). These studies suggested that maintenance of egg rejection in these species is cost-free (but this remains to be tested in most species apparently retaining egg rejection).

However, this may represent only a misleading inference resulting from overlooking alternative selection agents unrelated to defences against brood parasites (Avilés and Parejo, 2011; Avilés *et al.*, 2014). In traits with multiple functions, a previously secondary function may maintain a trait after the primary function is lost (Lahti *et al.*, 2009). Trnka and Grim (2014) showed that in great reed warbler, *Acrocephalus arundinaceus*, females, aggression towards adult cuckoos strongly correlates with aggression against human observers ('self-defence') and nest guarding (but not with probability of and latency to egg rejection). If the warbler's anti-cuckoo aggression is an integrated part of a suite of correlated behavioural characters (behavioural syndrome), then selection on other components of behaviour may indirectly lead to the maintenance of high anti-cuckoo aggression even in the absence of cuckoos in a population that is no longer parasitized. By the same logic, egg discrimination may be theoretically maintained by other selection pressures unrelated to interspecific parasitism, e.g. conspecific parasitism (Samas *et al.*, 2014a, b) or nest sanitation (Peer and Sealy, 2004). However, the latter might apply only to rejection of poorly mimetic interspecific eggs: rejection of mimetic eggs (especially conspecific eggs) requires extremely fine-tuned cognitive abilities that, in principle, cannot represent a by-product of very rough discrimination of objects needed in the context of nest cleaning (Samas *et al.*, 2014a). Importantly, nest sanitation is present in virtually all passerines and therefore cannot explain taxonomic variation in egg rejection (Peer and Sealy, 2004). No matter the mechanism, maintenance of anti-parasite defences after relaxation from parasite pressure would make it difficult for brood parasites to re-invade such hosts (Rothstein, 2001).

Limitations

To disentangle the causes responsible for observed patterns of covariation between host adaptations and parasite presence/absence, we would ideally need to know historical patterns of parasite–host contact. However, we know little about the length of the host–parasite interactions. Naturally, the (non)existence of long-term parasite–host contact, i.e. the scale of thousands and millions of years, is unknown in all cases. A currently parasitized population may have become parasitized only recently, thus being effectively allopatric at relevant evolutionary time-scales. By the same logic, a population that is currently not parasitized may have escaped from parasitism only recently, thus being effectively sympatric at evolutionary time-scales.

Most studies so far have concerned specific host and/or parasite populations at a specific time and place, representing 'snap-shots' of only a few breeding seasons. Such studies have resulted in important advances in our knowledge of coevolutionary interactions, and work adequately for addressing various questions related to brood parasitism, like variables affecting risks of parasitism, nest site selection, baseline levels of defences, egg mimicry, demography, and molecular analyses and dispersal patterns. However, it is impossible to track the temporal variation in coevolutionary interactions of specific populations based on just couple of years of study. Studies of long-term interactions are rare in general (Thompson, 2005), simply due to the work load required to undertake such investigations.

Assessing the length of the host–parasite interactions is further complicated considering that any host population we currently observe may have become locally extinct and recolonized its current geographical location multiple times. This effect is certain in northern areas: no current Scandinavian cuckoo or host population can be older than 10 000 years *in situ*. Pleistocene climate dynamics, with over 30 glaciation events during the last 2 million years, massively affected the ranges of all species globally, including tropics (e.g. Joseph *et al.*, 2002). Even today, 10 millennia after the end of the last glaciation, we witness range shifts that reflect Pleistocene climatic events (e.g. continuing westward range expansion of various East Asian species). With this large-scale persistent dynamic, it proves extremely difficult to assign a long-term sympatry/allopatry status to any host populations.

For example, Vikan *et al.* (2010) studied various native brambling, *Fringilla montifringilla*, populations and found the same level of host defences in both one parasitized and three currently non-parasitized populations. They explained this pattern by intense gene flow; under this scenario all studied populations would represent a single panmictic population. Indeed, gene flow might have been substantial historically: any of these populations could not have existed for longer than 10 000 years because of Pleistocene glaciations. The maintenance of strong anti-cuckoo defences in current allopatry thus may be explained by fixation of rejection alleles at the level of the whole species. Such species-fixed defences may be retained for longer than defences that are variable within species because selection has erased genetic variation in the defence traits (Foster and Endler, 1999). Alternatively, rejection alleles became fixed not in Scandinavia but during cuckoo–brambling coevolution in glacial refugia in more southern parts of Europe (or even earlier, in the Tertiary) and were retained in all populations that colonized Scandinavia after glaciers retreated 10 000 years ago. The same holds for hosts of brown-headed cowbirds in temperate America. Defences in allopatry may not necessarily reflect 'introgression of rejecter genes from sympatric populations' (Briskie *et al.*, 1992): after glacial retreat, host populations, e.g. American robins, *Turdus migratorius* (Figure 9.1), might have spread further north than cowbird populations and simply brought their defences from glacial refugia with no current gene flow needed.

The longest record of parasitism rates and host responses to standardized model eggs covers only 30 years. In reed warblers, *Acrocephalus scirpaceus*, decline in parasitism

Figure 9.1 Full diversity of types of hosts and parasites that should become the focus of future brood parasitism studies. Upper row shows examples of model introduced species: brood parasite (pin-tailed whydah), host (village weaver) and control non-host (goldfinch). Lower row shows examples of model species that expanded their ranges naturally: brood parasite (shiny cowbird), host (American robin) and control non-host (fieldfare). Note that these examples do not reflect parasite–host relationships between these particular species and come from geographically varied areas to exemplify research opportunities at global scales. See text for details on these and other potential model systems (photo credits: T. Grim).

rates caused by a population crash in cuckoos was accompanied by a parallel decline in host rejection rates (Thorogood and Davies, 2013). Imagine that we did not have the evidence accumulated over the three decades and were starting to study the population only now. We would observe low parasitism rates and low levels of host defences, inferring that parasitism is probably a recent phenomenon and hosts had not evolved strong defences yet. It is only due to the benefit of the existing evidence that we may firmly say that such a conclusion would be erroneous. Unfortunately, such historical evidence is lacking for almost all other sites where brood parasites and their hosts are studied (but see Igic *et al.*, 2012). Therefore, any studies about the evolution of parasite and host coevolutionary features that are not based on direct evidence of past parasitism might be inconclusive in principle.

Solutions

We envisage two possible research opportunities that may be particularly fruitful in overcoming these limitations. First, natural expansions or retractions of parasite or host population ranges allow direct measurement of the length of sympatry and allopatry, at least at short time-scales (inferences at long time-scales are impossible in principle, see above). Second, host or parasite populations introduced from their native ranges into novel geographically and genetically isolated populations represent invaluable 'natural experiments' (*sensu* Diamond, 1986) to study how coevolved traits may be affected by changing selective pressures. Avian invaders are often and productively studied to address various aspects of avian biology (e.g. Blackburn *et al.*, 2009), but their potential for understanding brood parasite–host coevolution has so far been poorly utilized. Such large-scale introduction experiments are ethically and legally unacceptable today. Fortunately, in a scientific sense, today's researchers can already harvest data from non-native populations that underwent decades or even centuries of evolution in isolation from their source populations.

Solution I: Range Shifts

Natural changes in host and parasite biogeography, including both expansions and retractions of their ranges, lead to novel host–parasite associations and new ecological and coevolutionary relationships (Morand and Krasnov, 2010). Therefore, natural range dynamics may help to elucidate parasite–host coevolution.

Expansions of novel parasites and hosts to areas both with and without native (indigenous) parasites may provide interesting study systems. If parasites invade locations where another parasite is already present (Cruz *et al.*, 1998), hosts may be pre-adapted to cope with the novel parasite (e.g. previously evolved egg discrimination), decrease parasite success, and slow down or prevent the new parasite's range expansion (Dinets *et al.*, 2015). Further, native and novel parasites might differ in their virulence. The presence of a new highly virulent parasite may increase overall parasite pressure, selecting for host defences which may, as a by-product, decrease fitness of a native less virulent parasite (cf. collateral damage hypothesis: Lyon and Eadie, 2004; Samas *et al.*, 2014a). Also host range expansion may bring the host into contact with new parasites; if the new parasites are sufficiently virulent they could constrain the host's range spread and prevent further coevolution.

Natural parasite or host population declines, local extinctions and consequent range retractions also represent an opportunity to study evolutionary dynamics of host–parasite coevolution, especially relaxed selection (Lahti *et al.*, 2009) when a parasite goes extinct and host switching (Morand and Krasnov, 2010) when a host goes extinct. The decline in the cuckoo population in the UK provided one such natural experiment showing that the resulting relaxed selection led to rapid decline of host defences, perhaps within the limits of phenotypic plasticity (Thorogood and Davies, 2013). In another study in the Czech Republic, a decline in the great reed warbler host population forced cuckoos to switch to an alternate host, the reed warbler, with a resulting evolutionary lag in adaptations of both cuckoos and the locally new host (Igic *et al.*, 2012).

Cuckoo females are specialist parasites, but other parasites are generalists parasitizing many hosts (see below). These differences may have large effects on successful establishment of naturally invading or introduced parasites. An invasive East Asian freshwater mussel, for instance, has spread throughout Central Europe because its parasitic larvae are able to complete its development on novel fish species (Douda et al., 2012). Similarly, invading or introduced generalist brood parasites may have the best odds in novel host communities, simply because their more flexible host choice increases the chances of finding appropriate fosterers. Indeed, the most famous case of range expansion with effects on native host populations entails an extremely generalist parasite, the brown-headed cowbird (Smith et al., 2000; Baker et al., 2014). This cowbird was in historical times closely associated with distribution of bison, *Bison bison*, on the Great Plains of North America. After the arrival of European settlers the cowbirds experienced a significant range expansion due to clearing of forests and introduction of domestic livestock. This range expansion continued well into the twentieth century. Cowbird parasitism is often most pronounced in forest fragments and edge areas, and may severely reduce host population viability, often in tandem with habitat destruction. Many hosts used today lack defences against the parasite while experiencing high parasitism rates (Smith et al., 2000). Nevertheless, the strength of the 'expanding range effect' on hosts is debated because cowbirds might have older historical contacts with various hosts (Rothstein and Peer, 2005) and many cowbird hosts retain their defences long after parasitism declines (Peer et al., 2011). Extreme Pleistocene dynamics of animal ranges, and consequently parasite–host contacts, makes most of the inferences about historical cowbird–host coevolution largely uncertain. In contrast, currently witnessed range changes may at least provide some certainty, although only at short time-scales, about parasite–host interactions and their consequences.

The shiny cowbird, *Molothrus bonariensis* (Figure 9.1), is another generalist brood parasite distributed throughout South America. From 1860 onwards, it colonized the islands of the West Indies, most probably with the aid of humans. In Puerto Rico, high rates of parasitism (~80%) led to a pronounced reduction in host reproductive output. Since novel hosts often show weak defences against brood parasitism coupled with the typical low population sizes in island endemics, concerns are raised about their ability to withstand the high costs of shiny cowbird parasitism (Woodworth, 1997). The shiny cowbird has continued its range expansion: the first individual was observed in the United States in 1985, and has since spread across the south-eastern part of the United States (Cruz et al., 1998). No direct evidence for parasitism has been documented in the country, but females in a state ready to lay eggs have been encountered (Post and Sykes, 2011). Shiny cowbirds have also expanded their range in South America due to human-induced habitat fragmentation, just as has happened in North America with brown-headed cowbirds (Cruz et al., 1998). The range expansion in parts of South America may have been boosted by introductions (Marín, 2000).

An additional level of complexity is seen in rare cases of multiple parasites colonizing the same geographic area. The south-eastern United States was recently invaded by three generalist parasites: shiny cowbirds (above), bronzed cowbirds, *Molothrus aeneus*, and brown-headed cowbirds, providing a unique three-parasite invasion system in Florida (Cruz *et al.*, 1998). Future research should address how previously cowbird-free host populations respond to the novel parasite pressure. On the other hand, it is important to stress that previous absence of cowbirds does not imply that the population was 'naïve' before the cowbirds arrived – gene flow can lead to an influx of rejecter alleles into phenotypically allopatric populations making them effectively genetically sympatric (see also Moskát *et al.*, 2008). Therefore, studies of 'naïve' native host population would benefit from quantifying gene flow among host populations sympatric and allopatric with brood parasites (Soler *et al.*, 1999).

Overall, local adaptation – including decay of traits that are no longer adaptive – requires that the spatial scale of selection is larger than scale of gene flow (Foster and Endler, 1999). In other words, restricted gene flow speeds up evolutionary change. Therefore, any study reporting maintenance of defences in native allopatric host populations that does not assess genetic similarities with sympatric populations is inconclusive – gene flow among parasitized and non-parasitized populations may have retarded a decline of host defences in currently allopatric populations (Soler *et al.*, 1999; Moskát *et al.*, 2008; Vikan *et al.*, 2010). This does not apply to introduced populations where gene flow is excluded by distances between source and introduced populations being extremely large compared to host dispersal distances (Lahti, 2006; Samas *et al.*, 2014a; Yang *et al.*, 2014). Gene flow also represents a potential statistical problem for source populations: if such populations are connected by gene flow they may not represent independent units; statistical analyses ignoring this then would be biased with overestimated degrees of freedom.

However, even populations subject to intense gene flow may express fine-tuned adaptations against parasites through phenotypic plasticity. Comparatively, populations isolated from each other (i.e. low dispersal and gene flow) tend to evolve non-plastic genetically 'rigid' defences. This paradigm would predict that perfectly isolated populations (e.g. introduced to distant geographical areas) should lose plasticity and/or evolve non-plastic defences towards novel parasites encountered in novel ranges (Foster and Endler, 1999; Tojo and Nakamura, 2014).

Despite all of the mentioned range expansions above, to our knowledge, no study has yet tested the same host population before and after a brood parasite colonized that host population (i.e. temporal allopatry followed by sympatry). Expansions of the three cowbird species provide an opportunity to glimpse at coevolution in action. This could be done by studying currently allopatric host populations that are expected to come into contact with the parasite(s) soon. Such populations should be tested with non-mimetic and mimetic foreign eggs, including conspecific eggs (Briskie *et al.*, 1992; Samas *et al.*, 2014a). Additionally, researchers need to quantify egg phenotypes of both hosts and arriving parasites (including those currently allopatric from the point of view of the focal population), and the population would need to be characterized genetically to quantify gene flow from additional conspecific populations, both allopatric and sympatric with parasites. Such procedures would need to be repeated across years and, more likely, decades, to document predicted changes in host and parasite egg phenotypes, host egg rejection rates, etc. Only such long-term studies will allow assessment of both phenotypic plasticity and genetic evolutionary change.

Range of Geographic Scales: Between Urbanization and Intercontinental Invasion

Most studies of native parasite and host populations performed so far have focused on an intermediate subcontinental scale where sympatric versus allopatric populations were distanced by dozens to hundreds of kilometres. We envisage that our understanding of enemy–victim coevolution will be greatly advanced if the research focus is extended from 'snap-shots' at intermediate scales to shifting ranges at both small and large scales.

Colonization of urban settlements by birds represents a special case of range shifts at the smallest geographic scale. Urban areas strikingly differ both abiotically and biotically from neighbouring rural landscapes. Yet some species, including various passerine hosts of brood parasites, termed 'urban exploiters', show strong preferences for such human-dominated environments, and other species, termed 'urban adapters' readily adapt to urban areas although they also commonly live in rural areas (Gil and Brumm, 2014). Similarly, some parasites, like brown-headed cowbirds, prefer disturbed urbanized areas (Barnagaud et al., 2015). Unlike these, most brood parasites are among the shyest birds and avoid human proximity (Erritzøe et al., 2012). Thus, host populations may become allopatric with parasites like common cuckoos through occupying urban areas (Grim et al., 2011) or even breeding indoors (Liang et al., 2013). The timing of the origin of urban populations (and, by implication, of allopatry) is often known (Evans et al., 2010), providing a crucial advantage over non-urban populations where the length of sympatry/allopatry is almost never known. Urban passerines represent largely ignored study systems where effects of sympatry versus micro-allopatry can be studied at finest distance scales (Samas et al., 2014a).

At the opposite spatial extreme, transoceanic expansions seem to be feasible in some brood parasites. Common cuckoos are increasingly often recorded in North America (including wintering and courting) and brown-headed cowbirds in Eurasia (Dinets et al., 2015). The first study that addressed the potential coevolution between native hosts and presumably invasive parasites (Dinets et al., 2015) suggested that interactions will be complex: specific outcomes of the first contact between new enemies and victims will depend on particular host species egg rejection reaction norms and, in the case of cuckoos, on particular invading gens (host-specific race) and its gens-specific egg phenotype. Future studies should test additional potential hosts and their populations located in the range extension zone of both brood parasites, both before and after the new contact.

A recent study showed that the successful establishment of a species in an exotic range is best predicted by its urbanization in the ancestral range (Møller et al., 2015). This provides an exciting opportunity to study the same species at all scales: in native range sympatric with parasites, native (urban) range allopatric with parasites, and exotic range allopatric with original parasites. So far this opportunity has been employed in only two model species, in blackbirds, *Turdus merula*, and song thrush, *T. philomelos* (Samas *et al.*, 2014a).

Solution II: Introductions

For any study of relaxed or novel selection, it is necessary to know the history of focal populations (Foster and Endler, 1999). Unfortunately, as we showed, historical sympatry/allopatry status of a host species/population is in most cases impossible to estimate. Therefore, the most reliable way to know the length of host–parasite contact is to manipulate it experimentally. This was essentially done by our ancestors who introduced many bird species, including hosts and even some parasites, outside their natural ranges (Blackburn *et al.*, 2009), creating large-scale natural experiments with many replicates (Diamond, 1986; Table 9.1). The human-assisted 'Great Escape' of hosts to novel areas without their parasites and related selection pressures may teach us much about changes in expression and evolution of host defensive traits within the framework of enemy release (Morand and Krasnov, 2010) and relaxed selection hypotheses (Lahti *et al.*, 2009).

Introduced populations show multiple advantages for studies on brood parasite–host coevolution which enable researchers to overcome some of the research and inferential constraints that are hard or impossible to overcome in native unmanipulated populations. Introduced host populations lack any recent experience with ancestral parasites. Thus, in the terms of geographic mosaic theory of coevolution, invasive populations effectively represent coevolutionary cold spots (Thompson, 2005). This is because their brood parasites were almost never co-introduced (but see Payne, 2010) and introduced hosts are only rarely parasitized by native parasites (Table 9.1; Tojo and Nakamura, 2014). Researchers also know accurately the length of allopatry (Hale and Briskie, 2007) which allows them to determine if time since isolation covaries with changes in host resistance traits across independent introduced populations (Lahti, 2006).

Introduced host populations have left behind not only their brood parasites but also a majority of their conspecifics. This is crucial, because in native populations, patterns of host defences incongruent with local parasitism pressure were often explained by possible (i.e. unknown) gene flow from other conspecific populations, but without actually quantifying it. Therefore, host populations introduced to remote oceanic islands (e.g. Hawaii) or old continental islands (e.g. New Zealand) provide fundamental advantages: chances of gene flow between source and recipient populations are typically zero. This is clear even without any molecular data: lifetime dispersal of hosts is often at the scale of kilometres, whereas distances between native and introduced populations including their migratory pathways are typically thousands of kilometres making any contact between the source and recipient populations effectively impossible.

Detectable evolutionary changes in traits related to parasite–host coevolution can be rapid, taking less than a century (egg colour: Lahti, 2005, 2008; Spottiswoode and Stevens, 2012; clutch size: Samas *et al.*, 2013). This is no surprise: various bird

Table 9.1 Overview of introduced passerine populations with information on their status as hosts of brood parasites in native and introduced range

Host species	Origin	Parasitism in native range	Number of introduced populations	Introduced ranges	Parasitism in introduced range
Alaudidae					
Alauda arvensis	Eurasia	CuCa (o), HiHy (r)	4	Australia, North America, Pacific islands	Not reported
Cardinalidae					
Cardinalis cardinalis	North America	MoAe (u), MoAt (f)	2	Pacific islands, West Indies	Not reported
Corvidae					
Corvus splendens	Asia	EuSc (f)	16	Africa, Arabia, Australia, Europe, Indian ocean islands, Middle East	Israel: ClGl (u)
Gymnorhina tibicen	Australia	ScNo (f)	2	Pacific islands	Not reported
Emberizidae					
Paroaria coronata	South America	MoBo (u)	4	Indian ocean islands, North America, Pacific islands	Not reported
Sicalis flaveola	South America	MoBo (o)	3	Pacific islands, West Indies	Not reported
Estrilidae					
Amandava amandava	Asia	Not reported	13	Africa, Arabia, Europe, Indian ocean islands, Pacific islands, West Indies	Not reported
Estrilda astrild	Africa	ViFu (r), ViMa (f), ViWi (r)	14	Atlantic islands, Europe, Indian ocean islands, Pacific islands, South America	Not reported
Estrilda melpoda	Africa	ViMa (u)	6	Europe, North America, Pacific islands, West Indies	Puerto Rico: ViMa (u)
Estrilda troglodytes	Africa	ViMa (u)	6	Atlantic islands, Europe, Pacific islands, West Indies	Not reported
Lonchura castaneothorax	Malay, Australia	Not reported	2	Pacific islands	Not reported
Lonchura cucullata	Africa	ViCh (r), ViMa (r)	2	Indian ocean islands, West Indies	Puerto Rico: MoBo (u)
Lonchura malabarica	Asia	Not reported	3	Europe, Pacific islands, West Indies	Not reported
Lonchura malacca	Asia	Not reported	6	Pacific islands, West Indies	Not reported

(cont.)

Table 9.1 (cont.)

Host species	Origin	Parasitism in native range	Number of introduced populations	Introduced ranges	Parasitism in introduced range
Lonchura oryzivora	Asia	Not reported	11	Africa, Atlantic islands, Australia, Indian ocean islands, North America, Pacific islands, West Indies	Not reported
Lonchura punctulata	Asia	Not reported	9	Australia, Indian ocean islands, Pacific islands, West Indies	Puerto Rico: MoBo (r)
Fringillidae					
Carduelis carduelis	Eurasia	CuCa (r)	5	Atlantic islands, Australia, Pacific islands, South America, West Indies	Australia: CaFl (u), CaPa (o), ChBa (r), ChLu (r)
Carduelis chloris	Eurasia	CuCa (o)	4	Atlantic islands, Australia, Pacific islands, South America	Australia: CaPa (r), ChBa (r). New Zealand: UrTa (u)
Carduelis flammea	Circumpolar	CuCa (r), MoAt (r)	2	Australia, Pacific islands	Not reported
Fringilla coelebs	Eurasia	CuCa (o)	2	Africa, Pacific islands	New Zealand: ChLu (u)
Serinus mozambicus	Africa	ChKl (u)	3	Indian ocean islands, Pacific islands	Not reported
Passeridae					
Passer domesticus	Eurasia	CuCa (r)	30	Africa, Atlantic islands, Australia, Indian ocean islands, Malay, North America, Pacific islands, South America, West Indies	Africa: ChCa (u). Australia: CaFl (u), CaPa (r), ChBa (r), ChLu (r). Brazil and Chile: MoBo (r/o). New Zealand: ChLu (r), UrTa (o). South Africa: ChKl (p). USA: MoAt (r/o)

Passer montanus	Eurasia	CuCa (r)	4	Australia, North America, Pacific islands	Not reported
Ploceidae					
Euplectes afer	Africa	Not reported	4	Europe, Pacific islands, West Indies	Not reported
Euplectes orix/franciscanus	Africa	ChCa (f), ChCu (u), ChKI (u), ViPa (p)	2	North America, West Indies	Not reported
Foudia madagascariensis	Indian ocean islands	Not reported	2	Arabia, Atlantic islands	Not reported
Ploceus cucullatus	Africa	ChCa (f), ChCu (f), ChKI (f), CuSo (p), ViMa (p)	4	Europe, Indian ocean islands, West Indies	Puerto Rico: MoBo (u)
Ploceus intermedius	Africa	ChCa (f), ChKI (u)	2	Arabia, Pacific islands	Not reported
Pycnonotidae					
Pycnonotus cafer	Asia	CaPs (p), ClJa (u), CuCa (u), HiSp (p), SuDi (u), SuLu (u)	8	Arabia, Pacific islands	Not reported
Pycnonotus jocosus	Asia	CaSo (u)	6	Arabia, Australia, Indian ocean islands, North America, Pacific islands	Australia: CaPa (r)
Sturnidae					
Acridotheres cristatellus	Asia	Not reported	4	North America, South America, Pacific islands	Not reported
Acridotheres fuscus	Asia	Not reported	2	Pacific islands	Not reported
Acridotheres tristis	Asia	EuSc (f)	23	Africa, Arabia, Atlantic islands, Australia, Europe, Indian ocean islands, North America, Pacific islands	South Africa: ClGl (u)
Gracula religiosa	Asia	EuSc (u)	3	North America, Pacific islands, West Indies	Not reported

(cont.)

Table 9.1 (*cont.*)

Host species	Origin	Parasitism in native range	Number of introduced populations	Introduced ranges	Parasitism in introduced range
Sturnus vulgaris	Eurasia	CuCa (r)	7	Africa, Australia, North America, Pacific islands, West Indies	USA: MoAt (r)
Timaliidae					
Garrulax canorus	Asia	ClCo (u), HiSp (u)	2	Pacific islands	Not reported
Leiothrix lutea	Asia	CuCa (u)	4	Europe, Pacific islands	Not reported
Turdidae					
Turdus merula	Eurasia	CuCa (o), CuMi (u)	2	Australia, Pacific islands	Australia: CaPa (r). New Zealand: ChLu (r)
Turdus philomelos	Eurasia	CuCa (r)	2	Australia, Pacific islands	New Zealand: UrTa (u)

Only cases with successful establishment in two or more geographical entities were extracted from Sol *et al.* (2012). According to Sol *et al.* (2012), only two obligate brood parasites were successfully introduced outside their native range (*Vidua paradisaea*, *V. macroura*). Pacific islands include New Zealand.

Data on parasite distribution and host use from Friedmann (1963, 1971), Friedmann *et al.* (1977), Friedmann and Kiff (1985), Moksnes and Røskaft (1995), Johnsgard (1997), Payne (2005), Yang *et al.* (2012) and Lowther (2014). Parasitism categorized into 'frequent' (f, commonly used host), 'occasional' (o, irregularly used host), 'rare' (r, only a few cases of parasitism), 'used' (u, no information on regularity of host use, but probably rare or occasional in most cases) and 'possible' (p, uncertain host status).

Brood parasite, abbreviations:

Cacomantis flabelliformis (CaFl); *Cacomantis pallidus* (CaPa); *Cacomantis passerines* (CaPs); *Cacomantis sonneratii* (CaSo); *Chrysococcyx basalis* (ChBa); *Chrysococcyx caprius* (ChCa); *Chrysococcyx cupreus* (ChCu); *Chrysococcyx klaas* (ChKl); *Chrysococcyx lucidus* (ChLu); *Clamator coromandus* (ClCo); *Clamator glandarius* (ClGl); *Clamator jacobinus* (ClJa); *Cuculus canorus* (CuCa); *Cuculus micropterus* (CuMi); *Cuculus solitarius* (CuSo); *Eudynamys scolopaceus* (EuSc); *Hierococcyx hyperythrus* (HiHy); *Hierococcyx sparverioides* (HiSp); *Molothrus aeneus* (MoAe); *Molothrus ater* (MoAt); *Molothrus bonariensis* (MoBo); *Scythrops novaehollandiae* (ScNo); *Surniculus dicruroides* (SuDi); *Surniculus lugubris* (SuLu); *Urodynamis taitensis* (UrTa); *Vidua chalybeata* (ViCh); *Vidua funerea* (ViFu); *Vidua macroura* (ViMa); *Vidua paradisaea* (ViPa); *Vidua wilsoni* (ViWi).

species, such as those that have colonized urban environments at time-scales equal to those of introduced populations (from decades up to two centuries), have shown micro-evolutionary changes (Miranda *et al.*, 2013), genetic divergence (Evans *et al.*, 2010) and even micro-geographical differentiation within urban populations (Björklund *et al.*, 2010). Similarly, some introduced bird populations diverged from their ancestors, sometimes even to the magnitude of subspecific differences (Baker and Moeed, 1987). Within a century and half from being introduced, two *Turdus* thrushes naturalized in New Zealand converged to local species life histories (e.g. smaller clutch sizes and no seasonal clutch size trends) and, consequently, fit large-scale macro-ecological patterns (Samas *et al.*, 2013). This indicates that one-and-a-half centuries was sufficient to allow for an apparent micro-evolutionary change in host phenotypes (actually the clutch size changes stabilized within the first century after introductions and did not change afterwards; Samas *et al.*, 2013). This directly confirms that introduced populations had enough time to respond to relaxed/new selection and thus are useful for testing coevolutionary hypotheses (Thompson, 1998). However, these studies of changes in expression of traits should also be accompanied by molecular studies to disentangle effects of plasticity and genetic changes.

Importantly, even if there was not enough time for an evolutionary (i.e. genetic) change, the same direction of changes between native and introduced populations is predicted from phenotypic plasticity. Lowered realized or perceived parasitism risk should produce relaxed expression of anti-parasite adaptations. Thus, both genetic and developmental change in allopatry should lead to lowered levels of defence. Crucially, increased levels of defences in allopatry would be unambiguous evidence against a presumed function of host behaviour in the context of interspecific parasitism: decreased parasitism pressure cannot select for increased host defences in principle. Alternative explanations (e.g. conspecific parasitism, see below) then need to be addressed (Samas *et al.*, 2014a, b).

Searching the literature, we found only a few publications investigating introduced populations in light of avian brood parasitism (see also Baker *et al.*, 2014). In the following sections, we will discuss the model systems that have been studied so far and possibilities for future research on introduced populations.

Introduced Hosts: Current State of Knowledge

The village weaver, *Ploceus cucullatus* (Figure 9.1), is a species native to sub-Saharan Africa, but was introduced to Hispaniola, West Indies, in the eighteenth century (Cruz and Wiley, 1989). In Africa, it is known to be a host of the diederik cuckoo, *Chrysococcyx caprius*, which lays eggs mimicking at least some of the weaver egg types (Erritzøe *et al.*, 2012). Weaver hosts have well-developed egg rejection skills, and show sophisticated defensive mechanisms like intricate egg signatures, extreme interclutch variation and remarkably low intraclutch variation in egg appearance (Cruz and Wiley, 1989; Lahti and Lahti, 2002). Weaver eggs may have a white, light green or dark blue-green ground colour, either with or without spots. A high interclutch variation between different females makes it more difficult for the cuckoo to mimic any clutch, and a low intraclutch variation may make it easier for the host to recognize even a mimetic egg (Feeney *et al.*, 2014). Intricate egg signatures enhance the effect of both these traits, making successful parasitism even more difficult for the parasite. Village weavers base their rejection decisions on differences in colour and spots between own and foreign eggs (Lahti and Lahti, 2002). These mechanisms work well against cuckoo parasitism, but also against parasitism by conspecifics, which is an additional selective pressure for the evolution of egg rejection, especially in colonially nesting birds like weavers. However, Lahti (2006) did not find any evidence for conspecific parasitism in both native and introduced study populations (see also Cruz *et al.*, 2008).

On arrival in Hispaniola, the weavers faced no brood parasite. If interspecific parasites are responsible for evolution of host defences, defensive mechanisms should deteriorate with time, especially if there are costs related to rejection behaviour. Indeed, rejection of foreign eggs has been found to be lower in Hispaniola than in Africa (Cruz and Wiley, 1989). Furthermore, intraclutch variation has increased and interclutch variation has decreased in the population compared to the native one, just as in another introduced population in a parasite-free environment in Mauritius (Lahti, 2005). An interesting twist to this story is the arrival of brood parasitic shiny cowbirds to Hispaniola in the 1970s, about 200 years after the introduction of village weavers on the island. Unlike diederik cuckoos, this parasite is a generalist using many hosts and laying eggs that resemble weaver eggs in size but not in colour. Cowbird parasitism on village weavers increased from about 1% in the 1970s to about 16% in 1982, and costs of parasitism to weavers were high (Cruz and Wiley, 1989). Due to this new selective pressure, Hispaniolan weavers were therefore likely to evolve better defences in the future. A study in 1998 confirmed this expectation, as high weaver rejection rates of experimental eggs were reported. The authors suggested that the rapid change in egg rejection rates may be due to both genetic processes and learning (Robert and Sorci, 1999). Lahti (2006), however, came to a different conclusion. He suggested that egg rejection abilities had not deteriorated, but were hampered due to changes in egg phenotypes. Yet, Cruz *et al.* (2008) found that rejection of experimental eggs were highest in areas where shiny cowbirds were present, suggesting phenotypic plasticity in the egg rejection behaviour due to costs of rejection.

Interestingly, a recent comparison of red-billed leiothrix, *Leiothrix lutea*, defences in a native and introduced population showed similar patterns as in the weaver case above (Yang *et al.*, 2014). Leiothrixes are parasitized by common cuckoos in Asia, but an introduced population in Hawaii has been living in a parasite-free environment for 100 years. Introduced leiothrixes have retained their egg rejection ability. Intraclutch variation was lower and interclutch variation was higher in the source than in the introduced population.

Chaffinches, *Fringilla coelebs*, introduced to New Zealand from Europe have kept their advanced egg rejection abilities even after separation from their European counterparts for more than 100 years (Hale and Briskie, 2007). Song thrush and blackbirds even rejected eggs at higher rates than in Europe (Hale and Briskie, 2007); however, model types used in native and introduced populations differed and therefore are not quantitatively comparable. Samas *et al.* (2011, 2012, 2013, 2014a) studied both above-mentioned thrushes and greenfinches, *Carduelis chloris,* in New Zealand. Using identical model types across both species of thrushes and all populations, they found that thrush responses to non-mimetic eggs did not decline in New Zealand under the presumed relaxed selection from cuckoos. Also individual egg rejection repeatability remained virtually identical between native (Grim *et al.,* 2014) and introduced populations (Samas *et al.,* 2011). In contrast, responses to natural conspecific eggs even increased in New Zealand compared to native populations, leading to one of the highest conspecific egg rejection rates revealed in passerines (~60%). Thrushes are only accidentally prone to interspecific parasitism at present, both in source and New Zealand populations. Maintenance of defences in both native and introduced populations cannot be explained by selective neutrality: both thrushes suffered substantial rejection costs and errors in all populations (Hale and Briskie, 2007; Samas *et al.,* 2014a). Thus, even though we cannot rule out other possibilities (just like in any study that infers historical causes of current host anti-parasite adaptations; Samas *et al.,* 2014b), the most likely explanation is that thrush egg discrimination evolved and was maintained due to conspecific parasitism; this is in line with current unsuitability of thrushes as cuckoo hosts (Grim *et al.,* 2011).

Phenotypic plasticity predicts immediate changes in host responses to changed parasitism pressure: some cuckoo hosts adjust their responses across years, and even during a single breeding season (Thorogood and Davies, 2013). Therefore, virtually identical responses to foreign eggs between Europe and New Zealand after dozens of generations clearly reject the phenotypic plasticity hypothesis in the context of cuckoo parasitism. Still, it remains to be shown whether the elevated rejection of conspecific eggs in New Zealand reflects micro-evolutionary change or phenotypic plasticity, both due to elevated breeding densities in New Zealand which increase risks of conspecific parasitism (Samas *et al.,* 2013).

In Africa, 20 species of the Viduidae family (whydahs and indigobirds) are obligate brood parasites (Payne, 2010). The pin-tailed whydah, *Vidua macroura* (Figure 9.1), for example, exploits several species of waxbills *Estrilda* spp., where the parasitic chick is reared together with host young. The *Vidua* species and their hosts show similar egg appearance (due to shared phylogeny; Payne, 2010) and similar gape patterns (due to convergence; Payne, 2010). Because of their colourful breeding plumages and voice, the whydahs and indigobirds have been popular in aviculture as cagebirds for centuries, similarly to many of their host species. One host of the pin-tailed whydah, the orange-cheeked waxbill, *Estrilda melpoda,* has been introduced to Puerto Rico several times, most recently in the 1950s. Here it established a free-ranging population. Similarly, pin-tailed whydahs were also introduced here about 10 years later and started to use the waxbills as hosts. As far as we know, both species are still present there representing an exciting yet unused research opportunity.

Introduced Hosts: Limitations

Although introduced host and parasite populations provide important heuristic advantages compared to native, expanding (including urbanized) or retreating populations (see above), they also show some limitations. Out of the total diversity of hosts and parasites, only a small subset has been introduced (Table 9.1). We note, however, that even among native hosts and parasites only a minor proportion of them is studied anyway. At a finer scale, only a subset of individuals (and hence genetic variation and adaptations) have been introduced. Thus, genetic constitution of introduced populations may not be representative of native ones leading to founder effects. However, many introduced host populations resulted from massive propagule pressure that prevented non-adaptive confounding effects (Briskie and Mackintosh, 2004). Variation in the propagule pressure provides opportunities to test how founder effects or genetic drift affect altered coevolutionary relationships.

There is also a possibility that only generalist parasites will choose exotic hosts, which would reduce the opportunity for coevolution. However, this will not necessarily be so because parasites can change their host specificity in invaded ranges (Douda et al., 2012). Rather than as a limitation, we see this as an opportunity to test what particular general life history and specific parasitism-related traits affect host selection and coevolution (see Grim et al., 2011).

Virtually all introductions that are of interest to students of brood parasitism were done within several last centuries. This may raise concerns whether this time frame is sufficient to detect noticeable evolutionary responses and whether studies of such populations would not be limited in their focus on early stages of evolution of the coevolutionary system. Ample evidence suggests that animal behaviour may evolve remarkably fast (Thompson, 1998; Miranda et al., 2013) while other case studies suggest limited evolutionary change even in the long term (Lahti et al., 2009; Peer et al., 2011). Whether anti-parasite defence declines after pressure from interspecific parasitism relaxes depends on multiple factors, including standing genetic variation (Vikan et al., 2010), gene flow (Soler et al., 1999), trait covariation (Avilés and Parejo, 2011), secondary trait function (Lahti et al., 2009), alternative selection pressures (Lyon and Eadie, 2004), or costs and errors associated with the behaviour (Samas et al., 2014a). These and additional parameters are typically idiosyncratic, preventing unambiguous general conclusions on how the length of human-assisted allopatry of host and parasite populations limits detection of coevolutionary changes. Anyway, the few pioneering studies of such populations (above) persuade us to be optimistic.

A problem common to both expansions and introductions is that the newly colonized host ranges differ from ancestral ranges in multiple factors, not only presence/absence of parasites. For example, adaptations involved in coevolutionary interactions may also be under selection from abiotic factors. Village weavers freed from diederik cuckoo parasitism on Hispaniola and Mauritius have, in general, more blue-green eggshell pigments (biliverdin) resulting in lower interclutch variation, which may be an adaptation to protect embryos against solar radiation (Lahti, 2008). Parasitized populations in Africa have moved away from this optimum due to selection on high interclutch variation among females as a defence against egg mimicry by diederik cuckoos. However, even between the source populations, the one experiencing more exposure to sun had more intense blue-green colours than the one experiencing less sun exposure (Lahti, 2008). Hence, care should be taken in concluding that specific traits have evolved solely as a response to coevolution without considering alternative explanations.

The same cautionary note applies to biotic factors: also these vary between native versus exotic ranges. There are new predators, competitors or non-brood parasites. These lead to changes in selective pressures on traits, including those related to parasitism. Without appropriate controls (see next section), resulting changes might be incorrectly interpreted as if they were causally related to changed parasite pressure. Further, new adaptations to these novel selection pressures may, via trade-offs (Ricklefs and Wikelski, 2002) or trait covariances (Trnka and Grim, 2014), also affect focal anti-brood parasite adaptations. However, such confounding effects are measurable and thus can be taken into account in carefully planned studies. In our view, these complexities therefore do not detract from the huge research potential of introduced host and parasite populations.

Conclusions and Future Avenues

As we have seen, studies of introduced populations offer a unique opportunity to study how expression of traits varies according to specific selective pressures, and also how parasites may affect host populations. To avoid comparing only one population to another (i.e. pseudoreplication: Hurlbert, 1984), future studies should ideally focus on hosts introduced to several isolated locations (Table 9.1). We acknowledge that such multiple-population studies are logistically demanding, but we stress that they are doable (Lahti, 2006; Samas et al., 2014a). Ideally, researchers should compare parasitized and unparasitized introduced populations. For example, the orange-cheeked waxbill was introduced to several parasite-free islands (Hawaii and Oahu in the Hawaiian Islands, Bermuda, and Saipan in the Northern Mariana Islands). Its parasite, the pin-tailed whydah was co-introduced with waxbills but only to several places (Puerto Rico, Guadeloupe, Martinique; Payne, 2010). Thus, this unique model system provides multiple replicates of both parasitized and non-parasitized populations (see also red-billed leiothrix; Tojo and Nakamura, 2014; Yang et al., 2014). Whydah and allies are famous for the intricate similarity of their gape patterns to those of the host young, although it is unclear who mimics whom (Hauber and Kilner, 2007). Comparison of introduced parasitized and non-parasitized waxbill populations may help resolve this enigma.

Studies on species not parasitized in either native, shifted or introduced ranges are also important (Table 9.1). This is because potential interpopulation differences in adaptations would reflect variation in selective pressures unrelated to parasitism (Lahti, 2008). Such species can therefore serve as controls (if the same direction and magnitude of genetic and phenotypic changes in novel ranges is recorded in both former host and control species, then such changes in hosts cannot be interpreted as a response to relaxed selection from parasite). The controls should be those species that are primarily unsuitable (sensu Grim et al., 2011) as hosts of interspecific parasites and, thus, the microevolution of their traits would not be confounded by parasite–host coevolution. Such control species should include both those that expanded their ranges naturally (e.g. fieldfares, Turdus pilaris, that recently colonized Iceland and Greenland from Europe; Figure 9.1), and those that were made allopatric from potential native parasites by human transport (e.g. goldfinches, Carduelis carduelis, introduced from Europe to New Zealand; Figure 9.1). Just like in the case of focal host species, the best control species allopatric populations are those that became established via long-distance dispersal or introduction to distant and well-isolated places (i.e. no additional gene flow from native populations). In the ideal case, conditions in introduced ranges should differ from those in native ranges as little as possible in traits other than the presence of brood parasite(s), to avoid confounding effects of multiple changed selection pressures.

Most studies of brood parasitism have concerned a specific host and/or parasite population at a specific time and place. Work based on more representative sampling across time (longitudinal studies) or space (multiple populations) is rare, simply due to the workload required to undertake such investigations. However, because a traditional single host population approach cannot answer some relevant questions in principle, we argue for a change in the focus from such studies (even though they might cover more host species) to larger-scale studies (even though they are demanding and fewer species can be addressed).

Such a metareplication approach (Kelly, 2006) is not only relevant for the study of introduced populations, it is fundamental for the study of native populations too (Soler et al., 1999; Vikan et al., 2010; Grim et al., 2011). For example, if sympatry/allopatry affects host responses and mostly parasitized populations are studied, then this biased sampling may affect interspecific trends. Further, intraspecific variation can be extreme (rejection rates varying from 5 to 69% in a single species: Stokke et al., 2008), highlighting that no single population can be representative of a 'species-typical' behaviour (Foster and Endler, 1999). Instead, we suggest that species may show species-specific reaction norms (e.g. species-specific patterns of covariation between parasitism and egg rejection rates), which can be revealed only by sampling across multiple populations subject to varying parasitism risks. This highlights the crucial importance of metareplication for any studies in ecology (Kelly, 2006).

In science, results are always determined by methods. For example, even seemingly subtle differences in model eggs design can have profound consequences for host responses. Any comparison in science needs to be based on experimental design that is consistent across all units (individuals, populations, species) that are being compared. This methodological aspect is crucial in studies of introduced or expanding populations because data are often collected by different researchers in different locations. Therefore, we strongly recommend that methods are not similar but *identical* across all spatial replicates.

Any defence in the absence of a threat is a wasteful investment. All previous work (but see Samas et al., 2014a) made a simplistic assumption that only rejection costs and errors at non-parasitized nests select against retention of anti-parasitic adaptations after parasites no longer use a particular host. For an adaptation to decay, rejection costs and errors are not necessary because multiple other mechanisms erase the adaptation (mutation pressure, genetic drift, costs of maintenance of neural networks, trade-offs with currently useful adaptations). This also highlights a necessity to consider alternative hypotheses, namely conspecific brood parasitism as a viable alternative to the interspecific parasitism hypothesis (Lyon and Eadie, 2004). We note that in contrast to interspecific parasitism, in the case of conspecific parasitism rejection costs and errors at all nests (i.e. not only non-parasitized ones) select against host defences, because conspecific parasites do not evict host progeny and are raised jointly (Samas *et al.*, 2014a).

Generally in biology, the most rapid micro-evolutionary changes, i.e. those happening at the scale of decades or centuries (Thompson, 1998), are typically found in populations introduced to novel environments (Blackburn *et al.*, 2009) and in urbanized populations (Gil and Brumm, 2014). Therefore students of brood parasite–host interactions should capitalize on many such systems that are currently available, yet remain unexamined. The major advantage of such an approach, especially in the case of introduced populations, is that it is best suited to answer some big unanswered questions in the study of brood parasitism. For example (Rothstein, 2001), does host–parasite coevolution follow a coevolutionary cycles scenario (host adaptations decline after parasitism pressure ceases, allowing parasites to re-invade) or a single-trajectory model (host adaptations are retained after parasitism pressure ceases, preventing parasites from re-invading)?

This central question is hardly possible to answer through the study of native populations because of confounding effects of gene flow and always unknown long-term length (time) and extent (ancient parasitism rates and costs) of parasite–host contact. Introduced populations are free from these two fundamental problems. Indeed, a few pioneering studies have already employed the framework that we detailed in this chapter, addressed the central question and changed our long-held views of parasite–host coevolution. We believe that future studies following the above listed conceptual and methodological framework, especially (a) metareplication across phylogeny, space and time, (b) standardized methods, (c) inclusion of control species, (d) attention to alternative hypotheses, and (e) realistic consideration and quantification of all costs and benefits, will bring novel, robust, and exciting evidence that will fundamentally deepen our understanding of enemy–victim coevolution.

Acknowledgements

We thank J. Briskie, D. Hanley, D. Lahti, W. Liang, B. Matysiokova, P. Samas, V. Remes, A. Trnka and especially D. Sol and an anonymous referee for perceptive comments on previous versions.

References

Avilés, J.M. and Møller, A.P. (2003). Meadow pipit (*Anthus pratensis*) egg appearance in cuckoo (*Cuculus canorus*) sympatric and allopatric populations. *Biological Journal of the Linnean Society*, 79, 543–549.

Avilés, J.M. and Parejo, D. (2011). Host personalities and the evolution of behavioural adaptations in brood parasitic–host systems. *Animal Behaviour*, 82, 613–618.

Avilés, J.M., Bootello, E.M., Molina-Morales, M. and Martínez, J.G. (2014). The multidimensionality of behavioural defences against brood parasites: evidence for a behavioural syndrome in magpies? *Behavioral Ecology and Sociobiology*, 68, 1287–1298.

Baker, A.J. and Moeed, A. (1987). Rapid genetic differentiation and founder effect in colonizing populations of common mynas (*Acridotheres tristis*). *Evolution*, 41, 525–538.

Baker, J., Harvey, K.J. and French, K. (2014). Threats from introduced birds to native birds. *Emu*, 114, 1–12.

Barnagaud, J.-Y., Papaix, J., Gimenez, O. and Svenning, J.-C. (2015). Dynamic spatial interactions between the native invader brown-headed cowbird and its hosts. *Diversity and Distributions*, 21, 511–522.

Björklund, M., Ruiz I. and Senar, J.C. (2010). Genetic differentiation in the urban habitat: the great tits (*Parus major*) of the parks of Barcelona city. *Biological Journal of the Linnean Society*, 99, 9–19.

Blackburn, T., Lockwood, J. and Cassey, P. (2009). *Avian Invasions*. Oxford: Oxford University Press.

Briskie, J.V. and Mackintosh, M. (2004). Hatching failure increases with severity of population bottlenecks in birds. *Proceedings of the National Academy of Sciences, USA*, 101, 558–561.

Briskie, J.V., Sealy, S.G. and Hobson, K.A. (1992). Behavioral defenses against avian brood parasitism in sympatric and allopatric host populations. *Evolution*, 46, 334–340.

Cruz, A. and Wiley, J.W. (1989). The decline of an adaptation in the absence of a presumed selection pressure. *Evolution*, 43, 55–62.

Cruz, A., Post, W., Wiley, J.W., *et al.* (1998). Potential impacts of cowbird range expansion in Florida. In *Parasitic Birds and their Hosts*, ed. Rothstein, S.I. and Robinson, S.K. New York: Oxford University Press, pp. 313–336.

Cruz, A., Prather, J.W., Wiley, J.W. and Weaver, P.F. (2008). Egg rejection behavior in a population exposed to parasitism: village weavers on Hispaniola. *Behavioral Ecology*, 19, 398–403.

Davies, N.B. and Brooke, M.L. (1989). An experimental study of co-evolution between the cuckoo, *Cuculus canorus*, and its hosts. I. Host egg discrimination. *Journal of Animal Ecology*, 58, 207–224.

Diamond, J.M. (1986). Overview: laboratory experiments, field experiments, and natural experiments. In *Community Ecology*, ed. Diamond, J.M. and Case, T.J. New York: Harper and Row, pp. 3–22.

Dinets, V., Samas, P., Croston, R., Grim, T. and Hauber, M.E. (2015). Predicting the responses of native birds to transoceanic invasions by avian brood parasites. *Journal of Field Ornithology*, 86, 244–251.

Douda, K., Vrtílek, M., Slavík, O. and Reichard, M. (2012). The role of host specificity in explaining the invasion success of the freshwater mussel *Anodonta woodiana* in Europe. *Biological Invasions*, 14, 127–137.

Erritzøe, J., Mann, C.F., Brammer, F.P. and Fuller, R.A. (2012). *Cuckoos of the World*. Helm Identification Guides. London: Christopher Helm.

Evans, K.L., Hatchwell, B.J., Parnell, M. and Gaston, K.J. (2010). A conceptual framework for the colonisation of urban areas: the blackbird *Turdus merula* as a case study. *Biological Reviews*, 85, 643–667.

Feeney, W.E., Welbergen, J.A. and Langmore, N.E. (2014). Advances in the study of coevolution between avian brood parasites and their hosts. *Annual Review of Ecology, Evolution, and Systematics*, 45, 227–246.

Foster, S.A. and Endler, J.A. (1999). *Geographic Variation in Behavior*. New York: Oxford University Press.

Friedmann, H. (1963). *Host Relations of the Parasitic Cowbirds*. United States National Museum Bulletin No. 233. Washington, DC: Smithsonian Institution.

Friedmann, H. (1971). Further information on the host relations of the parasitic cowbirds. *Auk*, 88, 239–255.

Friedmann, H. and Kiff, L.F. (1985). The parasitic cowbirds and their hosts. *Proceedings of the Western Foundation of Vertebrate Zoology*, 2, 225–302.

Friedmann, H., Kiff, L.F. and Rothstein, S.I. (1977). *A Further Contribution to Knowledge of the Host Relations of the Parasitic Cowbirds*. Smithsonian Contributions to Zoology No. 235. City of Washington: Smithsonian Institution Press.

Gil, D. and Brumm, H. (2014). *Avian Urban Ecology*, Oxford, UK: Oxford University Press.

Grim, T., Samas, P., Moskát, C., *et al.* (2011). Constraints on host choice: why do parasitic birds rarely exploit some common potential hosts? *Journal of Animal Ecology*, 80, 508–518.

Grim, T., Samas, P. and Hauber, M.E. (2014). The repeatability of avian egg ejection behaviors across different temporal scales, breeding stages, female ages and experiences. *Behavioral Ecology and Sociobiology*, 68, 749–759.

Hale, K. and Briskie, J.V. (2007). Response of introduced European birds in New Zealand to experimental brood parasitism. *Journal of Avian Biology*, 38, 198–204.

Hauber, M.E. and Kilner, R.M. (2007). Coevolution, communication, and host-chick mimicry in parasitic finches: who mimics whom? *Behavioral Ecology and Sociobiology*, 61, 497–503.

Honza, M., Procházka, P., Stokke, B.G., *et al.* (2004). Are blackcaps current winners in the evolutionary struggle against the common cuckoo? *Journal of Ethology*, 22, 175–180.

Hurlbert, S.H. (1984). Pseudoreplication and the design of ecological field experiments. *Ecological Monographs*, 54, 187–211.

Igic, B., Cassey, P., Grim, T., *et al.* (2012). A shared chemical basis of avian host–parasite egg colour mimicry. *Proceedings of the Royal Society B: Biological Sciences*, 279, 1068–1076.

Johnsgard, P.A. (1997). *The Avian Brood Parasites*. New York: Oxford University Press.

Joseph, L., Wilke, T. and Alpers, D. (2002). Reconciling genetic expectations from host specificity with historical population dynamics in an avian brood parasite, Horsfield's bronze-cuckoo *Chalcites basalis* of Australia. *Molecular Ecology*, 11, 829–837.

Kelly, C.D. (2006). Replicating empirical research in behavioral ecology: how and why it should be done but rarely ever is. *Quarterly Review in Biology*, 81, 221–236.

Kuehn, M.J., Peer, B.D. and Rothstein, S.I. (2014). Variation in host response to brood parasitism reflects evolutionary differences and not phenotypic plasticity. *Animal Behaviour*, 88, 21–28.

Lahti, D.C. (2005). Evolution of bird eggs in the absence of cuckoo parasitism. *Proceedings of the National Academy of Sciences, USA*, 102, 18057–18062.

Lahti, D.C. (2006). Persistence of egg recognition in the absence of cuckoo brood parasitism: pattern and mechanism. *Evolution*, 60, 157–168.

Lahti, D.C. (2008). Population differentiation and rapid evolution of egg color in accordance with solar radiation. *Auk*, 125, 796–802.

Lahti, D.C. and Lahti, A.R. (2002). How precise is egg discrimination in weaverbirds? *Animal Behaviour*, 63, 1135–1142.

Lahti, D.C., Johnson, N.A., Ajie, B.C., *et al.* (2009). Relaxed selection in the wild. *Trends in Ecology and Evolution*, 24, 487–496.

Liang, W., Yang, C., Wang, L. and Møller, A.P. (2013). Avoiding parasitism by breeding indoors: cuckoo parasitism of hirundines and rejection of eggs. *Behavioral Ecology and Sociobiology*, 67, 913–918.

Lowther, P. (2014). Brood parasitism: host lists. Available at: http://www.fieldmuseum.org/science/blog/brood-parasitism-host-lists, accessed 9 November 2014.

Lyon, B.E. and Eadie, J.M.A. (2004). An obligate brood parasite trapped in the intraspecific arms race of its hosts. *Nature*, 432, 390–393.

Marín, M. (2000). The shiny cowbird (*Molothrus bonariensis*) in Chile: introduction or dispersion? *Ornitologia Neotropical*, 11, 285–296.

Martín-Vivaldi, M., Soler, J.J., Møller, A. P., Pérez-Contreras, T. and Soler, M. (2013). The importance of nest-site and habitat in egg recognition ability of potential hosts of the common cuckoo *Cuculus canorus*. *Ibis*, 155, 140–155.

Miranda, A.C., Schielzeth, H., Sonntag, T. and Partecke, J. (2013). Urbanization and its effects on personality traits: a result of microevolution or phenotypic plasticity? *Global Change Biology*, 19, 2634–2644.

Moksnes, A. and Røskaft, E. (1995). Egg-morphs and host preference in the common cuckoo (*Cuculus canorus*): an analysis of cuckoo and host eggs from European museum collections. *Journal of Zoology*, 236, 625–648.

Moksnes, A., Røskaft, E., Braa, A.T., *et al.* (1991). Behavioural responses of potential hosts towards artificial cuckoo eggs and dummies. *Behaviour*, 116, 64–89.

Møller, A.P., Díaz, M., Flensted-Jensen, E., *et al.* (2015). Urbanized birds have superior establishment success in novel environments. *Oecologia*, 178, 943–950.

Morand, S. and Krasnov, B.R. (2010). *The Biogeography of Host–Parasite Interaction*. Oxford, UK: Oxford University Press.

Moskát, C., Szentpéteri, J. and Barta, Z. (2002). Adaptations by great reed warblers to brood parasitism: a comparison of populations in sympatry and allopatry with the common cuckoo. *Behaviour*, 139, 1313–1329.

Moskát, C., Hansson, B., Barabás, L., Bártol, I. and Karcza, Z. (2008). Common cuckoo *Cuculus canorus* parasitism, antiparasite defence and gene flow in closely located populations of great reed warblers *Acrocephalus arundinaceus*. *Journal of Avian Biology*, 39, 663–671.

Nakamura, H., Kubota, S. and Suzuki, R. (1998). Coevolution between the common cuckoo and its major hosts in Japan. In *Parasitic Birds and Their Hosts*, ed. Rothstein, S.I. and Robinson, S.K. New York: Oxford University Press, pp. 94–112.

Payne, R.B. (2005). *The Cuckoos*. Oxford, UK: Oxford University Press.

Payne, R. B. (2010). Family Viduidae (whydahs and indigobirds). In *Handbook of the Birds of the World, Vol. 15. Weavers to New World Warblers*, ed. del Hoyo, J., Elliott, A. and Christie, D.A. Barcelona, Spain: Lynx Edicions, pp. 198–232.

Peer, B.D. and Sealy, S.G. (2004). Correlates of egg rejection in hosts of the brown-headed cowbird. *Condor*, 106, 580–599.

Peer, B.D., Kuehn, M.J., Rothstein, S.I. and Fleischer, R.C. (2011). Persistence of host defence behaviour in the absence of avian brood parasitism. *Biology Letters*, 7, 670–673.

Post, W. and Sykes, P.W. (2011). Reproductive status of the shiny cowbird in North America. *Wilson Journal of Ornithology*, 123, 151–154.

Ricklefs, R.E. and Wikelski, M. (2002). The physiology/life-history nexus. *Trends in Ecology and Evolution*, 17, 462–168.

Robert, M. and Sorci, G. (1999). Rapid increase of host defence against brood parasites in a recently parasitized area: the case of village weavers in Hispaniola. *Proceedings of the Royal Society B: Biological Sciences*, 266, 941–946.

Rothstein, S.I. (1990). A model system for coevolution: avian brood parasitism. *Annual Review of Ecology and Systematics*, 21, 481–508.

Rothstein, S.I. (2001). Relic behaviours, coevolution and the retention versus loss of host defences after episodes of avian brood parasitism. *Animal Behaviour*, 61, 95–107.

Rothstein, S.I. and Peer, B.D. (2005). Conservation solutions for threatened and endangered cowbird (*Molothrus* spp.) hosts: separating fact from fiction. *Ornithological Monographs*, 57, 98–114.

Samas, P., Hauber, M.E., Cassey, P. and Grim, T. (2011). Repeatability of foreign egg rejection: testing the assumptions of co-evolutionary theory. *Ethology*, 117, 606–619.

Samas, P., Polacikova, L., Hauber, M.E., Cassey P. and Grim T. (2012). Egg rejection behavior and clutch characteristics of the European greenfinch introduced to New Zealand. *Chinese Birds*, 3, 330–338.

Samas, P., Grim, T., Hauber, M.E., *et al.* (2013). Ecological predictors of reduced avian reproductive investment in the southern hemisphere. *Ecography*, 36, 809–818.

Samas, P., Hauber, M.E., Cassey, P. and Grim, T. (2014a). Host responses to interspecific brood parasitism: a by-product of adaptations to conspecific parasitism? *Frontiers in Zoology*, 11, 34.

Samas P., Hauber M.E., Cassey P. and Grim T. (2014b). The evolutionary causes of egg rejection in European thrushes (*Turdus* spp.): a reply to M. Soler. *Frontiers in Zoology*, 11, 72.

Smith, J.N.M., Cook, T.L., Rothstein, S.I., *et al.* (2000). *Ecology and Management of Cowbirds and Their Hosts*. Austin, TX: University of Texas Press.

Sol, D., Maspons, J., Vall-Ilosera, M., *et al.* (2012). Unraveling the life history of successful invaders. *Science*, 337, 580–583.

Soler, J.J., Martínez, J.G., Soler, M. and Møller, A.P. (1999). Genetic and geographic variation in rejection behavior of cuckoo eggs by European magpie populations: an experimental test of rejecter-gene flow. *Evolution*, 53, 947–956.

Spottiswoode, C.N. and Stevens, M. (2012). Host–parasite arms races and rapid changes in bird egg appearance. *American Naturalist*, 179, 633–648.

Stokke, B.G., Hafstad, I., Rudolfsen, G., *et al.* (2008). Predictors of resistance to brood parasitism within and among reed warbler populations. *Behavioral Ecology*, 19, 612–620.

Thompson, J.N. (1998). Rapid evolution as an ecological process. *Trends in Ecology and Evolution*, 13, 329–332.

Thompson, J.N. (2005). *The Geographic Mosaic of Coevolution*, Chicago, IL: University of Chicago Press.

Thorogood, R. and Davies, N.B. (2013). Reed warbler hosts fine-tune their defences to track three decades of cuckoo decline. *Evolution*, 67, 3545–3555.

Tojo, H. and Nakamura, S. (2014). The first record of brood parasitism on the introduced red-billed leiothrix in Japan. *Ornithological Science*, 13, 47–52.

Trnka, A. and Grim, T. (2014). Testing for correlations between behaviours in a cuckoo host: why do host defences not covary? *Animal Behaviour*, 92, 185–193.

Vikan, J.R., Stokke, B.G., Rutila, R., *et al.* (2010). Evolution of defences against cuckoo (*Cuculus canorus*) parasitism in bramblings (*Fringilla montifringilla*): a comparison of four populations in Fennoscandia. *Evolutionary Ecology*, 24, 1141–1157.

Woodworth, B.L. (1997). Brood parasitism, nest predation, and season-long reproductive success of a tropical island endemic. *Condor*, 99, 605–621.

Yang, C., Liang, W., Antonov, A., *et al.* (2012). Diversity of parasitic cuckoos and their hosts in China. *Chinese Birds*, 3, 9–32.

Yang, C., Liu, Y., Zeng, L. and Liang, W. (2014). Egg color variation, but not egg rejection behavior, changes in a cuckoo host breeding in the absence of brood parasitism. *Ecology and Evolution*, 4, 2239–2246.

10 Flight Behaviour of an Introduced Parasite Affects its Galápagos Island Hosts: *Philornis downsi* and Darwin's Finches

Sonia Kleindorfer, Katharina J. Peters, Leon Hohl and Frank J. Sulloway

Introduction

Disease and parasite outbreaks have become more frequent and more rapid as the consequence of increasing global interconnectedness, trade and travel (Kilpatrick, 2011). Ecological theory predicts that a parasite epidemic will end when the hosts have died or evolved defences against the pathogen (Duffy and Sivars-Becker, 2007). Given that island populations are especially prone to extinction from introduced disease, islands warrant particular conservation scrutiny (Wikelski *et al.*, 2004). There is growing information about the molecular ecology of virulence, which is useful in predicting the likelihood of acquired defences such as immunity (Alizon and Van Baalen, 2008; McCallum, 2008; Kovaliski *et al.*, 2014). Populations not only evolve, but they are also composed of individuals with diverse behaviours. Because behaviour can shape evolutionary trajectories, understanding the behaviour of novel pathogens and naïve hosts can inform insights into evolutionary scenarios that may usefully be applied to conservation and epidemiological modelling (Nelson, 2014).

Linking insights from animal behaviour with conservation policy seems logical, but has been slow to take off in practice (Caro, 1999; Nelson, 2014). It is easy to imagine how an understanding of a species' dispersal, diet and mating system would usefully inform conservation practice. In addition to these life history behaviours, there are many untapped approaches to integrating behaviour with conservation science that have not been sufficiently promoted by ethologists. This lack of integration has motivated some researchers to champion the 'conservation behaviour framework' (Berger-Tal *et al.*, 2011; Caro and Sherman, 2011), which identifies three major linkages between animal behaviour and conservation practice: (i) anthropogenic impacts on behaviour that impact biodiversity; (ii) behaviour-based species management; and (iii) behavioural indicators of other processes of conservation concern (Berger-Tal *et al.*, 2011; Daly and Johnson, 2011; Brearley *et al.*, 2013; Caro and Riggio, 2014; Kleindorfer *et al.*, 2014a; Palestis, 2014). The proposed linkages between animal behaviour and conservation practice can

Biological Invasions and Animal Behaviour, eds J.S. Weis and D. Sol. Published by Cambridge University Press. © Cambridge University Press 2016.

be used to identify relevant thematically structured frameworks to generate research approaches that test ideas about the role of behaviour for species persistence. As Nelson (2014) summarizes, the challenge is for behavioural biologists to 'demonstrate how behavioural knowledge can make a difference to conservation management problems in light of the far more dominating effects of anthropogenic threats' (Caro and Riggio, 2014).

The Galápagos Islands offer a timely case study to understand the behaviour of a novel host–parasite system. Endemic land birds including naïve Darwin's finch hosts (Passeriformes: Thraupidae) are experiencing extensive malformation and/or mortality from the introduced fly *Philornis downsi* (Diptera: Muscidae) (Causton *et al.*, 2013). Larvae of *P. downsi* were first discovered in Darwin's finch nests on Santa Cruz Island in 1997 (Fessl *et al.*, 2001), although the fly was present on the island by 1964 (Causton *et al.*, 2006). The abundance of *P. downsi* in nests differs across Galápagos Islands (Wiedenfeld *et al.*, 2007), and gene flow is largely unrestricted across three of the major islands (Dudaniec *et al.*, 2008). The adult fly is a vegetarian, but its larvae feed on the blood and tissue of developing chicks (Fessl *et al.*, 2006a; Huber, 2008; Koop *et al.*, 2011; Kleindorfer and Dudaniec, 2016). The intensity of *P. downsi* ('intensity' refers to the number of parasites per nest) has generally been lower during years with lower rainfall (Dudaniec *et al.*, 2007; but see Koop *et al.*, 2013). Since 2008, under conditions of higher rainfall, studies on Darwin's finch nesting success have reported 50–100% annual chick mortality due to *P. downsi* (O'Connor *et al.*, 2010b, d; Koop *et al.*, 2011; Kleindorfer *et al.*, 2014a). The few chicks that survive may persist as adults with malformed beaks (Galligan and Kleindorfer, 2009). *Philornis downsi* parasites are considered the biggest threat to all Galápagos land birds (Causton *et al.*, 2006), and understanding the behaviour of the fly and Galápagos land birds is a top management and research priority (Causton *et al.*, 2013).

Since the initial discovery of *P. downsi* in Darwin's finch nests in 1997, *P. downsi* intensity in finch nests has doubled on both Santa Cruz and Floreana Islands. Between 2000 and 2002 on Santa Cruz Island, small tree finch (*Camarhynchus parvulus*) nests with chicks had ~15 larvae per nest (Kleindorfer, unpublished data), but between 2010 and 2012 they had ~30 larvae per nest (Cimadom *et al.*, 2014). On Floreana Island between 2004 and 2013, *P. downsi* intensity nearly doubled (~28 to ~48 parasites per nest), in-nest mortality nearly doubled (~50% to ~90%), and chicks died in half the time (~11 versus only ~5 days after hatching) (Kleindorfer *et al.*, 2014a). On both islands, finch populations have been declining (O'Connor *et al.*, 2010c, d; Dvorak *et al.*, 2012). Kleindorfer *et al.* (2014a) applied the conservation behaviour framework to understand how change in *P. downsi* oviposition behaviour was driving Darwin's finch mortality patterns on Floreana Island. Kleindorfer *et al.* (2014a) showed that, across the decade, female *P. downsi* laid eggs earlier in the Darwin's finch nesting cycle and that the size and age of larvae in host nests was more synchronous. The combination of higher *P. downsi* abundance per nest, earlier egg-laying, and a more synchronous age class of parasites has meant that Darwin's finch hosts were exposed to older (and hence larger) parasites, which were consuming their blood and tissue from an earlier age. Within the conservation behaviour framework, the findings from Kleindorfer *et al.* (2014a) provided evidence for behavioural indicators (early fly oviposition behaviour, synchronous larval cohorts) that signal processes of conservation concern (early and elevated host death).

The evolutionary theory of ecological specialization predicts that parasites should become locally adapted (Kawecki and Ebert, 2004). Parasites that are locally adapted will have a fitness advantage in sympatric hosts as opposed to allopatric hosts, because locally adapted parasites cannot be invaded by non-adapted parasites from different populations (Kaltz and Shykoff, 1998). Parasite and host behaviour may contribute to the process of local parasite adaptedness. Given slight initial differences in host populations (including nesting behaviour), the selection pressures exerted by novel parasites (including search behaviour) should lead to different evolutionary trajectories across populations. In the case of *P. downsi* and Darwin's finches, Kleindorfer *et al.* (2014a) found preliminary evidence for local parasite adaptation: there were more live *P. downsi* pupae in the nests of small ground finches (*Geospiza fuliginosa*) than in the nests of small tree finches (*C. parvulus*) or medium tree finches (*C. pauper*) – a difference that needs further study. From the hosts' perspective, some species consistently had more *P. downsi* per nest than others. In particular, the critically endangered medium tree finch consistently had more *P. downsi* larvae per nest compared with small tree finches, small ground finches, or hybrid tree finches on Floreana Island (Kleindorfer *et al.*, 2014b). High *P. downsi* intensity is considered the primary cause of decline in the medium tree finch (O'Connor *et al.*, 2010d) and is considered a possible causal factor in the recent local extinction of the large tree finch (*C. psittacula*) and warbler finch (*Certhidea fusca*) on Floreana Island (Grant *et al.*, 2005; Kleindorfer *et al.*, 2014b). Thus there are urgent conservation imperatives to explain why *P. downsi* intensity, Darwin's finch host mortality and *P. downsi* pupation success differ among host species that inhabit the same patch of forest and breed at the same time (Kleindorfer *et al.*, 2014a, b). In particular, we need to understand what attributes make certain hosts more attractive to *P. downsi*. Viewed from the perspective of the parasite, do some aspects of *P. downsi* behaviour make this parasite more likely to encounter particular hosts?

This study examines vertical flight behaviour in *P. downsi* and nesting height in Darwin's finch hosts on Floreana Island. We use an observational approach to compare *P. downsi* intensity in relation to nesting height in three Darwin's finch host species, and we measure *P. downsi* abundance and sex ratio at different forest heights. If we find significant differences in fly abundance according to forest height, we predict that Darwin's finch nests at the height at which *P. downsi* are most commonly encountered will have the highest *P. downsi* intensity. In this case, data would support the view that parasite intensity is shaped by the probability of encountering a host nest. The conservation management implication of this work is to place fly traps at the height preferred by *P. downsi* flies (to remove the greatest number of parasites from the study site), and to place artificial nest boxes below mean fly height (to manage host reproductive success in threatened populations). If neither fly abundance nor host nesting height show a consistent pattern of association, then the data would support the view that host attributes (and not nest attributes) predict parasite intensity. The conservation management suggestion would be to place fly traps at different forest heights (by varying trap height, one would remove more flies from the study site) and to apply pyrethrum to nests of threatened populations to manage host reproductive success until the fly population is reduced (Knutie *et al.*, 2014). This chapter is a case study demonstrating how insights from host and parasite behaviour can be harnessed to inform conservation management.

Methods

Study Site

This study was conducted during February (2004, 2005) and during February to April (2006, 2008, 2010, 2012, 2013, 2014) on Floreana Island, Galápagos Archipelago (described in Kleindorfer *et al.*, 2014a, b). The study site was in *Scalesia* forest at the base of Cerro Pajas volcano (1°17′46″S, 90°27′06″W) at an elevation of 300–400 m; the area is the stronghold of the tree finch population on Floreana Island (O'Connor *et al.*, 2010c). We sampled *P. downsi* flies and Darwin's finch nesting outcome from four 100 m × 200 m study plots, as previously described (Kleindorfer *et al.*, 2014b; O'Connor *et al.*, 2010c). The preferred habitat of Darwin's tree finches is dominated by endemic *Scalesia pedunculata* trees, which are endangered on Floreana Island and only remain in fragmented patches totalling less than 3 km²; the highland *Scalesia* forest overlaps with agricultural land (O'Connor *et al.*, 2010c).

Host Species

We examined nesting height and parasite intensity in three Darwin's finch species: the common small tree finch (*Camarhynchus parvulus*), the critically endangered medium tree finch (*C. pauper*) and the common small ground finch (*G. fuliginosa*) (Grant and Grant, 2008; O'Connor *et al.*, 2010d; Sulloway and Kleindorfer, 2013). Based on surveys in 2013, the estimated maximum highland male population size on Floreana Island was ~6000 small tree finches, ~2500 medium tree finches and ~3800 small ground finches (Peters, 2016).

Nesting in Darwin's finches begins with the onset of the rains that usually occur in January or February. Males build a display nest and sing to attract females (Kleindorfer, 2007a). Females visit the singing male and inspect the nest. If accepted, a female will subsequently lay a clutch size of 2–5 eggs per nest (Kleindorfer, 2007b); some nests contained six eggs in 2008 and 2010. In all three species, the female is the sole incubator and both parents provide food to chicks. The incubation and feeding phase are ~14 days each. In the lowlands, small ground finch males typically nest in *Opuntia* cacti or *Acacia* trees; but in the highlands, they generally nest in cat's claw (*Zanthoxylum fagara*) or *S. pedunculata* trees (Kleindorfer, 2007b; O'Connor *et al.*, 2010a). Highland tree finches mostly build nests in *S. pedunculata* and occasionally in *Z. fagara*.

Parasite Species

Philornis downsi is a parasitic Dipteran that has two temporally distinct feeding modes: first instar larvae feed internally on the nasal and body cavities of its avian nestling hosts, and second and third instar larvae mostly feed externally on the chicks (Fessl *et al.*, 2006b; Kleindorfer and Sulloway, 2016). Adult female *Philornis downsi* generally carry ~60 eggs; each female fly mates with an average of ~2 males (range of 1–5 males per female), and 1 to 6 females each contribute an average of 5 larvae per Darwin's finch nest (range = 1–24 eggs) (Dudaniec *et al.*, 2010). Larvae pupate in the nest base after feeding on chicks for 4–7 days, and they emerge as flies after 10–14 days (Fessl *et al.*, 2006b; Kleindorfer and Dudaniec, 2016). The instars can be identified based on the size and shape of the posterior spiracles: first instars are the smallest in body length and have no discernible spiracles; second instars have two light brown spiracles and vary in body length from 4–7 mm; third instars have two large black spiracles and vary in body length from 6–12 mm.

The genus *Philornis* has a neotropical distribution comprising ~50 species (Dudaniec and Kleindorfer, 2006; Quiroga *et al.*, 2012). It is not known how *P. downsi* arrived in the Galápagos Islands, but there are two likely scenarios: (i) introduction via known mainland hosts such as smooth-billed ani (*Crotophaga ani*) and/or rock pigeon (*Columbia livia*), which were both introduced to the Galápagos between 1962 and 1972 (Wikelski *et al.*, 2004; Thiel *et al.*, 2005; Santiago-Alarcon *et al.*, 2006); and (ii) given that adult *P. downsi* feeds on fruit, the fly may have arrived via cargo boats from mainland Ecuador carrying produce (Dudaniec and Kleindorfer, 2006; Causton *et al.*, 2013). The latter possibility is rendered more plausible given the recent discovery of *P. downsi* in bird nests on mainland Ecuador (Bulgarella *et al.*, 2015).

Philornis downsi is the only parasite that causes measurable fitness costs in Darwin's finches. Avian poxvirus (*Poxvirus avium*) has existed on the Galápagos Islands since the 1890s (Parker *et al.*, 2011), and has increased sharply from 2000 to 2009 (Kleindorfer and Dudaniec, 2006; Gottdenker *et al.*, 2008; Zylberberg *et al.*, 2012). Both paramyxovirus and adenovirus have been found in Darwin's finches on Floreana Island (Deem *et al.*, 2011). Blood parasites have not been detected in Darwin's finches, and intestinal protozoan parasites are rare (Dudaniec *et al.*, 2005, 2006; Morales, 2013). Eight genera of feather mites have been found on Darwin's finches, and mite abundance increased with host body mass (Villa *et al.*, 2013).

Host Nesting Height and Nest Contents

Nesting height (m) was visually estimated, and the estimate was calibrated using a 6 m extendable video scope inserted into nests to check nest contents during routine nest status inspections. We used two methods to assess nesting status: 20-minute observations every 2 days to quantify parental activity at each nest, as well as nest inspection using a ladder (2004–2006) or mirror/camera on an extendable 6 m pole (2008–2014) (Kleindorfer *et al.*, 2014a). Within 2 days of the death or fledging of the last chick, we collected the nest and counted the number of *P. downsi*. The nesting material was dismantled and all *P. downsi* larvae, pupae and pupae cases were counted to calculate parasite intensity per nest. Chicks that had recently died were immersed in alcohol so that larvae within the body would float out and could be counted. We stored the pupae and larvae in ethanol within 24 h of collection from the host nest. All Darwin's finch nests with chicks in this study had *P. downsi* parasites.

We monitored nesting outcome at 582 active Darwin's finch nests between 2004 and 2014 on Floreana Island. The sample size per species was 150 small tree finch, 198 medium tree finch and 234 small ground finch nests. We analysed the following subsets of data for this study: nests with information about *P. downsi* intensity ($N = 254$), brood size ($N = 253$), percentage chick mortality ($N = 225$), nesting height ($N = 466$), and nests with information on both nesting height and *P. downsi* intensity ($N = 206$). The sample sizes per species for nests with information on both nesting height and *P. downsi* intensity were 40 small tree finch, 48 medium tree finch and 118 small ground finch.

Vertical Distribution of *P. downsi* in Fly Traps

From 15 March to 15 April 2014 we collected 365 *P. downsi* flies from 28 McPhail traps sampled four times each ($N = 112$ trapping events; mean $= 3.3 \pm 0.3$ flies per trap, maximum $= 19$). McPhail traps are ball-shaped plastic traps with a yellow bottom and a clear top with narrowing entrance in the middle; the traps are designed to hang in trees. The bottom contained a liquid lure (see details below) whose odour, together with the yellow colour, attracts insects to the trap; the insects subsequently drown in the fluid or fail to exit the trap. We placed seven fly traps every 15 m along each of 4 × 90 m transects in the *Scalesia* forest, for a total of 28 traps. The four transects (A–D) along which fly traps were placed were located within study plots used to monitor Darwin's finch nesting biology; the study plots are referred to as plot 1 (90°27′05.1″W, 01°17′50.5″S), plot 2 (90°27′09.8″W, 01°18′02.9″S), plot 3 (90°27′09.3″W, 01°18′05.9″S) and plot 4 (90°27′02.0″W, 01°17′54.0″S). The onset of each fly trap transect was separated by 100 m. Traps were allocated different heights that remained the same across the four-week sampling period, and sequential traps never had the same height. Table 10.1 shows the height of each fly trap within each transect. The number of replicates per trap height

Table 10.1 The height (m) of 28 McPhail traps to capture *Philornis downsi*. Each fly trap was placed at a distance of 15 m from the preceding fly trap along one of four straight-line transects (A–D) spanning 90 m in the highlands of Floreana Island

	Transect A (Plot 1)	Transect B (Plot 2)	Transect C (Plot 3)	Transect D (Plot 4)
Trap 1	2 m	5 m	6 m	2 m
Trap 2	6 m	3 m	4 m	6 m
Trap 3	4 m	6 m	3 m	2 m
Trap 4	7 m	2 m	4 m	6 m
Trap 5	5 m	4 m	6 m	4 m
Trap 6	2 m	6 m	5 m	7 m
Trap 7	3 m	4 m	2 m	5 m

was six traps at 2 m, three traps at 3 m, seven traps at 4 m, four traps at 5 m, six traps at 6 m, and two traps at 7 m.

To lure the flies to the trap, we filled them with 140 ml 'bait juice' made from one Hawaiian papaya (blended), 4 l of tap water, and 6 tablespoons of white sugar (using the trapping protocol developed by P. Lincango and C. Causton). Each fly trap was placed on a metal hook at a specific height on a *Scalesia* tree. The height of fly traps was either 2, 3, 4, 5, 6 or 7 m; the height was chosen randomly and allocated to each trap and transect before going to the field, but some heights needed to be changed in the field to suit the height of available trees. Following the trapping protocol of P. Lincango and C. Causton, insects were collected twice per week and stored in ethanol for later sorting at the field station. The bait lure was changed every 7 days. We repeated this procedure per trap across 4 weeks.

Identification of *P. downsi* from Fly Traps

The wing of *P. downsi* is very distinctive compared to other Galápagos Muscidae species. The R4+5 and M1 veins are sinuous or wavy and the distance separating them at the wing margin is greater than other Galápagos muscids (B. Sinclair, pers. comm.). *P. downsi* have dark abdomen and thorax; body length ~8 mm; wing length ~9 mm in females and ~10 mm in males. The flies collected from the McPhail traps were sexed at camp using a magnifying glass. Males have longer, pale yellowish legs, and the eyes are positioned differently compared to females. Male eyes are closely approximated and are dorsally separated by the width of the ocellar triangle (~0.2 mm); male eyes appear to be almost touching when viewed from above (Figure 10.1a). Females have shorter, darker legs, and female eyes are dorsally separated by ~0.5 mm; female eyes appear more parallel (Figure 10.1b).

Statistical Analysis

Data were analysed with SPSS 22 for Windows (SPSS Inc., Chicago, USA) and SAS 9.4 for Windows (Cary, North Carolina, USA). Before conducting statistical analyses, we examined the data to determine if they conformed to assumptions of normality and homogeneity of variance. For statistical analyses, nesting height data were log transformed and *P. downsi* data were square root transformed to satisfy requirements of normality. We found significant heterogeneity of variance for nesting height as assessed by Levene's test for equality of variances ($P = 0.034$). Therefore, we used a Welch ANOVA to test for differences in nesting height across host species. To test if higher nests had more *P. downsi* parasites, we used multiple regression analysis with *P. downsi* intensity (square root transformed) as the dependent variable and nesting height (log transformed) and year as independent variables. In a separate analysis, we examined the effects of host brood size, percentage chick mortality in nests and chick age at death in relation to nesting height, again using multiple regression analysis.

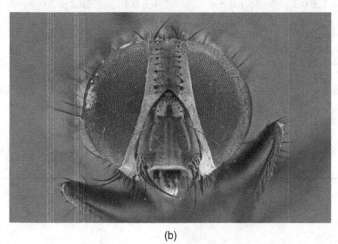

Figure 10.1 The frontal view of male (a) and female (b) *Philornis downsi* from Santa Cruz Island, Galápagos Archipelago. Male eyes are closely approximated and are dorsally separated by ~0.2 mm; female eyes appear more parallel and are dorsally separated by ~0.5 mm. Images provided by Bradley Sinclair.

We found significant heterogeneity of variance for male trap counts by height, for female trap counts by week, and for per cent male and female flies trapped by week. We also found excessive kurtosis (>3) in the data for male trap counts by height and by week. We therefore applied natural log transformations to normalize the data, which resolved these problems. The data by trapping events ($N = 112$) are not statistically independent. For this reason, we analysed these data using multilevel modelling in SAS (PROC MIXED). Our statistical design nested weekly capture data within collecting sites, and collecting sites within transects. Examination of the intraclass correlations between levels revealed that multilevel modelling was justified. Fifty per cent of the variation was found between transects and collection sites, and the remaining 50% of the variation occurred within individual traps.

Table 10.2 *Philornis downsi* intensity and nesting height (m) per species per year; data are shown as mean ± SE with sample size (N = number of nests). Nesting height differed significantly between species but not across years: lowest nesting height in small ground finch (3.5 ± 0.1), intermediate in small tree finch (4.8 ± 0.2), and highest in medium tree finch (6.8 ± 0.2). The intensity of *P. downsi* differed significantly between species: lower in small ground finch (31.7 ± 2.0) and small tree finch (31.1 ± 2.4), and higher in medium tree finch (55.2 ± 5.2) (ANOVA: species $F_{2,254} = 17.32, P < 0.001$).

	Small tree finch (*Camarhynchus parvulus*)		Medium tree finch (*C. pauper*)		Small ground finch (*Geospiza fuliginosa*)	
	P. downsi intensity (*N*)	Nesting height (m)	*P. downsi* intensity (*N*)	Nesting height (m)	*P. downsi* intensity (*N*)	Nesting height (m)
2004	30.9 ± 8.3 (13)	5.4 ± 0.5	65.5 ± 10.5 (5)	7.5 ± 0.5	19.9 ± 4.5 (14)	3.2 ± 0.4
2005	12.5 ± 0.5 (2)	4.0 ± 0.7	36.9 ± 3.4 (7)	4.7 ± 0.5	27.1 ± 5.7 (7)	3.3 ± 0.4
2006	20.4 ± 3.9 (10)	4.5 ± 0.3	36.1 ± 5.6 (35)	7.4 ± 0.4	28.2 ± 3.9 (68)	3.0 ± 0.2
2008	20.4 ± 4.2 (9)	4.9 ± 0.6	40.4 ± 7.6 (46)	5.6 ± 0.3	39.5 ± 4.1 (57)	3.6 ± 0.2
2010	31.5 ± 4.9 (42)	4.7 ± 0.3	64.3 ± 11.3 (53)	5.8 ± 0.3	21.3 ± 4.3 (58)	3.6 ± 0.3
2012	32.3 ± 3.8 (31)	4.2 ± 0.3	89.7 ± 31.4 (36)	6.8 ± 0.3		
2013	36.3 ± 5.2 (32)	4.7 ± 0.2	103.8 ± 28.7 (12)	6.8 ± 0.4	41.8 ± 5.2 (22)	4.4 ± 0.5
2014	51.9 ± 8.6 (11)	5.4 ± 0.5	68.0 ± 3.0 (3)	6.8 ± 0.6	43.0 ± 8.2 (8)	4.1 ± 0.6

Results

Philornis downsi Intensity and Nesting Height in Darwin's Finch Host Species

Nesting height differed significantly between the three species (Table 10.2). Nesting height was lowest in small ground finch (3.5 ± 0.1, N = 204), intermediate in small tree finch (4.8 ± 0.2, N = 90), and highest in medium tree finch (6.8 ± 0.2, N = 158) (Welch's ANOVA $F_{2,254.2} = 101.49, P < 0.0005$). We examined changes in nesting height per year from 2004–2014 in each species. Nesting height did not change significantly in small tree finch or medium tree finch (linear and quadratic regression analysis; all $P > 0.5$), but nesting height increased significantly across the decade in small ground finch ($r = 0.22, N = 206, P = 0.002$) (Table 10.2).

Host nesting height was positively correlated with the number of *P. downsi* parasites across Darwin's finch species (nesting height: $r_{partial} = 0.55, P < 0.001$; year: $r_{partial} = 0.24, P = 0.001, N = 208$): higher nests had more *P. downsi* parasites, and *P. downsi* intensity increased across the decade (Table 10.2). Within each species, the pattern between nesting height and *P. downsi* intensity was similar: higher nests had significantly more *P. downsi* in small tree finch (nesting height: $r_{partial} = 0.37, P = 0.018$;

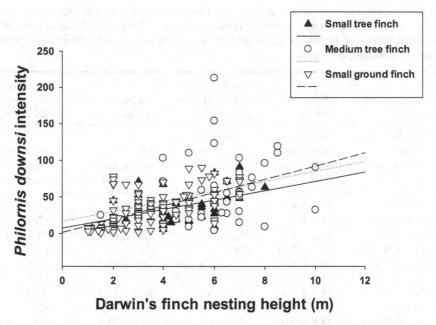

Figure 10.2 The significant linear association between host nesting height (m) and number of *Philornis downsi* parasites per nest on Floreana Island for data collected during the years 2004–14. The sample size per species was 40 small tree finch nests (*Camarhynchus parvulus*), 48 medium tree finch nests (*C. pauper*), and 118 small ground finch nests (*Geospiza fuliginosa*).

year: $r_{partial} = 0.22$, $P = 0.184$, $N = 40$) and small ground finch (nesting height: $r_{partial} = 0.62$, $P < 0.001$; year: $r_{partial} = 0.23$, $P = 0.01$, $N = 118$), and a marginally significant trend for more *P. downsi* in higher nests of medium tree finch (nesting height: $r_{partial} = 0.28$, $P = 0.056$; year: $r_{partial} = 0.45$, $P = 0.001$, $N = 48$) (Figure 10.2).

We used multiple regression analysis to test if other factors covaried with nesting height. In a single model, we tested host nesting height as the dependent variable against host brood size, percentage chick mortality in nests, and chick age at death. In small tree finch and medium tree finch, none of the variables was significantly associated with nesting height (small tree finch brood size: $r_{partial} = -0.37$, $P = 0.15$; percentage chick mortality: $r_{partial} = 0.10$, $P = 0.71$; chick age at death: $r_{partial} = 0.19$, $P = 0.46$, $N = 26$; medium tree finch brood size: $r_{partial} = -0.39$, $P = 0.34$; percentage chick mortality: $r_{partial} = 0.64$, $P = 0.09$; chick age at death: $r_{partial} = 0.19$, $P = 0.66$, $N = 27$). In small ground finch, higher nests had larger brood size (brood size: $r_{partial} = 0.39$, $P = 0.015$; percentage chick mortality: $r_{partial} = -0.06$, $P = 0.73$; chick age at death: $r_{partial} = -0.20$, $P = 0.22$, $N = 40$).

Philornis downsi Abundance at Different Heights

Table 10.3 and Figure 10.3 show the average number and percentage of male and female *P. downsi* flies caught in traps. To test specific hypotheses about fly behaviour we first computed linear and quadratic contrasts for trap height (Figure 10.4). Controlling the

Table 10.3 The percentage and number of male and female *Philornis downsi* per McPhail trap in relation to trap height (m) and the number of *P. downsi* per nest in relation to nesting height (m) in Darwin's finch species on Floreana Island. Data are shown as mean ± SE. Higher fly traps caught more female *P. downsi* and higher finch nests contained more *P. downsi* (statistical analyses in results)

Height (m)	Male *Philornis downsi* %*P. downsi* in fly traps (number of flies)	Female *P. downsi* %*P. downsi* in fly traps (number of flies)	Small tree finch (*Camarhynchus parvulus*) #*P. downsi* per nesting height (number of nests)	Medium tree finch (*C. pauper*) #*P. downsi* per nesting height (number of nests)	Small ground finch (*Geospiza fuliginosa*) #*P. downsi* per nesting height (number of nests)
2	30.0 ± 9.1 (1.1 ± 0.3)	28.3 ± 1.9 (1.0 ± 0.3)	27.0 ± 9.0 (5)	25.0 ± 0.0 (1)	17.7 ± 3.3 (76)
3	43.1 ± 3.7 (1.3 ± 0.1)	15.2 ± 5.0 (0.7 ± 0.2)	41.2 ± 8.2 (13)	27.0 ± 0.0 (1)	31.1 ± 4.3 (36)
4	50.5 ± 7.4 (2.1 ± 0.6)	35.3 ± 5.9 (1.6 ± 0.3)	21.0 ± 3.6 (23)	48.4 ± 10.7 (21)	29.5 ± 3.4 (38)
5	70.3 ± 8.1 (3.4 ± 1.3)	11.0 ± 3.9 (1.4 ± 0.5)	28.3 ± 6.3 (20)	38.0 ± 11.3 (21)	45.3 ± 4.0 (23)
6	36.3 ± 7.2 (1.8 ± 0.4)	38.8 ± 6.6 (1.5 ± 0.3)	37.2 ± 6.7 (12)	72.7 ± 21.5 (37)	59.2 ± 6.3 (21)
7	16.0 ± 7.8 (0.9 ± 0.7)	83.9 ± 7.9 (3.3 ± 0.5)	65.7 ± 9.1 (7)	52.0 ± 7.5 (27)	69.0 ± 5.1 (7)
8			63.0 ± 0.0 (1)	67.2 ± 13.1 (22)	None
9			None	87.8 ± 19.5 (15)	None

non-significant linear trend in height among males, we found that traps at an intermediate height caught a substantially greater number of male flies than traps placed at the very lowest and the very highest trap heights (for the linear trend: $r_{partial} = -0.03$, $t_{1,96.3} = 0.28$, $P = 0.79$; for the quadratic trend, $r_{partial} = 0.25$, $t_{1,105} = 2.67$, $P = 0.009$) (Figure 10.4a). In our analysis of female flies, we found a significant linear trend by height and, controlled for this trend, a non-significant trend for females to be trapped at the very lowest and the very highest (for the linear trend: $r_{partial} = 0.24$, $t_{1,94.8} = 2.36$, $P = 0.02$; for the quadratic trend: $r_{partial} = -0.15$, $t_{1,105} = -1.50$, $P = 0.14$, $N = 112$) (Figure 10.4b).

Analysing the data for both sexes collectively and controlling the non-significant linear and quadratic trends, we found a significant interaction effect between sex and the linear trend, as well as a significant interaction between sex and the quadratic trend (for the linear trend: $r_{partial} = 0.21$, $t_{1,112} = 2.35$, $P = 0.02$; for the quadratic trend interaction, $r_{partial} = -0.34$, $t_{1,112} = -3.84$, $P = 0.0002$). These findings show that males tended to be found at lower elevations than females, and especially at intermediate heights, whereas females tended to occur at the lowest and especially the highest elevations.

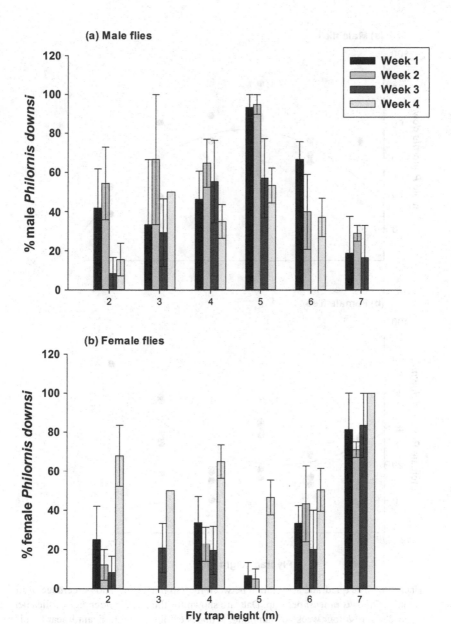

Figure 10.3 The percentage of (a) male and (b) female *Philornis downsi* caught in relation to the height (m) of McPhail traps on Floreana Island. The data are shown as means per week from 15 March to 15 April 2014 for each trap height. The sample size was 365 *P. downsi* caught in 112 trapping events using 28 McPhail traps.

Because of the relatively small number of male flies captured at 7 m, and because only two traps were placed at this height (versus an average of five traps placed at other heights), we repeated this last test by combining the data for flies trapped at 6 and 7 m and adjusting the model contrasts accordingly. The significant interaction between sex

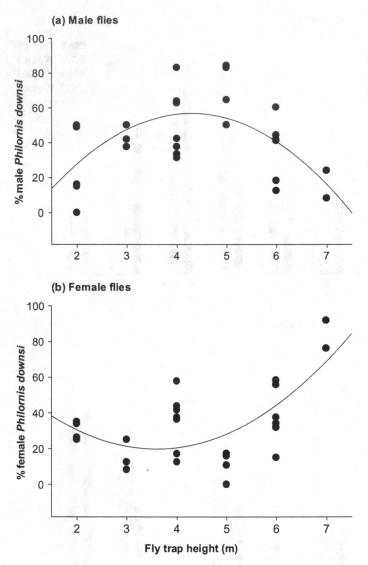

Figure 10.4 The quadratic association between fly trap height (m) and percentage of male and female *Philornis downsi* per trap. Data are shown for the mean percentage of flies per trap ($N = 28$) across four weeks of sampling. Females exhibited a significant linear trend, and a near-significant quadratic trend. The two quadratic trends were significantly different from each other: traps at a height of 4–5 m caught more male *P. downsi*, and traps at the lowest and highest elevations caught more female *P. downsi*.

and height as a linear trend was no longer significant, being replaced by a significant linear trend for the two sexes as a whole ($r_{partial} = 0.22$, $t_{1,94.3} = 2.16$, $P = 0.03$); however, the interaction between sex and the quadratic trend for height remained significant, confirming the previous finding that females are the least common where males are most common, namely, at intermediate heights ($r_{partial} = -0.25$, $t_{1,112} = -2.74$, $P = 0.007$).

Discussion

This study applied the conservation behaviour framework to a newly evolving host–parasite system on the Galápagos Islands, with the aim of identifying species-specific behaviour that could be used to alleviate the extremely high levels of virulence caused by *P. downsi* for its naïve Darwin's finch hosts. The study identified two behavioural traits that suggest new ways of thinking about *P. downsi* and Darwin's finch hosts. First, nesting height of Darwin's finches on Floreana Island was positively associated with *P. downsi* intensity: higher finch nests had more parasitic larvae. We found this pattern within and between host species using a substantial sample size (206 host nests sampled between 2004 and 2014). Second, fly traps placed higher in the canopy (7 m) caught significantly more female *P. downsi*, while fly traps placed at 4–5 m height caught more male *P. downsi*. Despite any limitations of the fly trap data, which pertain to a single year (2014), the findings are noteworthy for two reasons. First, they provide a plausible explanation for why we have consistently documented higher parasite intensity in the critically endangered medium tree finch (O'Connor *et al.*, 2010d; Kleindorfer *et al.*, 2014a, b). The average nesting height in medium tree finches was 6.8 m – and therefore the nests of this species appear to be more susceptible to being located by female *P. downsi* flies given that females were more common at 7 m. Second, these findings can be used to generate useful conservation management strategies to control *P. downsi* abundance. One obvious recommendation is to place fly traps at 6–8 m height to remove egg-laying female flies, and at 4–5 m height to remove male flies, from the habitats of critically endangered species such as medium tree finch on Floreana Island (O'Connor *et al.*, 2010d) and perhaps mangrove finch (*C. heliobates*) on Isabela Island (Fessl *et al.*, 2010) should fly trapping data reveal similar patterns of capture in relation to trap height. The results of this study suggest that *P. downsi* flight behaviour makes this parasite more likely to encounter particular hosts, rather than hosts having specific attributes that make them particularly attractive to the parasite.

Different research groups have been monitoring the impact of *P. downsi* intensity for Darwin's finch chick mortality since 1998 (Fessl *et al.*, 2006a; Dudaniec *et al.*, 2007; Huber *et al.*, 2010; O'Connor *et al.*, 2010a–d; Koop *et al.*, 2011; Knutie *et al.*, 2013; Cimadom *et al.*, 2014; Kleindorfer *et al.*, 2014a). One finding that has intrigued researchers for over a decade is that some Darwin's finch species have higher *P. downsi* intensity than others. On Santa Cruz Island, between 1998 and 2006, the highest *P. downsi* intensity (57 ± 4) has been found in the large-bodied (22 g) woodpecker finch (*Camarhynchus pallidus*); and the second highest *P. downsi* intensity (41 ± 6) occurred in the small-bodied (9 g) warbler finch (*C. olivacea*) (Kleindorfer and Dudaniec, 2009). Both species have had the strongest population declines from 2000 to 2010 (Dvorak *et al.*, 2012). In contrast, the medium-sized (13 g) small tree finch (*C. parvulus*) generally had fewer *P. downsi* (23 ± 3) (Dudaniec *et al.*, 2007), and its population on Santa Cruz Island has remained stable from 2000 to 2010 (Dvorak *et al.*, 2012). Hints suggesting that body size could be important for parasite intensity come from the finding that larger-bodied Darwin's finches build larger nests, and larger nests had more *P. downsi* (Kleindorfer and Dudaniec, 2009). Despite the appeal of host body size as a predictor of parasite intensity, the evidence has so far been inconclusive for *P. downsi* and Darwin's finches (Cimadom *et al.*, 2014). Rather, evidence is growing that nest attributes (nest size and location) predict parasite intensity. In a previous study, Kleindorfer and Dudaniec (2009) showed that nests that were larger in size and nests with many close neighbours had higher *P. downsi* intensity. It remains to be tested if host nesting density on Santa Cruz Island has changed across the past decade, and to what degree the height or abundance of *Scalesia* trees used for nesting may be associated with host nesting height. The conservation implications of managing vertical and horizontal forest attributes are important for biodiversity (e.g. DeVries *et al.*, 1997) but are only beginning to be explored for host–parasite systems (Peters and Kleindorfer, 2015). If the vertical forest is associated with particular patterns of parasite community, then forest height can become a target of tailored conservation management approaches.

As on Santa Cruz Island, there were different interspecific patterns of *P. downsi* intensity in Darwin's finch hosts on Floreana. The critically endangered medium tree finch had higher parasite intensity (55 ± 5) than the common small tree finch (31 ± 2) and small ground finch (31 ± 2) (Table 10.2). Since 2012, all medium tree finch chicks have died in the nest due to *P. downsi* parasites – a rather alarming finding (Kleindorfer *et al.*, 2014a). In addition to low nesting success and declining populations, tree finches on Floreana Island are hybridizing. In particular, female medium tree finches have increasingly paired with male small tree finches and have produced hybrid offspring. As a result, the proportion of hybrid tree finches has increased from 19% in 2005 to 41% in 2010 (Kleindorfer *et al.*, 2014b). Intriguingly, the hybrid finch nests had 50–79% fewer *P. downsi* compared with the nests of the parental tree finch (Kleindorfer *et al.*, 2014b). Therefore, the initial evidence suggests that hybrid offspring are being favoured by selection, perhaps because they are less likely to be parasitized by *P. downsi*. From another study, we know that hybrid tree finches foraged at ~4 m in the *Scalesia* canopy while medium tree finches foraged at ~6 m (Peters and Kleindorfer, 2015), evidence that supports the findings presented here: birds that foraged lower in the canopy also had fewer *P. downsi* parasites. Nesting height remains to be tested in hybrid finches, and we await a larger sample size to draw any firm conclusions. The increase in mean *P. downsi* intensity on Floreana Island coincides with species collapse via hybridization in Darwin's finches, the bird group that is known as a classic textbook example of speciation (Grant and Grant, 2008). Because *P. downsi* was likely introduced as the result of human activity, there are urgent conservation imperatives to explain why *P. downsi* intensity and Darwin's finch chick mortality differ between host species, and to do so before the populations collapse into a single hybrid swarm or become extinct (Kleindorfer *et al.*, 2014b).

This study suggests that the flight behaviour of *P. downsi* could be a key factor that predicts its frequency of occurrence in Darwin's finch nests. One major difference in *P. downsi* flight height is sex: we found significantly more female flies at 2 m and 7 m, and male flies at 4–5 m. This pattern suggests that gravid females may be avoiding males. Although we acknowledge the limits of our study design and its small sample size, we note that our findings are consistent with evidence from other study systems of female behavioural tactics designed to reduce the cost of mating and to avoid males once a female is gravid. *Drosophila melanogaster* is a model system to study the costs of reproduction because the mating frequency that maximizes male reproductive success is higher than that which maximizes female reproductive success (Bateman, 1948). Female *D. melanogaster* that mate frequently have a shorter lifespan and lower reproductive success (Fowler and Partridge, 1989). Wigby and Chapman (2005) identified a sex peptide in male *D. melanogaster* ejaculate that stimulates female egg production but also lowers female reproductive success. Clearly, sexual selection should favour females that avoid costly matings. In damselflies (*Ischnura elegans*) males will attempt to mate multiply with females, leading to male harassment and selection on the frequency of cryptic or male-type colour polymorphisms among females (Van Gossum *et al.*, 2001). Svensson and colleagues have shown that male species recognition in banded demoiselle (*Calopteryx splendens*) is fixed at emergence, but females learn to recognize heterospecifics to reduce costly hybrid matings (Svensson *et al.*, 2014). Male harassment has been shown to reduce reproductive success in damselflies (*I. senegalensis*), and females accordingly avoid oviposition sites having many males (Takahashi and Watanabe, 2010). Thus, there is evidence across taxonomic groups for behavioural tactics to avoid the high costs of multiple mating in female insects. From molecular data on family groups of *P. downsi* in nests of Darwin's finches, we know that female *P. downsi* remate between one and five times (an average of 1.91 ± 0.08 times) (Dudaniec *et al.*, 2010); previous studies have not tested for potential fitness costs of multiple mating by *P. downsi* females – a topic of potentially fruitful future research.

Little is known about the reproductive behaviour of *P. downsi* in the wild. We believe the morphology of *P. downsi* male and female eyes provides indirect evidence concerning reproductive behaviour as it relates to flight height in this system. Although this matter was not an explicit focus of this study, we used eye morphology (and other morphological traits) as a means of sexing the adult flies caught in traps. In *P. downsi*, male eyes are close together, whereas female eyes are wider apart (see Figure 10.1). From previous study of other flies, aspects of mate search behaviour can be predicted from eye morphology and vision. For example, Zeil (1986) recorded the flight paths of male house flies (*Fannia canicularis*) patrolling the airspace below indoor landmarks, such as lampshades. Male house flies approached these landmarks from below and defended the airspace immediately below the landmark. If a patrol area was occupied, the next arriving male occupied an area below that of the first male. Zeil (1986) hypothesized that female flies might approach landmarks from the side, and not from below as the males did, and that female flies would therefore pass through the dorsal visual field of the males. The sexual dimorphism in compound eye organization in *P. downsi* may provide indirect evidence that male vision and female flight behaviour are related. As is the case in many fly species (e.g. Collett and Land, 1975; Zeil, 1983; Land and Eckert, 1985), the eyes of male *P. downsi* extend more medially in the frontal and dorsal visual field, compared with female eyes, which is an indication of a fronto-dorsal acute zone that may be involved in detecting and chasing females. We therefore predict that male *P. downsi* at lower heights than females are better able to detect a female flying above – a hypothesis that remains to be tested.

Studies of differences in flight height in Diptera have found significantly more female flies higher in the canopy compared with males (Aluja *et al.*, 1989; Herczeg *et al.*, 2014), and more females closer to the ground compared with males (Gersabeck and Merritt,

1983; Birtele and Hardersen, 2012). In addition to avoiding costly reproduction with males, gravid female flies move to foraging areas to increase their foraging efficiency for egg production (Irvin *et al.*, 1999; Mavoungou *et al.*, 2013; Maguire *et al.*, 2014). These findings show that the vertical distribution of flies must be considered in relation to reproductive behaviour. Adult *P. downsi* feed on fruit (such as papaya) and decaying organic matter (Fessl *et al.*, 2001), and gravid *P. downsi* females in search for food may therefore fly at both lower and higher elevations, as we found in this study. Clearly, many different factors could influence the vertical distribution of flies, including meteorological conditions, vegetation type and cover, host location, and oviposition habits (Roberts, 1985; Van Hennekeler *et al.*, 2011; Birtele and Hardersen, 2012; Swanson *et al.*, 2012; Mavoungou *et al.*, 2013; Maguire *et al.*, 2014). One limitation of this study is that fly abundance per trap height was sampled within a single year and across a single month, and therefore that the findings could reflect other unexamined factors, such as microclimate and seasonality. Ideally, fly trapping should be carried out across the year and under varied environmental conditions.

The conservation behaviour framework encourages synergistic discourse between ethology and conservation. Accordingly, a fruitful approach to manage the impacts of invasive species is to identify behavioural traits that can be targeted for maximal efficacy of limited human and financial resources. Here we showed that both host nesting height and parasite flight height could be useful targets of conservation intervention. Maintaining a broad range of vertical forest is likely to improve host survival by creating stratified nesting areas for the avoidance of airborne parasites that preferentially occur at particular heights. The findings can also be applied to nest-box studies: if nesting height is associated with high parasite intensity then researchers should alter nest-box height. Finally, when attempting to remove airborne parasites from the population, trap heights that target female flies would remove more eggs and hence would maximize conservation dollars for trapping efforts. In these ways, one can imagine how linking insights from ethology and conservation management should generate more biologically relevant and cost-effective approaches to managing threatened species.

Acknowledgements

We are grateful to the Galápagos National Park and Charles Darwin Research Station for continued support to carry out our collaborative research on the impacts of *Philornis downsi* for Darwin's finch nesting success and population dynamics on the Galápagos Islands. We thank TAME airlines for reduced airfare to conduct the study. We greatly appreciate the generous support received from the community of Floreana Island. We thank Charlotte Causton and Piedad Lincango for sharing advice about fly trapping as well as preliminary patterns of fly height data from Santa Cruz Island, and Charlotte Causton, Francesca Cunningham, and Washington Tapia for organizing and hosting the International *P. downsi* Action Group. We are most grateful to Bradley Sinclair for answering numerous questions about *P. downsi* morphology and behaviour across the years and for providing the images used in this chapter. For funding, we thank the

Rufford Small Grant Foundation, Australian Research Council, Mohamed bin Zayed Species Conservation Fund, Max Planck Institute for Ornithology, Royal Society for the Protection of Birds/Birdfair, Club300 Bird Protection, the Ecological Society of Australia, and Earthwatch Institute. For assistance in the field, we thank Rachelle Bassi, Claire Charlton, Rebekah Christensen, Timothy Clark, Rachael Dudaniec, Christine Evans, James Forwood, Svenja Gantefoer, David Gaspard, Kathy Gavrilchuck, Marina Louter, Jody O'Connor, Jeremy Robertson, David Shimoda Arango Roldan, Tanya Seebacher, Matthias Schmidt, Robin Schubert, Santiago Torres and Carlos Vinueza. We thank Jody O'Connor and Rachael Dudaniec for providing critical comments on an earlier draft of this manuscript. This publication is contribution number 2127 of the Charles Darwin Foundation for the Galápagos Islands.

References

Alizon, S. and Van Baalen, M. (2008). Multiple infections, immune dynamics, and the evolution of virulence. *The American Naturalist*, 172, E150–E168.

Aluja, M., Cabrera, M., Guillen, J., Celedonio, H. and Ayora, F. (1989). Behaviour of *Anastrepha ludens*, *A. obliqua* and *A. serpentina* (Diptera: Tephritidae) on a wild mango tree (*Mangifera indica*) harbouring three McPhail traps. *International Journal of Tropical Insect Science*, 10, 309–318.

Bateman, A.J. (1948). Intra-sexual selection in *Drosophila*. *Heredity*, 2, 349–368.

Berger-Tal, O., Polak, T., Oron, A., *et al.* (2011). Integrating animal behavior and conservation biology: a conceptual framework. *Behavioral Ecology*, arq224.

Birtele, D. and Hardersen, S. (2012). Analysis of vertical stratification of Syrphidae (Diptera) in an oak-hornbeam forest in northern Italy. *Ecological Research*, 27, 755–763.

Brearley, G., Rhodes, J., Bradley, A., *et al.* (2013). Wildlife disease prevalence in human-modified landscapes. *Biological Reviews*, 88, 427–442.

Bulgarella, M., Quiroga, M.A., Dregni, J.S., *et al.* (2015). *Philornis downsi* (Diptera: Muscidae), an avian nest parasite invasive to the Galápagos Islands, in mainland Ecuador. *Annals of the Entomological Society of America*, sav026.

Caro, T. (1999). The behaviour–conservation interface. *Trends in Ecology and Evolution*, 14, 366–369.

Caro, T. and Riggio, J. (2014). Conservation and behavior of Africa's 'Big Five'. *Current Zoology*, 60(4), 486–499.

Caro, T. and Sherman, J. (2011). Endangered species and a threatened discipline: behavioural ecology. *Trends in Ecology and Evolution*, 26, 111–118.

Causton, C.E., Peck, S.B., Sinclair, B.J., *et al.* (2006). Alien insects: threats and implications for the conservation of the Galápagos Islands. *Annals of the Entomological Society of America*, 99, 121–143.

Causton, C., Cunninghame, F. and Tapia, W. (2013). Management of the avian parasite *Philornis downsi* in the Galápagos Islands: a collaborative and strategic action plan. In *Galápagos Report 2011–2012*. Puerto Ayora, Galapagos, Ecuador: GNPS, GCREG, CDF and GC, pp. 167–173.

Cimadom, A., Ulloa, A., Meidl, P., *et al.* (2014). Invasive parasites, habitat change and heavy rainfall reduce breeding success in Darwin's finches. *PLoS ONE*, 9, e107518.

Collett, T. and Land, M. (1975). Visual control of flight behaviour in the hoverfly *Syritta pipiens* L. *Journal of Comparative Physiology*, 99, 1–66.

Daly, E.W. and Johnson, P.T. (2011). Beyond immunity: quantifying the effects of host anti-parasite behavior on parasite transmission. *Oecologia*, 165, 1043–1050.

Deem, S., Jiménez-Uzcátegui, G. and Ziemmeck, F. (2011). CDF checklist of Galapagos zoopathogens and parasites. In *Galápagos Report 2011–2012*. Puerto Ayora, Galapagos, Ecuador: GNPS, GCREG, CDF and GC.

DeVries, P.J., Murray, D. and Lande, R. (1997). Species diversity in vertical, horizontal, and temporal dimensions of a fruit-feeding butterfly community in an Ecuadorian rainforest. *Biological Journal of the Linnean Society*, 62, 343–364.

Dudaniec, R.Y. and Kleindorfer, S. (2006). The effects of the parasitic flies *Philornis* (Diptera, Muscidae) on birds. *EMU*, 106, 13–20.

Dudaniec, R.Y., Hallas, G. and Kleindorfer, S. (2005). Blood and intestinal parasitism in Darwin's finches: negative and positive findings. *Acta Zoologica Sinica*, 51, 507–512.

Dudaniec, R.Y., Kleindorfer, S. and Fessl, B. (2006). Effects of the introduced ectoparasite *Philornis downsi* on haemoglobin level and nestling survival in Darwin's small ground finch (*Geospiza fuliginosa*). *Austral Ecology*, 31, 88–94.

Dudaniec, R.Y., Fessl, B. and Kleindorfer, S. (2007). Interannual and interspecific variation on intensity of the parasitic fly, *Philornis downsi*, in Darwin's finches. *Biological Conservation*, 139, 325–332.

Dudaniec, R.Y., Gardner, M.G., Donellan, S. and Kleindorfer, S. (2008). Genetic variation in the invasive avian parasite, *Philornis downsi* (Diptera, Muscidae) on the Galápagos archipelago. *BMC Ecology*, 8, 13.

Dudaniec, R.Y., Gardner, M.G. and Kleindorfer, S. (2010). Offspring genetic structure reveals mating and nest infestation behaviour of an invasive parasitic fly (*Philornis downsi*) of Galápagos birds. *Biological Invasions*, 12, 581–592.

Duffy, M.A. and Sivars-Becker, L. (2007). Rapid evolution and ecological host–parasite dynamics. *Ecology Letters*, 10, 44–53.

Dvorak, M., Fessl, B., Nemeth, E., Kleindorfer, S. and Tebbich, S. (2012). Distribution and abundance of Darwin's finches and other land birds on Santa Cruz Island, Galápagos: evidence for declining populations. *Oryx*, 46, 78–86.

Fessl, B., Couri, M. and Tebbich, S. (2001). *Philornis downsi* Dodge and Aitken, new to the Galápagos Islands, (Diptera, Muscidae). *Studia Dipterologica*, 8, 317–322.

Fessl, B., Kleindorfer, S. and Tebbich, S. (2006a). An experimental study on the effects of an introduced parasite in Darwin's finches. *Biological Conservation*, 127, 55–61.

Fessl, B., Sinclair, B.J. and Kleindorfer, S. (2006b). The life cycle of *Philornis downsi* (Diptera: Muscidae) parasitizing Darwin's finches and its impacts on nestling survival. *Parasitology*, 133, 739–747.

Fessl, B., Young, H.G., Young, R.P., *et al.* (2010). How to save the rarest Darwin's finch from extinction: the mangrove finch on Isabela Island. *Philosophical Transactions of the Royal Society B: Biological Sciences*, 365, 1019–1030.

Fowler, K. and Partridge, L. (1989). A cost of mating in female fruitflies. *Nature*, 338, 760–761.

Galligan, T.H. and Kleindorfer, S. (2009). Naris and beak malformation caused by the parasitic fly, *Philornis downsi* (Diptera: Muscidae), in Darwin's small ground finch, *Geospiza fuliginosa* (Passeriformes: Emberizidae). *Biological Journal of the Linnean Society*, 98, 9.

Gersabeck, E.F. and Merritt, R.W. (1983). Vertical and temporal aspects of Alsynite® panel sampling for adult *Stomoxys calcitrans* (L.)(Diptera: Muscidae). *Florida Entomologist*, 66, 222–227.

Gottdenker, N.L., Walsh, T., Jiménez-Uzcátegui, G., *et al.* (2008). Causes of mortality of wild birds submitted to the Charles Darwin Research Station, Santa Cruz, Galápagos, Ecuador from 2002–2004. *Journal of Wildlife Diseases*, 44, 1024–1031.

Grant, P.R. and Grant, B.R. (2008). *How and Why Species Multiply: The Radiation of Darwin's Finches*. Princeton, NJ: Princeton University Press.

Grant, P.R., Grant, B.R., Petren, K. and Keller, L.F. (2005). Extinction behind our backs: the possible fate of one of the Darwin's finch species on Isla Floreana, Galápagos. *Biological Conservation*, 122, 499–503.

Herczeg, T., Blahó, M., Száz, D., *et al.* (2014). Seasonality and daily activity of male and female tabanid flies monitored in a Hungarian hill-country pasture by new polarization traps and traditional canopy traps. *Parasitology Research*, 113, 1–10.

Huber, S.K. (2008). Effects of the introduced parasite *Philornis downsi* on nestling growth and mortality in the medium ground finch (*Geospiza fortis*). *Biological Conservation*, 141, 601–609.

Huber, S.K., Owen, J.P., Koop, J.A., *et al.* (2010). Ecoimmunity in Darwin's finches: Invasive parasites trigger acquired immunity in the medium ground finch (*Geospiza fortis*). *PLoS ONE*, 5, e8605.

Irvin, N., Wratten, S., Frampton, C., *et al.* (1999). The phenology and pollen feeding of three hover fly (Diptera: Syrphidae) species in Canterbury, New Zealand. *New Zealand Journal of Zoology*, 26, 105–115.

Kaltz, O. and Shykoff, J. A. (1998). Local adaptation in host–parasite systems. *Heredity*, 81, 361–370.

Kawecki, T.J. and Ebert, D. (2004). Conceptual issues in local adaptation. *Ecology Letters*, 7, 1225–1241.

Kilpatrick, A.M. (2011). Globalization, land use, and the invasion of West Nile virus. *Science*, 334, 323–327.

Kleindorfer, S. (2007a). Nesting success in Darwin's small tree finch (*Camarhynchus parvulus*): Evidence of female preference for older males and more concealed nests. *Animal Behaviour*, 74, 795–804.

Kleindorfer, S. (2007b). The ecology of clutch size variation in Darwin's small ground finch *Geospiza fuliginosa*: comparison between lowland and highland habitats. *Ibis*, 149, 730–741.

Kleindorfer, S. and Dudaniec, R.Y. (2006). Increasing prevalence of avian poxvirus in Darwin's finches and its effect on male pairing success. *Journal of Avian Biology*, 37, 69–76.

Kleindorfer, S. and Dudaniec, R.Y. (2009). Love thy neighbour? Social nesting pattern, host mass and nest size affect ectoparasite intensity in Darwin's tree finches. *Behavioural Ecology and Sociobiology*, 63, 731–739.

Kleindorfer, S. and Dudaniec, R.Y. (2016). Host-parasite ecology, behavior and genetics: a review of the introduced fly parasite *Philornis downsi* and its Darwin's finch hosts. *BMC Zoology*, 1:1.

Kleindorfer, S., Peters, K.J., Custance, G., Dudaniec, R.Y. and O'Connor, J.A. (2014a). Changes in *Philornis* infestation behavior threaten Darwin's finch survival. *Current Zoology*, 60, 542–550.

Kleindorfer, S., O'Connor, J.A., Dudaniec, R.Y., *et al.* (2014b). Species collapse via hybridization in Darwin's tree finches. *The American Naturalist*, 183, 325–341.

Kleindorfer, S. and Sulloway, F.J. (2016). Naris deformation in Darwin's Finches: Experimental and historical evidence for a post-1960s arrival of the parasite *Philornis downsi*. *Global Ecology and Conservation*, 7, 122–131.

Knutie, S.A., Koop, J.A., French, S.S. and Clayton, D.H. (2013). Experimental test of the effect of introduced hematophagous flies on corticosterone levels of breeding Darwin's finches. *General and Comparative Endocrinology*, 193, 68–71.

Knutie, S.A., Mcnew, S.M., Bartlow, A.W., Vargas, D.A. and Clayton, D.H. (2014). Darwin's finches combat introduced nest parasites with fumigated cotton. *Current Biology*, 24, R355–R356.

Koop, J.a.H., Huber, S.K., Laverty, S.M. and Clayton, D.H. (2011). Experimental demonstration of the fitness consequences of an introduced parasite of Darwin's finches. *PLoS ONE*, 6, e19706.

Koop, J.A., Le Bohec, C. and Clayton, D.H. (2013). Dry year does not reduce invasive parasitic fly prevalence or abundance in Darwin's finch nests. *Reports Parasitology*, 3, 11–17.

Kovaliski, J., Sinclair, R., Mutze, G., *et al.* (2014). Molecular epidemiology of rabbit haemorrhagic disease virus in Australia: when one became many. *Molecular Ecology*, 23, 408–420.

Land, M. and Eckert, H. (1985). Maps of the acute zones of fly eyes. *Journal of Comparative Physiology A*, 156, 525–538.

Maguire, D.Y., Robert, K., Brochu, K., *et al.* (2014). Vertical stratification of beetles (Coleoptera) and flies (Diptera) in temperate forest canopies. *Environmental Entomology*, 43, 9–17.

Mavoungou, J.F., Kohagne, T.L., Acapovi-Yao, G.L., *et al.* (2013). Vertical distribution of *Stomoxys* spp. (Diptera: Muscidae) in a rainforest area of Gabon. *African Journal of Ecology*, 51, 147–153.

McCallum, H. (2008). Tasmanian devil facial tumour disease: lessons for conservation biology. *Trends in Ecology and Evolution*, 23, 631–637.

Morales, V. (2013). Endoparásitos en varios pinzones de Darwin e cautiverio y pinzones silvestres en la isla Santa Cruz, Provincia Insular Galápagos, Ecuador-2008. Repositorio Digital Universidad Politecnica Salesiana.

Nelson, X.J. (2014). Animal behavior can inform conservation policy, we just need to get on with the job – or can it? *Current Zoology*, 60, 479–485.

O'Connor, J.A., Dudaniec, R.Y. and Kleindorfer, S. (2010a). Parasite infestation in Galápagos birds: contrasting two elevational habitats between islands. *Journal of Tropical Ecology*, 26, 285–292.

O'Connor, J.A., Robertson, J. and Kleindorfer, S. (2010b). Video analysis of host–parasite interactions in Darwin's finch nests. *Oryx*, 44, 588–594.

O'Connor, J.A., Sulloway, F.J. and Kleindorfer, S. (2010c). Avian population survey in the Floreana Highlands: Is the medium tree finch declining in remnant patches of *Scalesia* forest? *Bird Conservation International*, 20, 343–353.

O'Connor, J.A., Sulloway, F.J., Robertson, J. and Kleindorfer, S. (2010d). *Philornis downsi* parasitism is the primary cause of nestling mortality in the critically endangered Darwin's medium tree finch (*Camarhynchus pauper*). *Biodiversity and Conservation*, 19, 853–866.

O'Connor, J.A., Robertson, J. and Kleindorfer, S. (2014). Darwin's finch begging intensity does not honestly signal need in parasitised nests. *Ethology*, 120, 228–237.

Palestis, B.G. (2014). The role of behavior in tern conservation. *Current Zoology*, 60, 500–514.

Parker, P.G., Buckles, E.L., Farrington, H., *et al.* (2011). 110 years of avipoxvirus in the Galápagos Islands. *PLoS ONE*, 6, e15989.

Peters, K.J. (2016). *Unravelling the Dynamics of Hybridisation and its Implications for Ecology and Conservation of Darwin's Tree Finches.* Adelaide: Flinders University, School of Biological Sciences, p. 207.

Peters, K.J. and Kleindorfer, S. (2015). Divergent foraging behavior in a hybrid zone: Darwin's tree finches (*Camarhynchus* spp.) on Floreana Island. *Current Zoology*, 61, 181–190.

Quiroga, M.A., Reboreda, J.C. and Beltzer, A.H. (2012). Host use by *Philornis* sp. in a passerine community in central Argentina. *Revista Mexicana de Biodiversidad*, 83, 110–116.

Roberts, D. (1985). Vertical distribution of flying black-flies (Diptera: Simuliidae) in Central Nigeria. *Tropical Medicine and Parasitology: Official Organ of Deutsche Tropenmedizinische Gesellschaft and of Deutsche Gesellschaft fur Technische Zusammenarbeit (GTZ)*, 36, 102–104.

Santiago-Alarcon, D., Tanksley, S.M. and Parker, P.G. (2006). Morphological variation and genetic structure of Galápagos Dove (*Zenaida galapagoensis*) populations: issues in conservation for the Galápagos bird fauna. *The Wilson Journal of Ornithology*, 118, 194–207.

Sulloway, F.J. and Kleindorfer, S. (2013). Adaptive divergence in Darwin's small ground finch (*Geospiza fuliginosa*): divergent selection along a cline. *Biological Journal of the Linnean Society*, 110, 45–59.

Svensson, E.I., Runemark, A., Verzijden, M.N. and Wellenreuther, M. (2014). Sex differences in developmental plasticity and canalization shape population divergence in mate preferences. *Proceedings of the Royal Society B: Biological Sciences*, 281, 20141636.

Swanson, D., Adler, P. and Malmqvist, B. (2012). Spatial stratification of host-seeking Diptera in boreal forests of northern Europe. *Medical and Veterinary Entomology*, 26, 56–62.

Takahashi, Y. and Watanabe, M. (2010). Female reproductive success is affected by selective male harassment in the damselfly *Ischnura senegalensis*. *Animal Behaviour*, 79, 211–216.

Thiel, T., Whiteman, N.K., Tirapé, A., *et al.* (2005). Characterization of canarypox-like viruses infecting endemic birds in the Galápagos Islands. *Journal of Wildlife Diseases*, 41, 342–353.

Van Gossum, H., Stoks, R. and De Bruyn, L. (2001). Frequency-dependent male mate harassment and intra-specific variation in its avoidance by females of the damselfly *Ischnura elegans*. *Behavioral Ecology and Sociobiology*, 51, 69–75.

Van Hennekeler, K., Jones, R., Skerratt, L., Muzari, M. and Fitzpatrick, L. (2011). Meteorological effects on the daily activity patterns of tabanid biting flies in northern Queensland, Australia. *Medical and Veterinary Entomology*, 25, 17–24.

Villa, S.M., Le Bohec, C., Koop, J.A., Proctor, H.C. and Clayton, D.H. (2013). Diversity of feather mites (Acari: Astigmata) on Darwin's finches. *The Journal of Parasitology*, 99, 756–762.

Wiedenfeld, D.A., Jiménez, G., Fessl, B., Kleindorfer, S. and Valerezo, J.C. (2007). Distribution of the introduced parasitic fly *Philornis downsi* (Diptera, Muscidae) in the Galápagos Islands. *Pacific Conservation Biology*, 13, 14–19.

Wigby, S. and Chapman, T. (2005). Sex peptide causes mating costs in female *Drosophila melanogaster*. *Current Biology*, 15, 316–321.

Wikelski, M., Foufopoulos, J., Vargas, H. and Snell, H. (2004). Galápagos birds and diseases: invasive pathogens as threats for island species. *Ecology and Society*, 9, 5.

Zeil, J. (1983). Sexual dimorphism in the visual system of flies: the free flight behaviour of male Bibionidae (Diptera). *Journal of Comparative Physiology*, 150, 395–412.

Zeil, J. (1986). The territorial flight of male houseflies (*Fannia canicularis* L.). *Behavioral Ecology and Sociobiology*, 19, 213–219.

Zylberberg, M., Lee, K.A., Klasing, K.C. and Wikelski, M. (2012). Increasing avian pox prevalence varies by species, and with immune function, in Galápagos finches. *Biological Conservation*, 153, 72–79.

11   Eat or be Eaten: Invasion and Predation in Aquatic Ecosystems

Judith S. Weis

Predator–prey interactions are of great importance in ecology and evolution. Behaviours associated with prey capture or predator avoidance are a link between the underlying neurological processes and the community and ecosystem levels (Weis *et al.*, 2000). Predators must have effective capture behaviours in order to acquire the energy for growth and survival. Effective behaviours involve motivation to feed, detection of prey and coordination of the attack. For prey species, it is even more critical – they must be able to detect the presence of predators and respond appropriately in order to survive and stay in the gene pool. While response to predators may involve morphological changes such as larger spines or thicker shells (phenotypic plasticity), behaviour is a key component of predator avoidance. One important component is recognition of predator odours, which trigger responses such as decreased activity, cessation of foraging, feeding and grooming, and shifts to refuges or other safe habitats. Other important behavioural components of predator avoidance are camouflage and quick escape.

The introduction of a species into an ecosystem can have major impacts on potential prey, potential competitors for prey and potential predators. Introduced species can become successful by outcompeting native species for resources including food. Top-down effects of species at higher trophic levels on the abundance, diversity, productivity or community structure of species at lower trophic levels are well documented (Paine, 1966; Strong, 1992; Baum and Worm, 2009). Most studies of top-down effects experimentally exclude the top predator or consumer and determine the effects on the rest of community. In some instances, native consumers can keep introduced species under control. Higher trophic level invaders tend to have more severe, broader impacts than lower trophic level invaders (Sax and Gaines, 2008).

In the absence of a predator, prey species can become more abundant. An introduced species without a natural predator in the new environment similarly can become very abundant. An introduced predator whose potential prey do not recognize it as such also has an advantage over native predators which are recognized by their prey. This chapter focuses on trophic interactions and the roles of feeding and of predator avoidance in the success of selected aquatic invasive species. It also discusses cases in which native

Biological Invasions and Animal Behaviour, eds J.S. Weis and D. Sol. Published by Cambridge University Press. © Cambridge University Press 2016.

predators can keep the introduced species in check, and how, in the absence of predators, humans can step into that role (for some species). The chapter reviews findings on animals from a variety of taxa, ecological roles, and geographic regions.

Decapod Crustaceans: Crayfish and Crabs

There are a number of documented cases of invasive species outcompeting native crustaceans for the same food, generally in laboratory studies. Gherardi *et al.* (2001) compared the predation by the invasive red swamp crayfish, *Procambarus clarkii*, on different species of tadpoles with that of a native crayfish, *Austropotamobius pallipes*, and found that the invasive crayfish caught them more rapidly and also appeared to be faster in switching to different prey than the native species (Barbaresi and Gherardi, 2000). More information on this topic is provided in Chapter 16.

Invading rusty crayfish, *Orconectes rusticus*, are replacing the native northern crayfish, *O. virilis*, as well as a previous invader, the northern clearwater crayfish (*O. propinquus*) in the mid-Western United States. Hill and Lodge (1999) investigated interspecific competition for food and non-consumptive effects of predation by largemouth bass, *Micropterus salmoides*, on growth and mortality of the three species. In competition experiments, northern crayfish had reduced growth and northern clearwater crayfish had increased mortality in the presence of invading rusty crayfish. Rusty crayfish, on the other hand, were unaffected by the presence of the other two species. It appears that competitive interactions favour the invader. Tests of non-consumptive effects of fish predation in the presence of largemouth bass showed that northern crayfish growth was considerably reduced, rusty crayfish growth declined slightly and northern clearwater crayfish growth was unchanged. Mortality of all three species increased in the presence of largemouth bass, with northern crayfish experiencing the greatest and rusty crayfish, the invasive species, the least mortality, indicating again an advantage of the invasive species.

There are not only competitive differences between invasive versus native crayfish species, there are also differences between native and introduced populations of the same species. In competition experiments in the laboratory, rusty crayfish from introduced populations were better foragers and bolder to forage under predation risk than native populations of the same species. The invasive populations were better at stealing food from native crayfish species, something that native populations did not seem to do (Pintor and Sih, 2008). Thus, it would appear that individuals who are better foragers are the ones that are most likely to survive when introduced into a new habitat; this may exert selection pressure on the newly arrived animals. Introduction into a new environment can select for certain traits. Suites of behaviours (animal personalities or 'behavioural syndromes') is a growing field of study that has major relevance to invasion success (Chapple *et al.*, 2011); other chapters in this volume focus on this topic. It is also likely, however, that natives are more often better adapted than new arrivals to their specific environment, which could prevent introduced species from becoming invasive.

Invasive species may be more effective in avoiding predators than some native species. Invasive rusty crayfish had more effective anti-predator behaviour against fish than the native northern crayfish (Garvey *et al.*, 1994). While the latter swam away and was easily captured by the fish, the former approached fish aggressively with claws extended and generally escaped predation. Gherardi *et al.* (2011) studied the invasive North American red swamp crayfish, *Procambarus clarkii*, to investigate responses to odours of fish predators compared with conspecific alarm odour. They studied introduced populations of red swamp crayfish in two sites with different fish assemblages. Although crayfish from both populations responded to conspecific alarm odour with a greater reduction in feeding than their response to predator odours, red swamp crayfish seem to perceive a general fish odour that alerts them to possible predation risk. Where they coexist with fish, they can distinguish among fish species, adjusting the intensity of their response to the risk of predation by that species. These results indicate a high capacity for learning and predator recognition in this very successful invader.

The green crab, or shore crab, native to Europe, is a very successful invader in the United States, among other places. Competition was studied between invasive green crabs, *Carcinus maenas*, and native blue crabs, *Callinectes sapidus*, in which size-matched pairs were placed in competition for a mussel. Green crabs were far better competitors and ate the food more often than the blue crabs, despite the greater aggressiveness exhibited by the latter. Typically, blue crabs placed in the experimental tank with green crabs extended their claws in an aggressive display, while the green crabs scuttled in and took the food (MacDonald *et al.*, 2007). When fights took place over the food, the blue crab was the 'loser' a disproportionate number of times, despite their well-known aggression. These results suggest that juvenile blue crabs are at a disadvantage compared with green crabs; in areas where green crabs are common, juvenile blue crabs may spend more energy in conflict and may be outcompeted for desired food. Another advantage exhibited by the green crab was superior ability to learn. Roudez *et al.* (2008) found that they were able to learn and remember the location of hidden food much more rapidly than blue crabs. This could partially contribute to the success of green crabs.

However, since blue crabs eventually become larger than green crabs, the larger blue crabs may be able to provide 'biotic resistance' to the green crab invasion. Biotic resistance is the ability of resident species in a community to reduce the success of invasions (Elton, 1958). In Chesapeake Bay (Maryland), where blue crabs are more common than further north, green crabs have not successfully expanded their range. Predation rates of large blue crabs on tethered green crabs were much higher in these areas (DeRivera *et al.*, 2005).

A competition was set up between green crabs versus juvenile American lobsters, *Homarus americanus*, in which the green crabs also prevailed. They were the first to get to the food, fed in more trials than the lobsters, and spent more time with the food than did the lobsters (Rossong *et al.*, 2006). The green crabs also captured and consumed some of the lobsters. In a shelter experiment, in which a refuge for hiding was provided and using crabs of 53–76 mm carapace width and lobsters of 28–57 mm carapace length, the green crabs captured and consumed lobsters in 6 of 11 trials. The lobsters that survived spent more time in the shelter than those that were captured. Contrary to expectations, the lobsters that were larger in relation to the green crabs were more likely to be eaten, which authors believed to be due to more conspicuous activity and less use of the shelter. The authors concluded that green crabs have the potential to negatively impact juvenile lobsters by both competition and predation.

On the West Coast of North America, the invasive green crab is a predator of the native crab *Hemigrapsus oregonensis*, which showed 5-fold to 10-fold declines within 3 years of their arrival. Field and laboratory experiments indicated that green crab predation caused these declines (Grosholz *et al.*, 2000). The green crab also preferentially preys on native clams in the genus *Nutricola* (Grosholz, 2005) along the Pacific coast. The major decline of these clams following the green crab invasion allowed an exotic clam, *Gemma gemma*, which had been present in low numbers, to increase their population and become invasive. This phenomenon was referred to as an 'invasional meltdown,' in which invasion by one exotic species facilitates subsequent invasions by other species (Ricciardi, 2001). The concept is that ecosystems become more easily invaded as the cumulative number of species introductions increases, and that facilitative interactions among invaders can exacerbate their impacts.

However, there is an example where green crab predation seems to result in a general benefit to the ecosystem. In New England salt marshes, the (native) herbivorous marsh crab, *Sesarma reticulatum*, has been released from predator control (due to overfishing), which has led to extensive herbivory on cordgrass on salt marsh creek banks, causing a marsh die-back (Altieri *et al.*, 2012). However, in some locations the cordgrass is recovering, coincident with the invasion of green crabs that occupy marsh crab burrows and consume the smaller marsh crabs (Coverdale *et al.*, 2013). Thus, the green crabs are playing a positive role in assisting the recovery of these damaged salt marshes.

Lohrer and Whitlatch (2002) found that along the Atlantic Coast of the United States, the Japanese (or Asian) shore crab, *Hemigrapsus sanguineus*, had negative effects on the prior invader, the green crab. Asian shore crabs are replacing green crabs in rocky intertidal habitats. Asian shore crabs significantly reduced the recruitment of green crabs during field experiments by 50–75%, but when Asian shore crabs were immobilized in mesh bags or had both chelae removed, they did not affect the recruitment of green crabs, possibly suggesting direct predation on the juvenile green crabs. The rising densities of Asian shore crabs threaten green crabs because Asian shore crabs prey upon young green crabs (0 year). Less effective anti-predator behaviour by green crabs was found to facilitate their replacement by Asian shore crabs along the United States Atlantic Coast. While juvenile shore crabs tend to move away from larger crabs to escape attack, juvenile green crabs tend to remain inactive and rely on protective colouration, which is less effective against crab predators that are tactile hunters (Lohrer and Whitlatch, 2002). MacDonald *et al.* (2007) compared shell strength of the carapaces of green crabs and Asian shore crabs with size-matched native blue crabs. Green crabs and Asian shore crabs had thicker, heavier carapaces that did not break as easily as those of blue crabs, implying that this morphological, rather than behavioural, defence would make them less likely to be preyed upon.

Zooplankton and Small Crustaceans

The comb jelly, *Mnemiopsis leidyi*, was introduced into the Black Sea from the Western Atlantic in the 1980s and, by 1989, the Black Sea population had reached 400 specimens per cubic metre of water. Afterwards, because they had depleted their food, the population dropped somewhat. In 1999, the species was introduced into the Caspian Sea, where they depleted 75% of the zooplankton, affecting the entire food chain (Finenko *et al.*, 2006). *M. leidyi* eats eggs and larvae of pelagic fish and in the Black Sea caused a dramatic drop in fish, including the commercially important anchovy, *Engraulis encrasicholus*, by competing for the same food sources as well as by eating anchovy young and eggs. This invasion caused cascading effects. Bottom-up effects included the collapse of planktivorous fish and their predators, dolphins in the Black Sea and Caspian seals, *Pusa caspica*, in the Caspian Sea. Top-down effects included an increase in phytoplankton due to release from grazing pressure from other zooplankton, and increasing bacterioplankton populations, triggering increases in zooflagellate populations.

Purcell *et al.* (2001) reviewed the population distributions and compared population dynamics of *M. leidyi* in United States Atlantic waters, where it is native, and in the Black Sea region and examined effects of temperature and salinity, zooplankton availability and predator abundance on population size in both regions. In both regions, *M. leidyi* populations are restricted by low winter temperatures. In Chesapeake Bay, however, they are also limited by zooplanktivorous fish that compete with the ctenophores for food. In the Black Sea, no obvious predators of *M. leidyi* were present during the decade after its introduction when its populations flourished. Zooplanktivorous fish populations had already been severely reduced by overfishing in the Black Sea. Thus, reduced populations of potential predators and competitors for food enabled the *M. leidyi* populations to grow enormously in the new habitat. In Chesapeake Bay, *M. leidyi* also consumes substantial amounts of zooplankton daily, but reduces zooplankton populations significantly only when its own predators are rare. *M. leidyi* is an important predator of fish eggs in both locations. Purcell *et al.* (2001) concluded that the enormous impact of *M. leidyi* on the Black Sea ecosystem occurred because of the shortage of predators and competitors in the late 1980s and early 1990s. Subsequent accidental introduction of another comb jelly, the large *Beroe ovata*, into the Black Sea resulted in a dramatic decline in *M. leidyi* because *B. ovata* is a predator on the smaller *M. leidyi*. Shiganova *et al.* (2001) calculated that the *B. ovata* ingested up to 10% of the *M. leidyi* population daily. A marked decrease in *M. leidyi* density was recorded along with increased abundance of zooplankton (about 5-fold) and ichthyoplankton (about 20-fold) compared with years after the *M. leidyi* invasion but before the *B. ovata* invasion. The *B. ovata* population underwent an initial explosion until the numbers of both ctenophores stabilized. Both *M. leidyi* and *B. ovata* remain today in the Black Sea. Intentional introduction of *B. ovata* has been suggested for biological control of *M. leidyi* in the Caspian Sea.

The Eurasian spiny water flea, *Bythotrephes longimanus*, was first detected in 1984 in Lake Huron (Bur *et al.*, 1986) and is now established in all the Great Lakes and many other lakes in that region of the United States. *B. longimanus* consume small zooplankton such as smaller cladocerans, copepods and rotifers, and compete directly with planktivorous larval fish for food (Berg and Garton, 1988; Evans, 1988). The species has been implicated as a factor in the decline of alewife, *Alosa pseudoharengus*, throughout the Great Lakes (Evans, 1988). *B. longimanus* not only compete with, but also prey on, other cladocerans such as *Leptodora kindtii* and may be a causal factor in its decline, as the abundances of the two species are often negatively correlated (Branstrator, 1995). The invasion severely altered planktonic food webs. A number of cladoceran species have declined dramatically since the invasion, and overall species richness has decreased as a result (Barbiero and Tuchman, 2004). Their results indicate that this predator can have pronounced effects on zooplankton community structure. Boudreau and Yan (2003) compared the summer crustacean zooplankton communities of 17 lakes invaded by *B. longimanus* and 13 non-invaded (reference) lakes in Ontario. The communities of the two lake groups differed greatly. Average species richness was 30% higher in the reference compared to the invaded lakes. Total zooplankton biomass was significantly lower in the invaded lakes, mainly because of lower abundances of all common cladoceran species.

Non-lethal negative effects of predatory invaders have also been noted. In response to *B. longimanus*, daphniids move deeper in the water column to avoid predation, but this reduces their growth rate. In the presence of *B. longimanus* kairomones, daphniids migrated vertically, occupying a middle region by night and a low, cold region during the day (Pangle and Peacor, 2006). Over a 4-day experiment, the vertical migration induced by *B. longimanus* caused a 36% reduction in the somatic growth rate of the daphniids, a level sufficient to affect their population growth rate. Concentrations of *B. longimanus* kairomones in water taken directly from the field (Lake Michigan) were high enough to induce behavioural shifts that led to these large reductions in somatic growth rate.

As Bailey *et al.* (2006) demonstrated, the invasive amphipod, *Gammarus tigrinus*, in Ireland preys upon both adults and juveniles of the native opossum shrimp *Mysis relicta*, while adults of the latter prey only upon juveniles of the former. Furthermore, the presence of *G. tigrinus* alters habitat use by *M. relicta*, which then become more vulnerable to fish predation. In the Netherlands, another freshwater amphipod invader, *Dikerogammarus villosus*, preys on both the native *Gammarus duebeni* and on *G. tigrinus*, including both intermoult and recently moulted individuals (Dick and Platvoet, 2000). *D. villosus* has been nicknamed 'killer shrimp' because of its behaviour towards other invertebrates as well as fishes (Casellato *et al.*, 2007). In laboratory studies with various invertebrates, *D. villosus* consumed far more prey and a broader range of prey than other amphipods studied. Adults consumed about one-third of their body weight per day (Krisp and Maier, 2005), suggesting that they may contribute to the decline of its invertebrate prey in some European streams. *D. villosus* also exhibits better predator-avoidance behaviour by sheltering in (invasive) zebra mussels more than other invasive amphipods, *Pontogammarus robustoides*, as well as native European amphipods, *Gammarus fossarum*. In a related study, the different amphipods were exposed to predation of two fish species: the racer goby, *Babka gymnotrachelus* (Ponto-Caspian), and Amur sleeper, *Perccottus glenii* (Eastern Asian). Kobak *et al.* (2014) tested gammarid survival in the presence of fish and five different substrata: sand, macrophytes, stones, living zebra mussels (*Dreissena polymorpha*) and empty mussel valves. *D. villosus* had the highest survival on all substrates, and its survival was highest in living mussels, whereas survival of the other gammarids was similar on all substrates. Thus, *D. villosus* was better protected from fish predators than the other two amphipods, and living zebra mussels provided them the best protection.

The invasive amphipod, *Gammarus roeseli*, which has invaded the same areas in Eastern Europe as *D. villosus*, was compared with the native *G. pulex* for susceptibility to fish predators. Brown trout, *Salmo trutta*, preyed more on *G. pulex*, although there was no significant difference observed in anti-predator behaviour between the two amphipod species. The presence of spines on *G. roeseli* appeared to be responsible for the differential predation (Bollache *et al.*, 2006), a morphological rather than behavioural trait.

Stable isotope analysis of the two invasive gammarids *D. villosus* ('killer shrimp') and *G. roeselii* was done by Rothhaupt *et al.* (2014). They exhibited no significant differences in $\delta^{13}C$ and $\delta^{15}N$, indicating a considerable overlap of the diet of these sympatric invaders. Their similarity in diet, as well as voracious behaviour, could promote their invasion success, with serious consequences for local biodiversity.

In some cases, predation by resident species may keep invaders under control. The European predatory freshwater amphipods, *Gammarus pulex* and *Gammarus duebeni celticus*, have negative effects on and restrict the abundance of their prey, the invasive North American amphipod *Crangonyx pseudogracilis* (MacNeil *et al.*, 2013).

Paterson *et al.* (2015) investigated predatory functional responses of native *G. duebeni celticus* and invasive *G. pulex* amphipods in Ireland towards three prey species (*Asellus aquaticus*, *Simulium* spp., *Baetis rhodani*) and found that the invasive amphipod had a higher predatory impact (lower handling time) on two of three prey species, which agreed with impacts observed in the field. To incorporate even more complexity and ecological realism, they also investigated effects of parasites (the acanthocephalan *Echinorhynchus truttae*) and predators (brown trout, *Salmo trutta*) on the focal amphipods' predation and found opposite effects; parasites tended to reduce the predatory impact of the invasive amphipod and increase the predatory impact of the native amphipod on the prey, in the presence of the higher-order fish predator. However, when the fish was not there, parasitism alone did not alter the functional responses of parasitized amphipods. Use of different prey species and the more complex four-species community (higher-order fish predator; focal native or invasive amphipod predators; parasites of focal predators; native prey) revealed differences in the native and invasive predator's functional responses towards different prey species, which can be used to interpret impacts observed in the field.

Molluscs

One of the first aquatic invasive species to receive a lot of attention in the United States was the zebra mussel, *Dreissena polymorpha*, that arrived in the Great Lakes in the late 1980s, probably as a result of ballast water discharge from ships from Eastern Europe. In its new location, this bivalve had no natural predators and grew very densely, completely covering hard surfaces, clogging up intake pipes for water systems, and

causing considerable economic and ecological damage. On the other hand, their extensive filter feeding reduced plankton blooms and improved water clarity. Several years
later, the round goby, *Neogobius melanostomus* (or *Apollonia melanostomus*), arrived,
probably also in ballast water. These fish, unlike any native of the Great Lakes, eat
large numbers of zebra mussels, almost 80 a day (Ray and Corkum, 1997) and can
reduce population density of the zebra mussels (Barton *et al.*, 2005). Native species
like pumpkinseed sunfish, *Lepomis gibbosus*, and rusty crayfish, *Orconectes rusticus*,
eat the mussels, but only about one-fifth of the amount consumed by the round goby
(Naddafi and Rudstamb, 2014). Zebra mussels and subsequent invasive quagga mussels, *D. rostriformis*, exhibit anti-predator defences by increasing attachment strength,
increasing aggregation rates and use of refuges, and decreasing clearance rate. Zebra
mussels responded more strongly than the quagga mussels (Naddafi and Rudstamb,
2013). These anti-predator behaviours probably initially evolved in their native areas
that they share with round gobies.

In the late 1980s, the Asian clam, *Potamocorbula amurensis*, became established
in San Francisco Bay, replacing the native clam, *Macoma balthica*, as the dominant
benthic macroinvertebrate. It is a filter feeder, and its rapid spread and high density
resulted in reduced phytoplankton in the bay, which meant less food for zooplankton
and fish (Murrell and Hollibaugh, 1998). This clam has altered nutrient cycling in the
bay and caused major ecosystem changes. The invasive clam is eaten by diving ducks,
and since it lives closer to the surface of the mud than native clams, it is more available
to the ducks (Richman and Loworn, 2004). However, shells of Asian clams are thicker
and much harder to crush than shells of native clams, which could reduce predation on
them.

Veined rapa whelks, *Rapa venosa*, are carnivorous gastropods whose main diet consists of bivalves that they smother by wrapping around the hinged region of the shell;
they then feed between the opened valves. Native to the western Pacific, this species
favours compact sandy bottoms where it can readily burrow, and it tolerates low salinities, water pollution and low oxygen (Mann and Harding, 2003). Rapa whelks were
introduced into the Black Sea in the 1940s and spread along the Caucasian and Crimean
coasts and to the Sea of Azov, and the coastlines of Romania, Bulgaria and Turkey. It has
become established in the northern Adriatic and Aegean Seas. It was found in the United
States in Chesapeake Bay in 1998. It grows to over 150 mm shell length (SL) and has
a thick shell. Because of their predatory ability and lack of significant predators, they
have caused major changes in the benthic ecology in the Black Sea. Scientists are concerned about its potential impacts on Chesapeake Bay. However, juvenile rapa whelks
in Chesapeake Bay may be kept in check by predation by the blue crab, *C. sapidus*.
Feeding experiments were performed by Harding (2003) using three size classes of blue
crabs and a size range of rapa whelks. Blue crabs of all sizes consumed Age 1 whelks;
Age 2 whelks were consumed by medium and large crabs but not small crabs. The attack
methods of medium and large crabs changed with whelk age and size. Rapa whelks less
than approximately 35 mm SL were vulnerable to predation by all sizes of blue crabs,
which may be one reason that the whelk does not appear to be doing great damage in
the Chesapeake.

Fishes

Round gobies, *Neogobius melanostomus*, while able to reduce numbers of zebra mussels, also can displace native bottom-dwelling fish. Major reductions in populations of sculpins (*Cottus bairdi* and *C. cognatus*) occurred in areas where gobies became established because gobies compete with them for food or drive them from their habitat and spawning areas (Dubs and Corkum, 1996). Round gobies also eat darters and other small fishes and feed on eggs and young of lake trout (*Salvelinus namaycush*), which had already declined in the Great Lakes before the round goby arrived (Steinhart *et al.*, 2004).

In addition to eating native fish eggs, round gobies outcompete juvenile sport fish like walleye (*Sander vitreus*), yellow perch (*Perca flavescens*) and smallmouth bass (*Micropterus dolomieu*) for habitat. They also outcompete adult darters and sculpin, and prey on juveniles of these species, whose populations have decreased since gobies arrived. They outcompete native fish for food due partially to an ability to feed in darkness and to the presence of a ventral suction disc by which they can attach to rocks/substrates and stay on the bottom even in fast currents. However, smallmouth bass that survive long enough to reach about 5 cm in length can begin to eat gobies, which make up as much as 85% of an adult's diet. Smallmouth bass that eat gobies grow twice as fast as those that do not eat gobies (Winslow, 2010). But if competition with round gobies slows juvenile smallmouth bass growth, then the population may decline. Winslow (2010) found that the presence of gobies causes bottom-dwelling juvenile smallmouth bass to rise up from the bottom into the water column to find food. This not only forces them to eat tiny zooplankton (rather than the larger bottom-dwelling macroinvertebrates), but it also makes them more vulnerable to predators.

By changing food webs, round gobies have increased the potential of contaminants to reach humans. Zebra mussels accumulate pollutants, which can be transferred to gobies that eat them (Hogan *et al.*, 2007). This is of concern because gobies are eaten by larger smallmouth (*Micropterus dolomieu*) and rock bass (*Ambloplites rupestris*), walleyes (*Sander vitreus*), yellow perch (*Perca flavescens*) and brown trout (*Salmo trutta*), which accumulate the toxicants and transfer them to human consumers.

Invasive fishes in the Pacific Northwest pose a threat to native salmonids by predation. Sanderson *et al.* (2009) identified a number of non-indigenous freshwater fishes – channel catfish (*Ictalurus punctatus*), black and white crappie (*Pomoxis nigromaculatus* and *P. annularis*), largemouth bass (*Micropterus salmoides*), smallmouth bass (*Micropterus dolomieu*), walleye (*Sander vitreus*), and yellow perch (*Perca flavescens*) – that were consuming hundreds of thousands to millions of endangered juvenile salmonids, *Oncorhynchus* spp., at just a few sites. Salmon constituted a large fraction of the diet of some of these invasive species. This predation by invasive species appears to be a major source of salmon mortality, comparable to well-known threats such as fishing, hatcheries, dams and habitat alteration.

Salmonids themselves can be invasive species, due to authorized introductions of large numbers for recreational fishing, escapes from fish culture and illegal introductions by anglers. The rainbow trout, *Oncorhynchus mykiss*, has been introduced for food or sport in North America, Africa, Japan, Southeast Asia, most of South America and Central America, Australia and New Zealand, Europe and Hawaii. It has spread throughout the world and is included among the world's worst invasive species (Fausch, 2007). Invasive salmonids can have both direct effects on native salmonids and other fishes through biological interactions (i.e. predation, competition), and indirect effects by fragmenting their habitat and isolating native populations. In addition, they can alter aquatic food webs through cascading effects by displacing native trout and altering the invertebrate community. They are also a threat to frogs, which are already in decline throughout the world. Introduced trout in California have diminished the distribution and abundance of a native ranid frog, *Rana* (=*Lithobates*) *cascadae*, primarily via predation on frog larvae. Moreover, trout also feed on larval aquatic insects and compete for food with the declining native amphibian (Joseph *et al.*, 2011). Their impact on native fishes seems to be greatest in places that never had trout (e.g. Australia and New Zealand). In Southern Europe, Australia and South America, they have affected native freshwater fishes by eating them, outcompeting, transmitting diseases or hybridizing with closely related species (Crowl *et al.*, 1992). Hitt *et al.* (2003) found much hybridization between native west slope cutthroat trout, *Oncorhynchus clarki lewisi*, and non-native rainbow trout, *O. mykiss*, in streams of the Flathead River system in Montana. Seiler and Keeley (2007) found that the introduced rainbow trout and their hybrids with cutthroat trout have higher sustained swimming stamina than native cutthroat trout, and thus are likely to be better predators and better at avoiding predation than the native species.

Rainbow (*Oncorhynchus mykiss*) and brown trout (*Salmo trutta*) are widespread invasive salmonids with important effects as predators. Correa *et al.* (2012) investigated how density of invasive trout in 25 Patagonian lakes alters the trophic niche of a widespread native fish, *Galaxias platei*. The trophic height (level in the lake where they fed) of large *G. platei* declined with increasing trout density and the trophic height of large *S. trutta* decreased with decreasing *G. platei* density. Authors considered this reflected a depletion of galaxiid prey for both piscivorous *G. platei* and *S. trutta* in lakes with high trout density, and supported a management strategy of culling trout from lakes, which would both protect native fish and enhance a sport fishery for large trout. Young *et al.* (2010) compared effects of rainbow trout and brown trout in Chilean Patagonia and found that the rainbow trout colonized more streams and had a wider geographic range than brown trout. However, they had dramatically different effects on the native drift-feeding galaxiid *Aplochiton* spp., which were virtually absent from streams that had been invaded by brown trout but shared a broad sympatric range with rainbow trout. In New Zealand, Elkins and Grossman (2014) found that the presence of invasive rainbow trout produced a habitat shift by native warpaint shiners, *Luxilus coccogenis*, from pool habitats to shallower, higher velocity habitats with more variable substrata. The presence of rainbow trout also reduced prey capture success and feeding efficiency of the shiners.

Baxter *et al.* (2004) found complex effects of rainbow trout on both aquatic and nearby terrestrial food webs in Northern Japan. Rainbow trout ate terrestrial insects that fell into the stream, causing native Dolly Varden char, *Salvelinus malma*, to switch to forage on insects that graze algae from the stream bottom, which decreased the biomass of insects emerging from the stream to the forest. This, in turn, caused a major reduction in riparian-specialist spiders in the forest. Thus, this invasion altered predator–prey interactions and interrupted flows of resources between adjacent aquatic and terrestrial ecosystems. Such effects would be difficult to understand without in-depth knowledge of food web relationships.

Red lionfish, *Pterois volitans*, a venomous species from the western Pacific Ocean, appeared in the Atlantic off the coast of Florida around 1990 and have spread throughout the Caribbean, the Gulf of Mexico and the Southwest Atlantic. They were probably introduced by aquarium releases, perhaps also escaping from Florida aquaria during flooding from Hurricane Andrew in 1992. They are found from shallow mangrove areas to about 300 m deep. A tropical species, they tolerate temperatures as low as 16°C, so they can probably survive year-round as far north as North Carolina and have been found further north along the mid-Atlantic Coast. They compete with local fish for space and food (including the smaller fish that they eat). They are solitary ambush predators that use their fan-like pectoral fins to herd or corner prey against corals or ledges. They consume juveniles of many important commercial species, like snappers, groupers and shrimp. Albins and Hixon (2008) found lionfish caused significant reductions in recruitment of native fishes by an average of 79% over a 5-week experiment. This strong effect on a key life stage of coral reef fishes suggests that lionfish are already having substantial negative impacts on Atlantic coral reefs. Morris and Akins (2009) found that 21 families and 41 species of teleosts were represented in the lionfish diet in the Bahamas. A lionfish's stomach can expand to 30 times its normal size. This ability and its voracious appetite make it an effective predator. Côté and Maljkovi (2010) calculated rates of lionfish predation from field observations on Bahaman reefs. They consumed fish at a rate of over one kill per hour on clear days and over two kills per hour on overcast days, higher than its predation rate in its native range, probably due to naïveté of local prey species. Barbour *et al.* (2010) noted that in the Western Atlantic it occurs at higher densities and forages more successfully than in its native range. They compared its population size-structure and stomach contents in mangroves versus coral reefs in the Bahamas and found that lionfish stomachs contained similar prey with similar weight, demonstrating that they colonize and find equivalent food in mangrove habitats and coral reefs. (Personal note: Snorkelling in shallow water in the Turks and Caicos islands in the winter of 2012, the author observed more lionfish than all other fish combined.)

Amphibians

While, as a group, amphibians are on the decline and of great concern to conservationists, at least one invasive amphibian has become a major problem – the cane toad, *Rhinella marina*, in Australia. Trophic relations and negative impacts of this species on native biota are not as much on its prey, but on its predators. Because the cane toad is toxic, it has caused widespread mortality among its vertebrate predators. Mass mortality has been noted in native tadpoles of various species after consuming cane toad eggs (Crossland *et al.*, 2008). Cane toad larvae themselves are cannibals on conspecific eggs. While resistant to the toxin, they benefit by both nutrition and by reduction in competition from younger tadpoles (Crossland *et al.*, 2011). Mass mortality was seen in freshwater crocodiles coincident with the invasion of cane toads, causing the population density of the crocodiles, *Crocodylus johnstoni*, to decrease by as much as 77% following the invasion (Letnic *et al.*, 2008). Invertebrate aquatic predators, in contrast, such as coleopterans and hemipterans, appear to be resistant to the toxins and can consume cane toad tadpoles and reduce recruitment success in some water bodies (Cabrera-Guzman *et al.*, 2012). Predatory ants, *Iridomyrmex reurrus*, kill and consume large numbers of newly metamorphosed toads, which are more vulnerable to ant predators than are native frogs, due primarily to reduced predator detection and evasion, and slower locomotion by the cane toads (Ward-Fear *et al.*, 2009).

The vertebrate predators, however, may learn that the toad is toxic and learn to avoid it. Naïve individuals of the marsupial *Planigale maculata* in the Northern Territory would seize the first toad offered, but the majority did not repeat the attack on subsequent toads. On the other hand, planigales from Queensland, where the toads invaded 60 years earlier, have adapted to survive ingestion of toads, apparently by becoming resistant to the toxins over 60 generations. These marsupials all survived ingestion of toads and continued to attack them (Llewelyn *et al.*, 2010). This type of adaptation is, of course, more likely to reduce invasion success, since the toads are being consumed rather than avoided. Webb *et al.* (2008) found that after exposure to toads, planigales learned rapidly to avoid the toxic toads. However, they also avoided native frogs, which is a benefit to native fauna. Another carnivorous marsupial, the northern quoll, *Dasyurus hallucatus*, became locally extinct across northern Australia after the cane toad invasion. O'Donnell *et al.* (2010) investigated whether conditioned taste aversion could be used to mitigate toad impacts on the quolls in other parts of the country. They found that quolls that had been conditioned in the lab were much less likely to attack cane toads after release, and recommended that managers deploy taste-aversion baits in the field ahead of the invasion front as a way to improve survival of these endangered marsupials.

Raptors (black kites, *Milvus migrans*, and whistling kites, *Haliastur sphenurus*) consumed road-killed cane toads, although they preferred native frogs. They appeared to recognize and avoid the toxins by consuming only the toads' tongues (Beckmann and Shine 2011). Freshwater crocodiles, *Crocodylus johnstoni*, in the laboratory learned to avoid toads as prey after initially attacking them (Somaweera *et al.*, 2011). Even Australian marbled frogs, *Limnodynastes convexiusculus*, exhibit rapid avoidance learning through taste aversion. This learning allows them to persist in the presence of the toxic invader (Greenlees *et al.*, 2010) but also facilitates the invasion.

Conclusions

This chapter has shown that in freshwater and marine environments, many invasive species from a variety of taxa can succeed because of predator–prey interactions. They may be able to outcompete native species for food (such as red swamp and rusty crayfish, green crab, spiny waterflea and round goby). They may prey extensively on native species of concern and thus alter communities and food webs (such as spiny water flea, killer shrimp, Asian clam, round goby, rainbow trout and lionfish). They may be better equipped to escape predation by native species (such as rusty crayfish, red swamp crayfish and killer shrimp). Success in avoiding predation may be because predators do not recognize them as food, or because the invasive species has better sensory detection of predators and/or quicker escape responses. In the unusual case of the cane toad, predation can be fatal to the predator since the toad is toxic. Hosts learn to avoid eating the toads, whose populations can grow more rapidly due to reduced predation. In some cases, native predators can keep invasive species in check (biotic resistance) such as blue crab keeping green crab and rapa whelk in check in Chesapeake Bay. We have discussed how behavioural syndromes, or animal personality can affect the success of an invasion, and how complex trophic interactions and food webs can be disrupted by cascading top-down and bottom-up effects of feeding and/or predator avoidance of an invasive species.

Future studies of invaders competing with native species should include video analysis to investigate what specific behaviours allow one species to outcompete the other. Future studies of avoiding predation should also include analysis of videos to learn what aspects of predator avoidance are involved – is it quicker predator detection or quicker escape? Laboratory experiments on olfactory and visual systems could shed light on which sensory system is mainly involved in predator detection. It would also be of interest to investigate whether juvenile stages of invaders are similar to adults in terms of advantage over native species as predators or as prey. More studies such as Paterson *et al.* (2015) that include different prey as well as parasites and predators of the focal species and thus more closely resemble the complexity of nature, should be performed. Another field that should receive more attention is the learning ability of the invasive versus native species, and learning ability/speed of native species to respond appropriately to the new arrival as potential prey or predator. This also could play a significant role in invader success. Undoubtedly, in the future, studies of animal personality (discussed in greater detail in Chapters 2 and 7) will become much more frequent and we will learn of its involvement in many more invasive species interactions.

A Management Option: Eating Invasive Species

Top-down effects of predators can limit success of introduced species. The idea of humans eating invaders when other predators do not do so has been suggested as a way to control the spread of some species, although it probably cannot eliminate them entirely. Direct harvest of abundant invaders can reduce their numbers, and therefore their ecological and economic impacts. Pasco and Goldberg (2014) reviewed the field and concluded that, if done carefully, harvesting and marketing of invasive species represents a significant opportunity to improve ecosystems and provide economic incentives to local populations. They acknowledged potential problems that could arise and offered recommendations on how to implement such programmes properly and safely.

This idea is a controversial one. Those opposed to it argue that creation of a market for an invasive species provides an incentive to maintain the species in order to sustain the market, or even to spread the invader in order to improve profits (Nunez *et al.*, 2012). Illegally introducing fish in order to create a sport fishery has been done often. A number of fishes and crayfishes became invasive because of this practice.

This approach has been widely adopted for lionfish in the Atlantic and Caribbean. Efforts are underway to educate anglers about how to catch, clean, and eat them safely. Research indicates that targeted derbies and spearfishing contests can effectively reduce lionfish populations in local areas. Green *et al.* (2014) found that reducing lionfish below predicted threshold densities prevented native fish biomass from declining. Reductions in lionfish density of 25–92%, depending on the reef, were needed to keep lionfish below levels predicted to overconsume prey. On reefs where lionfish were kept below threshold densities, native prey fish biomass increased by 50–70%.

Pienkowski *et al.* (2015) reported that invasive Australian red claw crayfish, *Cherax quadricarinatus*, is an important source of income for fishermen in parts of Jamaica, who supplement their incomes during times when shrimp, *Macrobrachium* spp., catches decline. Full-time fishermen and those with no other occupations expend the greatest fishing effort; the least wealthy appear to be the most dependent on fishing, and consequently benefit the most from the invasive crayfish.

Asian carp are voracious eaters that can grow up to 1.8 m and 50 kg. These species were first imported and used in Louisiana catfish farms in the 1970s to control snails and vegetation. In the mid-1990s, flooding allowed them to escape from fish farms. They have spread to most of the Mississippi River watershed and the Missouri River, devastating food resources and habitats of native and sport fish. The bighead carp, *Hypophthalmichthys nobilis*, and silver carp, *H. molitrix*, now account for the majority of fish in the Missouri River. They are threatening to invade the Great Lakes.

An Asian Carp Marketing Summit in Illinois was convened in the fall of 2010 to identify opportunities to market the species as a way to reduce their numbers and recommended that eating the carp may be feasible. Representatives from restaurants, commercial fishing, processing businesses, government agencies and academic institutions agreed that Asian carp fillets marketed to restaurants and retailers could provide a financial incentive for extensive harvesting of these fish. Because they are filter feeders, they are generally low in contaminants, making them healthy to eat. Summit attendees also recommended that large numbers of harvested fish be exported to Asian markets, where these species are popular food fish, and that by-products be converted into pet food to eliminate waste and maximize efficiency and profit. However, the United States Fish and Wildlife Service published a rule that prohibits the importation and transportation of bighead carp. The efforts recommended by the marketing summit would appear to be an infraction of the rule prohibiting transportation of the species.

Chefs in New England are considering ways to cook and market green crabs, including broth and soft-shell crabs (Warner, 2015).

Other species that potentially could be included in this top-down management approach ('if you can't beat 'em, eat 'em') are other fishes and crabs, as well as mussels and whelks that are tasty seafood. There are, clearly, limits to what can be accomplished by eating invasive species. Eating them will not eradicate them, but will lessen their impacts and at the same time provide a new food source for this hungry world, in which many traditional seafood species are in short supply because of overfishing, habitat loss and pollution.

References

Albins, M.A. and Hixon, M.A. (2008). Invasive Indo-Pacific lionfish *Pterois volitans* reduce recruitment of Atlantic coral-reef fishes. *Marine Ecology Progress Series*, 367, 233–238.

Altieri, A.H., Bertness, M.D., Coverdale, T.C., Herrmann, N.C. and Angelini, C. (2012). A trophic cascade triggers collapse of a salt-marsh ecosystem with intensive recreational fishing. *Ecology*, 93, 1402–1410.

Bailey, R.J., Dick, J.T., Elwood, R.W. and MacNeil, C. (2006). Predatory interactions between the invasive amphipod *Gammarus tigrinus* and the native opossum shrimp, *Mysis relicta*. *Journal of North American Benthological Society*, 25, 393–405.

Barbaresi, S. and Gherardi, F. (2000). The invasion of the alien crayfish *Procambarus clarkii* in Europe, with particular reference to Italy. *Biological Invasions*, 2, 259–264.

Barbiero, R.P. and Tuchman, M.L. (2004). Changes in the crustacean communities of Lakes Michigan, Huron, and Erie following the invasion of the predatory cladoceran *Bythotrephes longimanus*. *Canadian Journal of Fisheries and Aquatic Sciences*, 61, 2111–2125.

Barbour, A.B., Montgomery, M.L., Adamson, A.A., Diaz-Ferguson, E. and Silliman B.R. (2010). Mangrove use by the invasive lionfish *Pterois volitans*. *Marine Ecology Progress Series*, 401, 291–294.

Barton, D., Johnson, R.A., Campbell, L., Petruniak, J. and Patterson, M. (2005). Effects of round gobies (*Neogobius melanostomus*) on dreissenid mussels and other invertebrates in Eastern Lake Erie, 2002–2004. *Journal of Great Lakes Research*, 31 Suppl. 2, 252–261.

Baum, J.K. and Worm, B. (2009). Cascading top-down effects of changing oceanic predator abundances. *Journal of Animal Ecology*, 78, 699–714.

Baxter, C.V., Fausch, K.D., Murakami, M. and Chapman, P.L. (2004). Fish invasion restructures stream and forest food webs by interrupting reciprocal prey subsidies. *Ecology*, 85, 2656–2663.

Beckman, C. and Shine, R. (2011). Toad's tongue for breakfast: exploitation of a novel prey type, the invasive cane toad, by scavenging raptors in tropical Australia. *Biological Invasions*, 13, 1447–1455.

Berg, D.J. and Garton, D.W. (1988). Seasonal abundance of the exotic predatory Cladoceran, *Bythotrephes cederstroemi*, in Western Lake Erie. *Journal of Great Lakes Research*, 14, 479–488.

Bollache, L., Kaldonski, N., Troussard, J.P., Lagrue, C. and Rigaud, T. (2006). Spines and behaviour as defences against fish predators in an invasive freshwater amphipod. *Animal Behaviour*, 72, 627–633.

Boudreau, S.A. and Yan, N.D. (2003). The differing crustacean zooplankton communities of Canadian Shield lakes with and without the nonindigenous zooplanktivore *Bythotrephes longimanus*. *Canadian Journal of Fisheries and Aquatic Sciences*, 60, 1307–1313.

Branstrator, D. (1995). Ecological interactions between *Bythotrephes cederstroemi* and *Leptodora kindtii* and the implications for species replacement in Lake Michigan. *Journal of Great Lakes Research*, 21, 670–679.

Bur, M., Klarer, D.M. and Kriege, K.A. (1986). First records of a European Cladoceran, *Bythotrephes cederstroemi*, in Lakes Erie and Huron. *Journal of Great Lakes Research*, 12, 144–146.

Cabrera-Guzman, E., Crossland, M. and Shine, R. (2012). Predation on the eggs and larvae of invasive cane toads (*Rhinella marina*) by native aquatic invertebrates in tropical Australia. *Biological Conservation*, 153, 109.

Carlton, J.T. (2001). *Introduced Species in US Coastal Waters: Environmental Impacts and Management Priorities*. Arlington, VA: Pew Oceans Commission.

Casellato, S., Visentin, A. and Piana, G. (2007). The predatory impact of *Dikerogammarus villosus* on fish. In *Biological Invaders in Inland Waters: Profiles, Distribution and Threats. Invading Nature – Springer Series in Invasion Ecology*, Volume 2, Part 5, ed. Gherardi, F. Berlin: Springer, pp. 495–506.

Chapple, D.G., Simmonds, S.M. and Wong, B.B. (2011). Can behavioral and personality traits influence the success of unintentional species introductions? *Trends in Ecology and Evolution*, 27, 57–64.

Correa, C., Bravo, A.P. and Hendry, A.P. (2012). Reciprocal trophic niche shifts in native and invasive fish: salmonids and galaxiids in Patagonian lakes. *Freshwater Biology*, 57, 1769–1781.

Côté, I.M. and Maljkovi, A. (2010). Predation rates of Indo-Pacific lionfish on Bahamian coral reefs. *Marine Ecology Progress Series*, 404, 219–225.

Coverdale, T.C., Axelman, E.E., Brisson, C.P., *et al.* (2013). New England salt marsh recovery: opportunistic colonization of an invasive species and its nonconsumptive effects. *PLoS ONE*, 8(8): e73823.

Crossland, M.R., Brown, G.P., Anstis, M., Shilton, C.M. and Shine, R. (2008). Mass mortality of native anuran tadpoles in tropical Australia due to invasive cane toad (*Bufo marinus*). *Biological Conservation*, 141, 2387–2394.

Crossland, M.R., Hearndon, M.N., Pizzatto, L., Alford, R.A. and Shine, R. (2011). Why be a cannibal? The benefits to cane toad, *Rhinella marina* (=*Bufo marinus*) tadpoles of consuming conspecific eggs. *Animal Behaviour*, 82, 775–782.

Crowl, T.A., Townsend, C.R. and McIntosh, A.R. (1992). The impact of introduced brown and rainbow trout on native fish: the case of Australasia. *Reviews of Fish Biology and Fisheries*, 2, 217–241.

deRivera, C.E., Ruiz, G.M., Hines, A.H. and Jivoff, P. (2005). Biotic resistance to invasion: native predator limits abundance and distribution of an introduced crab. *Ecology*, 86, 3364–3376.

Dick, J.T. and Platvoet, D. (2000). Invading predatory crustacean *Dikerogamarus villosus* eliminates both native and exotic species. *Proceedings of the Royal Society*, 267, 977–983.

Dubs, D. and Corkum, L. (1996). Behavioral interactions between round gobies (*Neogobius melanstomus*) and mottled sculpins (*Cottus bairdi*). *Journal of Great Lakes Research*, 22, 838–844.

Elkins, D. and Grossman, G. (2014). Invasive rainbow trout affect habitat use, feeding efficiency, and spatial organization of warpaint shiners. *Biological Invasions*, 16, 919–933.

Elton, C.S. (1958). *The Ecology of Invasions by Animals and Plants*. London: Methuen.

Evans, M.S. (1988). *Bythotrephes cederstroemi*: its new appearance in Lake Michigan. *Journal of Great Lakes Research*, 14, 234–240.

Fausch, K.D. (2007). Introduction, establishment and effects of non-native salmonids: Considering the risk of rainbow trout invasion in the United Kingdom. *Journal of Fish Biology*, 71, 1–32.

Finenko, G., Romanova, Z.A., Abolmasova, G.I., *et al.* (2006). Ctenophores-invaders and their role in the trophic dynamics of the planktonic community in the coastal regions off the Crimean coasts of the Black Sea (Sevastopol Bay). *Okeanologiya*, 46, 507–517.

Garvey, J.E., Stein, R.A. and Thomas, H.M. (1994). Assessing how fish predation and interspecific prey competition influence a crayfish assemblage. *Ecology*, 75, 532–547.

Gherardi, F., Renai, B. and Corti, C. (2001). Crayfish predation on tadpoles: a comparison between a native (*Austropotamobius pallipes*) and an alien species (*Procambarus clarkii*). *Bulletin Francais de Pêche Pisciculture*, 361, 659–668.

Gherardi, F., Mavuti, K.M., Pacini, N., Tricarico, E. and Harper, D.M. (2011). The smell of danger: chemical recognition of fish predators by the invasive crayfish *Procambarus clarkii*. *Freshwater Biology*, 56, 1567–1578.

Green, S.J., Dulvy, N.K., Brooks, A.M., *et al.* (2014). Linking removal targets to the ecological effects of invaders: a predictive model and field test. *Ecological Applications*, 24, 1311–1322.

Greenlees, M.J., Phillips, B.L. and Shine, R. (2010). Adjusting to a toxic invader: native Australian frogs learn not to prey on cane toads. *Behavioral Ecology*, 21, 966–971.

Grosholz, E.D. (2005). Recent biological invasions may hasten invasional meltdown by accelerating historical introductions. *Proceedings of the National Academy of Sciences, USA*, 102, 1088–1091.

Harding, J.M. (2003). Predation by blue crabs, *Callinectes sapidus*, on rapa whelks, *Rapana venosa*: possible natural controls for an invasive species? *Journal of Experimental Marine Biology and Ecology*, 297, 161–177.

Hill, A.M. and Lodge, D. (1999). Replacement of resident crayfishes by an exotic crayfish: the roles of competition and predation. *Ecological Applications*, 9, 678–690.

Hitt, N.P., Frissell, C.A., Muhlfeld, C.C. and Allendorf, F.W. (2003). Spread of hybridization between native westslope cutthroat trout, *Oncorhynchus clarki lewisi*, and nonnative rainbow trout, *Oncorhynchus mykiss*. *Canadian Journal of Fisheries and Aquatic Sciences*, 60, 1440–1451.

Hogan, L.S., Marschall, E., Folt, C. and Stein, R.A. (2007). How non-native species in Lake Erie influence trophic transfer of mercury and lead to top predators. *Journal of Great Lakes Research*, 33, 46–61.

Joseph, M.B., Piovia-Scott, J., Lawler, S.P. and Pope, K.L. (2011). Indirect effects of introduced trout on Cascades frogs (*Rana cascadae*) via shared aquatic prey. *Freshwater Biology*, 56, 828–838.

Kobak, J., Jermacz, Ł. and Płąchocki, D. (2014). Effectiveness of zebra mussels to act as shelters from fish predators differs between native and invasive amphipod prey. *Aquatic Ecology*, 48, 397–408.

Krisp, H. and Maier, G. (2005). Consumption of macroinvertebrates by invasive and native gammarids: a comparison. *Journal of Limnology*, 64, 55–59.

Letnic, M., Webb, J.K. and Shine, R. (2008). Invasive cane toads (*Bufo marinus*) cause mass mortality of freshwater crocodiles (*Crocodylus johnstoni*) in tropical Australia. *Biological Conservation*, 141, 1773–1782.

Llewelyn, J., Webb, J.K., Schwartzkopf, L., Alford, R. and Shine, R. (2010). Behavioural responses of carnivorous marsupials (*Planigale maculata*) to toxic invasive cane toads (*Bufo marinus*). *Austral Ecology*, 35, 560–567.

Lohrer, A.M. and Whitlatch, R.B. (2002). Interactions among aliens: apparent replacement of one exotic species by another. *Ecology*, 83, 710–732.

MacDonald, J.A., Roudez, R., Glover, T. and Weis, J.S. (2007). The invasive green crab and Japanese shore crab: behavioral interactions with a native crab species, the blue crab. *Biological Invasions*, 9, 837–848.

MacNeil, C., Dick, J., Alexander, M.E., Dodd, J.A. and Ricciardi, A. (2013). Predators vs. alien: differential biotic resistance to an invasive species by two resident predators. *Neobiota*, 19, 1–19.

Mann, R. and Harding, J.M. (2003). Salinity tolerance of larval *Rapana venosa*: Implications for dispersal and establishment of an invading predatory gastropod on the North American Atlantic Coast. *Biological Bulletin*, 204, 96–103.

Morris, J.A. and Akins, J.L. (2009). Feeding ecology of the invasive lionfish (*Pterois volitans*) in the Bahaman archipelago. *Environmental Biology of Fishes*, 86, 389–398.

Murrell, M.C. and Hollibaugh, J.T. (1998). Microzooplankton grazing in northern San Francisco Bay measured by the dilution method. *Aquatic Microbial Ecology*, 15, 53–63.

Naddafi, R. and Rudstam, L.G. (2013). Predator-induced behavioural defences in two competitive invasive species: the zebra mussel and the quagga mussel. *Animal Behaviour*, 86, 1275–1284.

Naddafi, R. and Rudstam, L.G. (2014). Predation on invasive zebra mussel, *Dreissena polymorpha*, by pumpkinseed sunfish, rusty crayfish, and round goby. *Hydrobiologia*, 721, 107–115.

Nunez, M.A., Kuebbing, S., Dimarco, R.D. and Simberloff, D. (2012). Invasive species: to eat or not to eat, that is the question. *Conservation Letters*, 5, 334–341.

O'Donnell, S., Webb, J.K. and Shine, R. (2010). Conditioned taste aversion enhances the survival of an endangered predator imperiled by a toxic invader. *Journal of Applied Ecology*, 47, 558–565.

Paine, R.T. (1966). Food web complexity and species diversity. *American Naturalist*, 100, 65–73.

Pangle, K.L. and Peacor, S.D. (2006). Non-lethal effect of the invasive predator *Bythotrephes longimanus* on *Daphnia mendotae*. *Freshwater Biology*, 51, 1070–1078.

Pasco, S. and Goldberg, J. (2014). Review of harvest incentives to control invasive species. *Management of Biological Invasions*, 5, 263–277.

Paterson, R.A., Dick, J.T., Pritchard, D.W., *et al.* (2015). Predicting invasive species impacts: a community module functional response approach reveals context dependencies. *Journal of Animal Ecology*, 84, 453–463.

Pienkowski, T., Williams, S., McLaren, K., Wilson, B. and Hockley, N. (2015). Alien invasions and livelihoods: Economic benefits of invasive Australian Red Claw crayfish in Jamaica. *Ecological Economics*, 112, 68–77.

Pintor, L.M. and Sih, A. (2008). Differences in growth and foraging behavior of native and introduced populations of an invasive crayfish. *Biological Invasions*, 11, 1895–1902.

Purcell, J.E., Shiganova, T.A., Decker, M.B. and Houde, E.D. (2001). The ctenophore *Mnemiopsis* in native and exotic habitats: US estuaries versus the Black Sea basin. *Hydrobiologia*, 451, 145–176.

Ray, W.J. and Corkum, L.D. (1997). Predation of zebra mussels by round gobies, *Neogobius melanostomus*. *Environmental Biology of Fishes*, 50, 267–273.

Ricciardi, A. (2001). Facilitative interactions among aquatic invaders: is an 'invasional meltdown' occurring in the Great Lakes? *Canadian Journal of Fisheries and Aquatic Sciences*, 58, 2513–2525.

Richman, S.E. and Loworn, J.R. (2004). Relative foraging value to lesser scaup ducks of native and exotic clams from San Francisco Bay. *Ecological Applications*, 14, 1217–1231.

Rossong, M.A., Williams, P.J., Coneau, M., Mitchell, S.C. and Apaloo, J. (2006). Agonistic interactions between the invasive green crab, *Carcinus maenas* (Linnaeus) and juvenile American lobster *Homarus americanus* (Milne Edwards). *Journal of Experimental Marine Biology and Ecology*, 329, 281–288.

Rothhaupt, K.-O., Hanselmann, A.J. and Yohannes, E. (2014). Niche differentiation between sympatric alien aquatic crustaceans: an isotopic evidence. *Basic and Applied Ecology*, 15, 453–463.

Roudez, R., Glover, T. and Weis, J.S. (2008). Learning in an invasive and a native predatory crab. *Biological Invasions*, 10, 1191–1196.

Sanderson, B.L., Barnas, K.A. and Rub, A.W. (2009). Nonindigenous species of the Pacific Northwest: an overlooked risk to endangered salmon? *Bioscience*, 59, 245–256.

Sax, D. and Gaines, S. (2008). Species invasions and extinction: the future of native biodiversity on islands. *Proceedings of the National Academy of Sciences*, 105, 11490–11497.

Seiler, S.M. and Keeley, E.R. (2007). Morphological and swimming stamina differences between Yellowstone cutthroat trout (*Oncorhynchus clarkii bouvieri*), rainbow trout (*Oncorhynchus mykiss*), and their hybrids. *Canadian Journal of Fisheries and Aquatic Sciences*, 64, 127–135.

Shiganova, T.A., Bulgakova, Y.V., Volovik, S.P., Mirzoyan, Z.A. and Dudkin, S.L. (2001). The new invader *Beroe ovata* Mayer 1912 and its effect on the ecosystem in the northeastern Black Sea. *Hydrobiologia*, 451, 187–197.

Somaweera, R., Webb, J.K., Brown, G.P. and Shine, R. (2011). Hatchling Australian freshwater crocodiles rapidly learn to avoid toxic invasive cane toads. *Behaviour*, 148, 501–517.

Steinhart, G., Marschall, E.A. and Stein, R.A. (2004). Round goby predation on smallmouth bass offspring in nests during simulated catch-and-release angling. *Transactions of the American Fisheries Society*, 133, 121–131.

Strong, D.R. (1992). Are trophic cascades all wet? Differentiation and donor control in a speciose system. *Ecology*, 73, 747–754.

Ward-Fear, G., Brown, G.P., Greenlees, M.J. and Shine, R. (2009). Maladaptive traits in invasive species: in Australia, cane toads are more vulnerable to predatory ants than are native frogs. *Functional Ecology* 23:559–568.

Warner, R. (2015). Green crabs are multiplying. Should we eat the enemy? *Boston Globe*, 12 February. Available at: http://www.bostonglobe.com/magazine/2015/02/12/the-green-crab-problem-shall-eat-enemy/Ahtg6L87Gpxs0RMKntYAoN/story.html, accessed 19 April 2016.

Webb, J.K., Brown, G.P., Child, T., *et al.* (2008). A native dasyurid predator (common planigale, *Planigale maculata*) rapidly learns to avoid a toxic invader. *Austral Ecology*, 33, 821–829.

Weis, J.S., Smith, G. and Santiago-Bass, C. (2000). Predator/prey interactions: a link between the individual level and both higher and lower level effects of toxicants in aquatic systems. *Journal of Aquatic Ecosystem Stress and Recovery*, 7, 145–153.

Winslow, C. (2010). Competitive interactions between young-of-the-year smallmouth bass (*Micropterus dolomieu*) and round goby (*Apollonia melanostomus*). PhD dissertation. Columbus, OH: Ohio State University 85 pp.

Young, K.A., Dunham, J.B., Stephenson, J.F., *et al.* (2010). A trial of two trouts: comparing the impacts of rainbow and brown trout on a native galaxiid. *Animal Conservation*, 13, 399–410.

12 Evolutionary Novelty and the Behaviour of Introduced Predators

Edwin D. Grosholz and Elizabeth H. Wells

Introduction

Much has been written about the ecological and economic costs of biological invasions (Chapin *et al.*, 2000; Pimentel *et al.*, 2005; Lodge *et al.*, 2006). Among the most important consequences of these ongoing invasions is the creation of communities composed of species that share dramatically different levels of evolutionary history. While some species have evolved together for centuries if not millennia, others share only a brief history as the result of human-mediated introductions ranging from years to decades in most cases. The consequences of this variable lack of shared history have a wide range of implications for a variety of species interactions (Verhoeven *et al.*, 2009). However, interactions involving interspecies recognition and behavioural response may be those most influenced by lack of familiarity (e.g. Sih *et al.*, 2010). In particular, the interactions of predators and prey often require recognition on at least one side, either a realization that a prey item is available for consumption or that a predator is present and avoidance is required.

Theories of Novel Predator–Prey Interactions

There has been a great deal of thought given to what might be expected between encounters of introduced or 'novel' predators and prey with counterparts native to the system. These include a wide range of hypotheses that predict disproportionate advantages for either predator or prey (Table 12.1). The Enemy Release Hypothesis (ERH) (Torchin and Mitchell, 2004) posits that introduced species benefit from a 'release' by leaving the competitors, parasites and predators and competitors in their native range behind. The ERH assumes that enemies in the new range are less able to impact the demography of the invader, either because these enemies fail to recognize the invader or are unable to overcome their defences (Elton, 1958; Crawley, 1987; Mack *et al.*, 2000; Keane and Crawley, 2002). A second hypothesis known as the Novel Weapons Hypothesis (NWH) suggests that invaders with weapons (morphological, chemical, etc.) that are evolutionarily novel are more likely to be successful against prey or competitors and, thus, be more successful invaders with larger impacts (Callaway and Ridenour, 2004; Salo *et al.*,

Biological Invasions and Animal Behaviour, eds J.S. Weis and D. Sol. Published by Cambridge University Press. © Cambridge University Press 2016.

Table 12.1 Evolutionary theories of novel predator–prey interactions and hypotheses about the prey choices of predators

	Experimental hypothesis	Theory supported
1.	A predator consumes more of an evolutionarily familiar prey species than of a novel one	
	1a. A native predator consumes more of a native prey species than an invasive prey species	Enemy release, Novel defence
	1b. An invasive predator consumes more of an evolutionarily familiar invasive prey species than a native or unfamiliar invasive prey species	Facilitation between invaders
2.	A predator consumes more of an evolutionarily novel prey species than of a familiar one	
	2a. An invasive predator consumes more of evolutionarily novel prey (either native or from another non-native region) than evolutionarily familiar prey	Novel weapons
	2b. A native predator consumes more of an evolutionarily novel non-native prey than familiar native prey	Biotic resistance
3.	A predator consumes the prey that is the easiest and most profitable, regardless of evolutionary or historical familiarity	Optimal foraging

2007; Kumar and Bais, 2010). A third hypothesis referred to as Biotic Resistance (Elton, 1958) reasons that native predators, competitors and pathogens can reduce the spread and impact of a new introduction even if it is unable to prevent the introduction initially (Mitchell and Power, 2003; Levine *et al.*, 2004; de Rivera *et al.*, 2005; Cheng and Hovel, 2010; Kimbro *et al.*, 2013).

Recent syntheses of these ideas have attempted not only to summarize these many hypotheses, but to develop a broader conceptual framework that could predict invasion success or failure, how predation risk varies as a function of familiarity (or conversely naïveté), and the role of novelty in dictating the relative strength of consumptive versus non-consumptive effects of predators on prey (Cox and Lima, 2006; Sih *et al.*, 2010; Carthey and Banks, 2014). Most of this theory has been developed from the perspective of prey avoiding consumers (Maron and Vilà, 2001; Sih *et al.*, 2010; Carthey and Banks, 2014), and consequently there has been less theory developed for the role of predators' recognition, selection, handling and consumption of novel prey, whether as native predators consuming novel prey, or as non-native predators consuming native prey or non-native prey from a different native source. Therefore, our primary focus for this chapter is to contribute to a broader conceptual framework describing the behaviour of non-native predators.

Prey Choice by Introduced Predators

To date, the studies that have examined the impacts made by introduced consumers on familiar (introduced from the same source) relative to unfamiliar (native or introduced from another source) prey are mixed in their conclusions. This scenario has real-world relevance and is fairly common, since several species are often introduced from the same source, e.g. pest herbivores accompanying agricultural shipments of grain or fruit, or marine predators hitchhiking along with bivalve aquaculture. Studies of plants and exotic herbivores show that introduced herbivores prefer novel native species (Parker *et al.*, 2006), whereas other studies have shown less significant patterns. The consequences of invasive predators consuming native versus non-native prey may have important consequences for native ecosystems, including the strength of biotic resistance and the spread and impacts of non-native consumers (Levine *et al.*, 2004; Parker *et al.*, 2006; Kimbro *et al.*, 2013).

Introduced predators clearly switch to use novel food sources in some cases, and fail to do so in other cases. Many introduced predators, particularly generalist predators, have successfully adapted to new prey including small Asian mongooses (*Herpestes javanicus*) in Hawaii and Caribbean islands (Simberloff *et al.*, 2000; Hays and Conant, 2007), cane toads (*Rhinella marina*) in Australia (Phillips and Shine, 2004), brown tree snakes (*Boiga irregularis*) in Guam (Fritts and Rodda, 1998), Nile perch (*Lates niloticus*) in Lake Victoria (Ogutuohwayo, 1990) and red lionfish (*Pterois volitans*) in the Caribbean (Albins and Hixon, 2008). California alone offers multiple examples of invasive predators which have adapted well to new prey, including striped bass (*Morone saxatilis*) (Bryant and Arnold, 2007), European green crabs (*Carcinus maenas*) (Grosholz, 2005), bull frogs (*Lithobates catesbeianus*) and mosquitofish (*Gambusia affinis*) (Lawler *et al.*, 1999). A recent study looking at foraging by populations of a native marine whelk on native but unfamiliar prey showed that the predators had no significant preference for prey commonly encountered by that population (McWilliam *et al.*, 2013). In another study involving a native lizard and fire ants, an unfamiliar non-native prey, the lizards showed little aversion to consuming invasive fire ants despite their toxicity (Robbins *et al.*, 2013). Examples also include intentional introductions of biological control agents which switched their prey choice away from the targeted pest and instead became an established predator of other native species (Simberloff and Stiling, 1996; Louda *et al.*, 2003). Failure of an introduced predator to switch to a novel food source is in fact a characteristic of a successful biocontrol agent (Simberloff and Stiling, 1996; Louda *et al.*, 2003; Stiling and Cornelisson, 2005). Alternatively, an invasion may fail to establish if a predator without a familiar food resource is unsuccessful at adapting to new ones (Sih *et al.*, 2010).

Predator Naïveté and Loss of Naïveté

There are certainly many reasons to assume that introduced predators in a new range are likely to be hesitant to consume novel, unfamiliar prey. There are many theories and observations from outside invasion biology that address the potential hesitance of predators including 'dietary conservatism', which has been shown for a diverse range of animals (McMahon *et al.*, 2014), although not necessarily invertebrates.

Although there are several examples of predators acting naïvely with regard to novel prey, there are many good case studies of predators dealing very effectively with novel prey from several systems. In addition to the examples listed above of invasive predators adapting to new prey, there are numerous documented cases of biotic resistance in which native predators actually limit invasive prey (Levine *et al.*, 2004; de Rivera *et al.*, 2005; Kimbro *et al.*, 2013).

Following the assumptions made about predators (including introduced ones) preferring familiar prey are additional theories about the role of time required for novel predators to become familiar with and exploit novel prey, which of course over time will become more familiar. Implicit in this thinking is that there will be time lags in recognition of and interaction of novel predators and prey, and that these will attenuate over time (Hastings *et al.*, 2005), although theory predicts that time lags for generalist predators would be shorter than for specialists (Cox and Lima, 2006; Ferrari *et al.*, 2008; Sih *et al.*, 2010). Again, most of the thinking in the case of invasive species revolves around time lags in the response of prey to consumers, and recent studies have demonstrated that the increased time spent with a novel predator can influence prey responses (Freeman and Byers, 2006; Freeman *et al.*, 2014).

The time required for a species to lose its naïveté is important in assessing the degree to which this naïveté influences the decisions of introduced predators. A critical test for measuring the loss of naïveté is to compare the preferences of introduced predators simultaneously from both the native and invasive ranges of that species. This will allow for testing the utility of evolutionary hypotheses such as ERH, NWH and the Evolution of Increased Competitive Ability (EICA) hypothesis (Blossey and Notzhold, 1995; Maron and Vilà, 2001; Hierro *et al.*, 2005). Again, most studies have focused on native and invasive plants rather than native versus invasive consumers. This may be due to the fact that if invasive grazers or predators are not successful in consuming unfamiliar species that are native in the introduced range, they may not become established and instead join the majority that are uncounted, unsuccessful invaders (Lodge, 1993; Williamson and Fitter, 1996; Jeschke and Strayer, 2005).

Experimental Tests of Novel Predator Choice

In the surprising absence of quantitative experimental studies of introduced predators on familiar and novel prey, we developed an experimental system using native and introduced prey and novel predators in the highly invaded benthic communities of central California estuaries. Our predators were invasive snails known as whelks (aka oyster drills) feeding on either native or introduced prey species from two types of bivalve prey, either mussels or oysters. Our intent was to test the existing theories of novel predator–prey dynamics in order to see which one had the most predictive power. However, as we show below, our results provided only mixed support for these current theories based on novelty, and in some cases provide more support for the older Optimal Diet Theory (ODT) (Emlen, 1966; MacArthur and Pianka, 1966; Werner and Hall, 1974; Pyke *et al.*, 1977; Sih and Christensen, 2001). ODT suggests that prey choice would maximize the ratio of energetic intake to handling and search time. Our results also showed little support for time lags in recognition in these predators and instead found that novel

Table 12.2 The origin and date of introduction for species of whelk predators and oyster and mussel prey described in the experiments

Class	Species	Origin	Date of first record in Tomales Bay
Whelk	*Acanthinucella spirata*	Northeastern Pacific	Native
	Urosalpinx cinerea	Northwestern Atlantic	1935
	Ocenebra inornata	Northwestern Pacific	1941
Oyster	*Ostrea lurida*	Northeastern Pacific	Native
	Crassostrea virginica	Northwestern Atlantic	Not established
	Crassostrea gigas	Northwestern Pacific	1941 (aquaculture)
Mussel	*Mytilus trossulus, Mytilus galloprovincialis*	Northeastern Pacific Mediterranean, Black Sea	Native 1880s
	Mytilus edulis	Northwestern Atlantic	Not established
	Musculista senhousia	Northwestern Pacific	1941

populations of the same species performed nearly the same as populations that had been introduced many decades earlier.

Introduced Predatory Whelks: A Case Study

This work focused primarily on the benthic communities in Tomales Bay, CA (38°9′1″N, 122°53′19″W), which is a well-studied estuary typical of many estuaries in western North America (for further description, see Hearn and Largier, 1997; Kimbro et al., 2009). We summarize the main results of this work and provide minimal details regarding the procedures used, but additional details and explanation can be found in Wells (2013).

The species chosen for this study are all established in many estuaries as the result of extensive introduction of oysters in the last century on the West Coast of North America (Carlton, 1992). The three whelks used in the interspecies predation studies are all common in Tomales Bay, are all in the family Muricidae, and are all generalist predators which feed largely on mussels and barnacles, but may also consume bryozoans, oysters, clams or scavenged dead tissue (Rittschof et al., 1984; Perry, 1987). We used the angular unicorn drill (*Acanthinucella spirata*), the Eastern oyster drill (*Urosalpinx cinerea*), and the Japanese or Asian oyster drill (*Ocenebra inornata*) (Table 12.2).

In addition to these generalist predatory whelks, we used native and non-native oysters and mussels, many species of which are also resident in Tomales Bay. This included the Pacific oyster (*Crassostrea gigas*), which is grown commercially but not naturalized in Tomales Bay, and the native Olympia oyster (*Ostrea lurida*). We also used the eastern oyster (*Crassostrea virginica*), which was introduced but never established in Tomales Bay. To compare mussel species, we used *Mytilus trossulus*, whose population is largely native *M. trossulus* but also likely includes some hybrid individuals (*Mytilus trossulus* × *M. galloprovincialis*; Braby and Somero, 2006), and *Musculista senhousia*, which is an

Asian mussel introduced into Tomales Bay (Carlton, 1992). We also used the eastern blue mussel (*Mytilus edulis*) from the Atlantic coast of the United States, which is not currently in Tomales Bay (Table 12.2).

All three whelks have inhabited Tomales Bay with two oyster and two mussel species for over 70 years (Carlton, 1992), so for a majority of the predator–prey pairings any preferences that were based in evolutionary history would have persisted through a period sufficient to allow within-lifetime individual learning or rapid post-invasion adaptation (Trussell, 1996; Freeman and Byers, 2006).

Sources of Native and Non-Native Species

For the experiments comparing whelks of different species, performed between May and August 2009, we collected *Acanthinucella spirata*, *Urosalpinx cinerea* and *Ocenebra inornata* from intertidal cobbles and mud in Tomales Bay. All whelks used were among the largest that could be found at the collection sites.

For oyster predation experiments using multiple whelk species, small *C. gigas* were obtained in 2009 in Tomales Bay. Similarly sized *C. virginica* were obtained in 2009 from the Virginia Institute of Marine Science's Aquaculture Genetics and Breeding Technology Center, in Gloucester Point, VA. Native Olympia oysters *O. lurida* were collected in 2009 from local populations in San Francisco Bay. For the mussel predation experiments comparing multiple whelk species, both *Mytilus trossulus* and *Musculista senhousia* were collected in 2009 in Tomales Bay. *Mytilus edulis* was collected in 2009 in Lewes, Delaware.

Comparing Predator Choice Among Native and Introduced Populations

We also sought to compare prey choice between native and introduced populations of the eastern United States whelk, *U. cinerea*. We collected *U. cinerea* from three native populations on the Atlantic coast of North America and from four invasive populations on the Pacific coast of North America.

Experimental Setup and Procedures

All specimens were housed and experiments were conducted at San Francisco State University's Romberg Tiburon Center for Environmental Studies on San Francisco Bay in Tiburon, CA. To examine predation on multiple prey species, we offered the whelks two-species prey combination treatments, which allowed comparison of each whelk's predation success directly between each pair of prey species rather than indirectly as in a buffet design. In the experiments comparing the three whelk species, an experimental block consisted of four different predator treatments (three whelk species and a no-predator control) crossed with three prey treatments (three different two-oyster combinations, or three different two-mussel combinations). In the experiments comparing the prey choice of native and introduced populations of *U. cinerea* populations, an experimental block consisted of eight different predator treatments (seven populations and a no-predator control) crossed with three different prey treatments (three different two-oyster combinations, or three different two-mussel combinations). Predator and prey numbers represented the low end of field densities. Bivalves that had been drilled through the shell or were substantially consumed without drilling were counted as successful predation events. Consumed or dead oysters and mussels were removed and replaced during daily checks through each experimental run so that whelks always had a choice between equal numbers of two-prey species.

We also conducted handling time experiments. The time required for each whelk to handle each oyster was measured in consumption rate observations of a single whelk and a single oyster for all nine predator–prey combinations (three whelk × three oyster species), which ran over the course of 4 months until sufficient replicates had been completed. We ran a similar handling time experiment for the three whelk species preying on the three mussel species over the course of 10 days. For all experiments, predator and prey length were measured, along with the mass of live bivalves, dead bivalves and clean shell mass. Handling time was determined as number of days (for oysters) or hours (for mussels) from when the whelk was first seen with a full foothold on the bivalve until it left the prey or the gaping shell of the dead bivalve was clearly empty. We calculated biomass/handling time ratio as (live prey mass – dead prey mass)(g)/handling time (hours or days).

We also approximated the energetic content of the prey by measuring live mass, dry tissue mass and ash-free dry weight for each prey species. We measured shell thickness at the drill location for a random subsample of 10 drilled shells of each prey species drilled by each whelk predator species from the predation choice experiments, and 10 of each prey species drilled by *U. cinerea* from each location.

We used a variety of statistical tests to analyse the experimental results. For all predation experiments including comparisons of the three whelk species, each whelk species in each two-prey combination was compared to the 50:50 proportion predicted by the null hypothesis with Wilcoxon signed rank test. To compare the native and invasive ranges of *U. cinerea*, the proportion of each prey species drilled or dead due to *U. cinerea* from either the native or invasive range was compared against the null 50:50 hypothesis with Wilcoxon signed tests, pooling populations in each range. To determine whether the ranges differed, the proportions of each prey drilled in each two-prey combination was compared for *U. cinerea* from each range, using analysis of variance (ANOVA) with a quasibinomial distribution and log link, again pooling the populations in each range.

We analysed biomass/handling time ratios as a function of prey species with Kruskal–Wallis tests and multiple pairwise Mann–Whitney U tests. We compared the three mussel and oyster species' available biomass by analysing the log transformed dry biomass as the function of prey species with a one-way fixed-effect generalized linear model (GLM) in a permutational ANOVA (PERMANOVA) (Clarke and Gorley, 2015). We analysed shell as the function of prey species, with prey length as a covariate, using a one-way fixed-effect PERMANOVA.

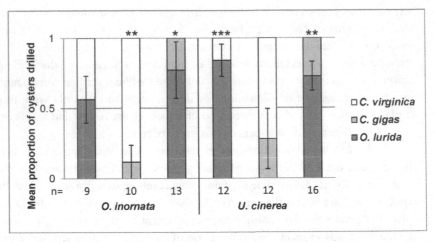

Figure 12.1 Mean proportion of oysters dead or drilled. Error bars are ±95% CI and significance tests show differences between predators: *$p < 0.05$, **$p < 0.01$, ***$p < 0.001$.

Predation Comparison

We found that *O. inornata* consumed significantly more *O. lurida* than *C. gigas* and more *C. virginica* than *C. gigas*. *Urosalpinx cinerea* consumed more *O. lurida* than either *C. gigas* or *C. virginica* (Figure 12.1). *Acanthinucella spirata* did not consume a significant number of oysters and was excluded from the analysis.

When we compared the different whelk species regarding predation patterns on mussels, we found that *A. spirata* consumed more *M. edulis* than *M. senhousia*. *Ocenebra inornata* consumed more *M. senhousia* than either *M. edulis* or *M. trossulus*, and more *M. edulis* than *M. trossulus*. *Urosalpinx cinerea* consumed more *M. senhousia* than either *M. edulis* or *M. trossulus*.

In comparing oyster prey choice between *U. cinerea* populations from the native and invasive ranges, *U. cinerea* from both ranges consumed more *C. virginica* than *C. gigas*. There were marginally significant trends for native range *U. cinerea* to consume more *C. virginica* than *O. lurida*, and for invasive range *U. cinerea* to consume more *C. virginica* than *O. lurida* and more *O. lurida* than *C. gigas*. The native and invasive range did not differ significantly from each other in oyster consumption patterns. (Figure 12.2a).

Urosalpinx cinerea from the native range consumed more *M. senhousia* than *M. edulis*, and more *M. senhousia* than *M. trossulus*. *Urosalpinx cinerea* from the invasive range consumed more *M. senhousia* than *M. edulis*, more *M. senhousia* than *M. trossulus*, and more *M. edulis* than *M. trossulus*. Compared to the native range, *U. cinerea* from the invasive range consumed a higher proportion of *M. senhousia* than either *M. edulis* or *M. trossulus*, but the trends of the two ranges were always in the same direction (Figure 12.2b).

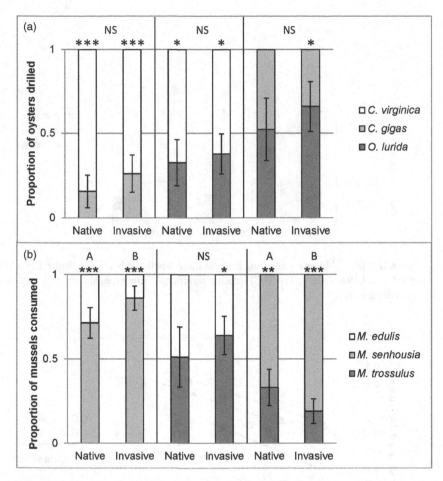

Figure 12.2 Mean proportion of bivalve prey dead or drilled by *U. cinerea* from its native and invasive range; (a) oysters, (b) mussels. Error bars are ±95% CI and significance tests show differences between predators (*$p < 0.05$, **$p < 0.01$, ***$p < 0.001$) and different letters (A,B) indicate significant differences between ranges for each prey combination.

Comparisons of Biomass and Handling Time

In comparing the different whelk species' biomass/handling time ratios, we found that neither *O. inornata* nor *U. cinerea* had any significant differences among the three oyster species. For mussel prey, *Acanthinucella spirata* had lower biomass consumed/handling time ratios for *M. trossulus* than for either *M. edulis* or *M. senhousia*. For mussel prey, *Ocenebra inornata* had a higher biomass/handling time ratio for *M. senhousia* than for either *M. edulis* or *M. trossulus* and a higher ratio for *M. edulis* than for *M. trossulus*. *Urosalpinx cinerea* had a lower biomass/handling time ratio for *M. trossulus* than for either *M. edulis* or *M. senhousia* (Figure 12.3).

From the assessments of available biomass, we found that *C. gigas* offered significantly higher dry biomass as a function of prey length than either *C. virginica* or

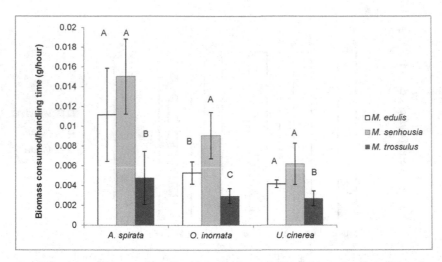

Figure 12.3 Mussel biomass/handling time for three whelk predators, in g/hour. Error bars are ±95% CI. Different letters (A,B,C) indicate significant differences among prey means within predator groupings.

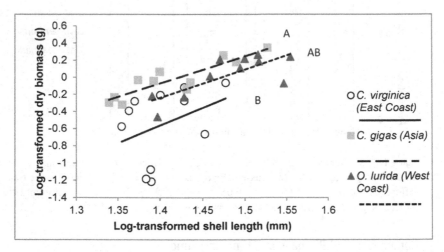

Figure 12.4 Least squares regression of log transformed dry biomass on log transformed shell length for each of the three oysters.

O. lurida. We also found that *M. senhousia* offered less dry biomass as a function of prey length than either *M. edulis* or *M. trossulus* (Figure 12.4).

Regarding oyster shell thickness, we found no significant difference in the shell thickness at the drill holes for *O. inornata* or *U. cinerea* (from pooled native and invasive ranges). For mussels, both *O. inornata* and *U. cinerea* (from pooled native and invasive ranges) drilled through thinner shells in *M. senhousia* than in either *M. trossulus* or *M. edulis* and through thinner shells in *M. edulis* than in *M. trossulus*. For *U. cinerea* from the native and invasive ranges, we found no effect of range on drill hole thickness in either oyster or mussel species. *Acanthinucella spirata*'s drill holes in mussels did not differ in shell thickness between mussel species.

Seeking Generality for Prey Choice by Introduced Predators

The results of our data suggest that the choices made by introduced predators when faced with novel versus native prey do not necessarily follow current predictions from naïveté theory or from general expectations based on evolutionary history outlined by enemy release, novel weapons and biotic resistance. Based on these ideas, we would expect that whelks would prefer to consume more familiar prey (ERH) or novel native prey, rather than novel invasive prey (NWH). While there are certainly data from other systems that support these ideas with regard to the dynamics of novel predator–prey interactions (Maron and Vilà, 2001; Callaway and Ridenour, 2004; Torchin and Mitchell, 2004), these theories may not be broadly predictive depending on the circumstances of the invasion. In our experiments we found that whelks were making at least some choices based on other criteria including ODT (Sih and Christensen, 2001). We also found that the basis for predator choice varied for our two sets of novel and familiar prey; therefore, the same set of novel predators can use different methods or tactics for selecting novel versus familiar prey depending on the traits of the prey.

Support for Optimal Diet Theory

The results in experiments with familiar and novel mussels consistently indicated that invasive whelks generally did not choose species based on evolutionary history or novelty, but rather the results are best explained as ODT. There was a significant pattern of whelks selecting *M. senhousia*, which was the most energetically optimal prey choice based on biomass/handling time for both *O. inornata* and *U. cinerea*. While *M. senhousia* offered the least amount of size-specific dry tissue, it also had a significantly thinner shell at the drill hole sites, requiring the least handling time and energy.

These results are not well explained by evolutionary history, since *M. senhousia* shares no evolutionary history with *U. cinerea*, although it does with *O. inornata*. Both whelks also share a more recent history with *M. senhousia* in Tomales Bay, so there could be a role for recent (i.e. within-lifetime) experience. However, the results from the experiments using both introduced and native populations of *U. cinerea* showed little difference in their preference for *M. senhousia*, indicating that there was little if any lag in this predator adjusting to a new, optimally favoured prey.

In the case of oysters, the choices were not consistent with ODT according to the energetic and time metrics we measured. Our results show that *C. gigas* provides the greatest amount of tissue per size, giving the biggest energetic payoff. There was a statistically insignificant but notable trend for *C. gigas* to have the highest biomass/handling time ratio for both *O. inornata* and *U. cinerea*. If whelks were choosing prey on the basis of ODT, we would therefore have expected the more-rewarding *C. gigas* to be the preferred prey, but this was not the case. Instead, in our experiments comparing whelk species there was a trend of selecting the novel native oyster rather than the oyster that was evolutionarily familiar to either oyster, which would not be predicted by most of the evolutionary hypotheses excepting NWH.

We found that evolutionary history could explain some part of the trend, although only for one species and in some circumstances. In the experiments comparing *U. cinerea* from different ranges, we found a trend towards favouring *C. virginica* with which *U. cinerea* shares significant evolutionary history in eastern North America. However, in the experiments comparing different whelk species, we found that both *U. cinerea* and *O. inornata* preferred *O. lurida*, which shares no evolutionary history with either species.

There appears to be no consistent pattern of evolutionary novelty predicting predation outcomes, and while we cannot point to ODT as being definitively supported by our oyster biomass/handling time results, the fact that *U. cinerea* favoured different species of oysters in different sets of very similar experiments indicates that their predation choices are not constrained by evolutionary history. This is supported by the lack of significant differences in predation trend direction between the native and invasive populations of *U. cinerea*.

The few studies that have investigated similar pairs of species have had different results from our work. For instance, studies from Willapa Bay show that *U. cinerea* preferentially consumes non-native *C. gigas* rather than the native *O. lurida* (Buhle and Ruesink, 2009). These results, which conflict with ours, suggest that predator preferences may be contingent upon more recent experience and may vary considerably even among bays in the same region. This would even further reduce the apparent importance of evolutionary history and theories of naïveté as important determinants of prey choice.

History may not Matter

Overall, our conclusions from experiments with native and introduced populations of *U. cinerea* showed that shared evolutionary history does not appear to be the primary mechanism underlying prey choice of either oysters or mussels, since we observed no consistent trend among categories of prey. The most preferred oyster was either the evolutionarily familiar *C. virginica* or the evolutionarily novel *O. lurida*, depending on circumstances, while the most preferred mussel was the evolutionarily novel *M. senhousia*.

In fact, naïve native range *U. cinerea* immediately preferred *M. senhousia* despite any prior experience, which is consistent with similar predator studies (Reusch, 1998; Crooks, 2005; Kushner and Hovel, 2006). Also, for the population of *U. cinerea* from the introduced range, the lack of any experience with *C. virginica* for several decades did not reduce their preference for *C. virginica* in the range-comparison experiments.

Considering the overall results, we conclude that even if we had a detailed understanding of the history of an invasion and with that the degree of novelty (either phylogenetically or trait based), this might not provide much guidance for predicting novel predator–prey interactions in this system. We would argue that this is likely to be true for many benthic invertebrate systems, since there is little about the species we have chosen that differs from many dozens of other similar invaders. It may be broadly true that if the cues used in these systems are very generalized and not species specific, there may be little functional naïveté, and if new species encountered are functionally similar to species the consumer has evolved with, the level of naïveté will be small and easily overcome by learning or behavioural plasticity (Suboski, 1992; Webb *et al.*, 2008). Darwin himself made the observation in his 'naturalization hypothesis' that historical experience may not be very important if a novel species is functionally similar to a familiar species (Darwin, 1872; Ferrari *et al.*, 2007; Turner, 2008; Pearse and Hipp, 2009). However, support for this hypothesis is equivocal (Ricciardi and Mottiar, 2006).

We also acknowledge that our results need to be interpreted in the light of several important caveats. First, these experiments are not conducted in the field and only include captive animals in aquaria. Therefore, search cues may be affected in unrealistic ways and we have avoided discussions of non-consumptive effects of predators on prey for this reason (Weissburg *et al.*, 2014). Second, we know from earlier studies that prey with limited mobility such as whelks can avoid predators by changing microhabitats in response to their own crab predators, which may affect their own predation impacts on oysters (Grabowski and Kimbro, 2005). Our experiments were not designed to measure those responses in the context of predator choice. Third, there are some important prey behaviours that sessile or nearly sessile prey can use to influence predation outcomes that we did not investigate. These include morphological changes like shell thickening or physiological options like induced changes in anti-predator chemistry.

Generalizations for Generalist Predators

We certainly acknowledge that evolutionary history may be more important for predators for which the rules of predation are more complicated and where more complex predatory behaviour is required both to recognize and subdue prey. Evolutionary history-based hypotheses have been shown to be useful and predictive in systems that involve invaders that in the novel range have left behind specialized predators or parasites (Torchin *et al.*, 2003; Müller-Schärer *et al.*, 2004; Torchin and Mitchell, 2004; Joshi and Vrieling, 2005), or introduced predators in on islands, in lakes or other systems lacking specific functional groups of predators or parasites (Ogutuohwayo, 1990; Fritts and Rodda, 1998; Knapp and Matthews, 2000; Blackburn *et al.*, 2004). It is also the case that prey morphology can differ significantly in the introduced range, such as shifts to increased size that often accompany introductions into many aquatic and marine habitats (Grosholz and Ruiz, 2003). Larger prey size could result in either less predation in the case of size refugia due to larger shell size or thickness. Alternatively, larger prey size could result in increased predation in the case of less well-defended prey that became energetically desirable. Unfortunately, we know little about how these changes in the introduced range ultimately influence novel predator–prey dynamics. Finally, evolutionary responses can include important morphological changes that can mediate predator behaviour such as the coevolved changes in claw morphology of introduced green crabs (Smith, 2004) and the subsequent changes in shell morphology of their snail prey (*Littorina obtusata*) (Trussell and Nicklin, 2002).

We conclude that for predators like marine invertebrates where diets are often broad and the cues used to distinguish predator and prey may be general enough for plasticity or adaptation to include new species relatively recently (Suboski, 1992; Trussell and Nicklin, 2002), detailed knowledge of the evolutionary history and the timing of introduction may not be very predictive. In these systems simpler rules based on ODT may be the best guide to predicting the prey choice of introduced predators and their impacts on native species. Introduced invertebrate predators may also be very quick to change behaviour after a handful of experiences, or after exposure to chemical cues regarding a host's vulnerability or makeup. This information needed for simple predator decisions may be 'learned' within a single generation or even less; even truly adaptive changes may be very rapid since, after all, any non-native predators that do not learn to eat new things will be unlikely to establish. We acknowledge that although prey choice may involve simple, relatively quickly learned behaviours, these may not be well explained by ODT. Many factors discussed above including chemical defences, previous feeding experiences, etc., may influence predator's decisions and may complicate explanations based on optimal diet considerations. In addition, there are many substantive criticisms of ODT in general (Pierce and Ollason, 1987), so this is not likely to be the basis for a general explanation for diet choice of introduced predators any more than it is for native predators. However, some of our data are consistent with ODT, which has been shown to more strongly predictive in systems involving largely sessile prey (Sih and Christensen, 2001).

Thoughts for the Future

Consequently, we may be equipped with much less interesting theory regarding novel predator–prey interactions, but instead with a simpler set of well-established theory and rules to help us predict novel predator–prey interactions between generalist predators and familiar types of prey. These rules may ultimately require less information about any particular invasion and allow greater likelihood of predicting the impacts of invasive species in highly invaded coastal habitats.

Perhaps the most striking aspect of our study is finding out how few studies have examined the basis of prey choice by introduced predators faced with native and novel prey. Future work would require investigating to what degree the results we present here apply to similar systems and different systems. We realize that this call to action is complicated by the circumstances we acknowledge above that introduced predators that are highly averse to novel prey are less likely to become successfully established in their new range. We may largely be faced with a world where the lists of successful introduced consumers are disproportionately made up of species that are relatively generalist in their prey choice and generally willing to consume unfamiliar prey that are native in the new range. This is likely the starting point for future study, but whether this is true or not, there is much to be learned about prey choice among novel predators.

Among the more tractable issues that could be addressed is temporal change in the willingness of novel predators to consume unfamiliar versus familiar prey. Whether this is due to learning, inherent plasticity or other mechanisms is largely unknown. We know very little about whether change takes place at all, as well as little about the mechanisms underlying change. To date, the studies that have addressed this question, including our own, have been inadequate, generally substituting space for time or using other proxies that do not provide much power or insight. We would suggest that new invasions should be identified and then predator preferences tested at regular intervals over meaningful periods of time. Understanding the mechanisms underlying temporal changes in prey responses may provide insights into expected changes for other invasions. For instance, it is often observed that many new invasions remain at low density for some period of time and that there is an observed 'lag time' before that species experiences exponential population growth and spread (Hastings *et al.*, 2005). For introduced predators, this lag time may be the result of either a learned or rapidly evolved response to use a novel prey species. If we could determine the mechanism underlying changing prey choice for particular introduced predators, this might help with prioritizing management actions for novel predators that present different levels of risk to native species.

Acknowledgements

We would like to thank the editors of this book, Judith Weis and Daniel Sol, for their kind invitation to contribute to this effort and two thoughtful reviewers for their comments on this manuscript. We would also like to thank the San Francisco Bay National Estuarine Research Reserve, which funded this work through a Graduate Student Fellowship, and in particular Dr Matt Ferner, the SF Bay NERR Research Coordinator. Finally, we would like to thank San Francisco State University's Romberg Tiburon Center for Environmental Studies and its entire community of researchers, which hosted this research.

References

Albins, M.A. and Hixon M.A. (2008). Invasive Indo-Pacific lionfish (*Pterois volitans*) reduce recruitment of Atlantic coral-reef fishes. *Marine Ecology Progress Series*, 367(1), 233–238.

Blackburn, T.M., Cassey, P., Duncan, R.P., Evans, K.L. and Gaston, K.J. (2004). Avian extinction and mammalian introductions on oceanic islands. *Science*, 305(5692), 1955–1958.

Blossey, B. and Notzold, R. (1995). Evolution of increased competitive ability in invasive non-indigenous plants: a hypothesis *Journal of Ecology*, 83(5), 887–889.

Braby, C.E. and Somero, G.N. (2006). Ecological gradients and relative abundance of native (*Mytilus trossulus*) and invasive (*Mytilus galloprovincialis*) blue mussels in the California hybrid zone. *Marine Biology*, 148(6), 1249–1262.

Bryant, M.E. and Arnold, J.D. (2007). Diets of age-0 striped bass in the San Francisco Estuary, 1973–(2002). *California Fish and Game*, 93(1), 1–22.

Buhle, E.R. and Ruesink, J.L. (2009). Impacts of invasive oyster drills on Olympia oyster (*Ostrea lurida* Carpenter 1864) recovery in Willapa Bay, Washington, United States. *Journal of Shellfish Research*, 28(1), 87–96.

Callaway, R.M. and Ridenour, W.M. (2004). Novel weapons: invasive success and the evolution of increased competitive ability. *Frontiers in Ecology and the Environment*, 2(8), 436–443.

Carlton, J.T. (1992). Introduced marine and estuarine mollusks of North America: An end-of-the-20th-century perspective. *Journal of Shellfish Research*, 11(2), 489–505.

Carthey, A.J.R. and Banks, P.B. (2014). Naïveté in novel ecological interactions: lessons from theory and experimental evidence. *Biological Reviews*, 89(4), 932–949.

Chapin, F.S., Zavaleta, E.S., Eviner, V.T., *et al.* (2000). Consequences of changing biodiversity. *Nature*, 405(6783), 234–242.

Cheng, B. and Hovel, K. (2010). Biotic resistance to invasion along an estuarine gradient. *Oecologia*, 164(4), 1049–1059.

Clarke, K. and Gorley, R. (2015). *PRIMER v7: User Manual/Tutorial*. Plymouth, UK: PRIMER-E.

Cox, J.G. and Lima, S.L. (2006). Naïveté and an aquatic-terrestrial dichotomy in the effects of introduced predators. *Trends in Ecology and Evolution*, 21(12), 674–680.

Crawley, M.J. (1987). What makes a community invasible? In *Colonization, Succession, and Stability: the 26th Symposium of the British Ecological Society held jointly with the Linnean Society of London*, ed. Gray, A.J., Crawley, M.J. and Edwards, P.J. Boston, UK: Blackwell Scientific Publications, pp. 429–453.

Crooks, J.A. (2005). Lag times and exotic species: the ecology and management of biological invasions in slow-motion. *Ecoscience*, 12(3), 316–329.

Darwin, C. (1872). *On the Origin of Species*. London: John Murray.

de Rivera, C.E., Ruiz, G.M., Hines, A.H. and Jivoff, P. (2005). Biotic resistance to invasion: native predator limits abundance and distribution of an introduced crab. *Ecology*, 86(12), 3364–3376.

Elton, C.S. (1958). *The Ecology of Invasions by Animals and Plants*. Chicago, IL: University of Chicago Press.

Emlen, J.M. (1966). Role of time and energy in food preference. *American Naturalist*, 100(916), 611–617.

Ferrari, M.C.O., Gonzalo, A., Messier, F. and Chivers, D.P. (2007). Generalization of learned predator recognition: an experimental test and framework for future studies. *Proceedings of the Royal Society B: Biological Sciences*, 274(1620), 1853–1859.

Ferrari, M.C.O., Messier, F., Chivers, D.P. and Messier, O. (2008). Can prey exhibit threat-sensitive generalization of predator recognition? Extending the predator recognition continuum hypothesis. *Proceedings of the Royal Society B: Biological Sciences*, 275(1644), 1811–1816.

Freeman, A.S. and Byers, J.E. (2006). Divergent induced responses to an invasive predator in marine mussel populations. *Science*, 313(5788), 831–833.

Freeman, A.S., Dernbach, E., Marcos, C. and Koob, E. (2014). Biogeographic contrast of *Nucella lapillus* responses to *Carcinus maenas*. *Journal of Experimental Marine Biology and Ecology*, 452, 1–8.

Fritts, T.H. and Rodda, G.H. (1998). The role of introduced species in the degradation of island ecosystems: a case history of Guam. *Annual Review of Ecology and Systematics*, 29, 113–140.

Grabowski, J.H. and Kimbro, D.L. (2005). Predator-avoidance behavior extends trophic cascades to refuge habitats. *Ecology*, 86(5), 1312–1319.

Grosholz, E.D. and Ruiz, G.M. (2003). Biological invasions drive size increases in marine and estuarine invertebrates. *Ecology Letters*, 6(8), 700–705.

Grosholz, E.D. (2005). Recent biological invasion may hasten invasional meltdown by accelerating historical introductions. *Proceedings of the National Academy of Sciences, USA*, 102(4), 1088–1091.

Hastings, A., Cuddington, K., Davies, K.F., *et al.* (2005). The spatial spread of invasions: new developments in theory and evidence. *Ecology Letters*, 8(1), 91–101.

Hays, W.S.T. and Conant, S. (2007). Biology and impacts of Pacific Island invasive species. 1. A worldwide review of effects of the small Indian mongoose, *Herpestes javanicus* (Carnivora: Herpestidae). *Pacific Science*, 61(1), 3–16.

Hearn, C.J. and Largier, J.L. (1997). The summer buoyancy dynamics of a shallow Mediterranean estuary and some effects of changing bathymetry: Tomales bay, California. *Estuarine Coastal and Shelf Science*, 45(4), 497–506.

Hierro, J.L., Maron, J.L. and Callaway, R.M. (2005). A biogeographical approach to plant invasions: the importance of studying exotics in their introduced and native range. *Journal of Ecology*, 93(1), 5–15.

Jeschke, J.M. and Strayer, D.L. (2005). Invasion success of vertebrates in Europe and North America. *Proceedings of the National Academy of Sciences, USA*, 102(20), 7198–7202.

Joshi, J. and Vrieling, K. (2005). The enemy release and EICA hypothesis revisited: incorporating the fundamental difference between specialist and generalist herbivores. *Ecology Letters*, 8(7), 704–714.

Keane, R.M. and Crawley, M.J. (2002). Exotic plant invasions and the enemy release hypothesis. *Trends in Ecology and Evolution*, 17(4), 164–170.

Kimbro, D.L., Largier, J. and Grosholz, E.D. (2009). Coastal oceanographic processes influence the growth and size of a key estuarine species, the Olympia oyster. *Limnology and Oceanography*, 54(5), 1425–1437.

Kimbro, D.L., Cheng, B.S. and Grosholz, E.D. (2013). Biotic resistance in marine environments. *Ecology Letters*, 16(6), 821–833.

Knapp, R.A. and Matthews, K.R. (2000). Non-native fish introductions and the decline of the mountain yellow-legged frog from within protected areas. *Conservation Biology*, 14(2), 428–438.

Kumar, A.S. and Bais, H.P. (2010). Allelopathy and exotic plant invasion. In *Plant Communication from an Ecological Perspective*, ed. Baluška, F. and Ninkovic, V. Berlin: Springer, pp. 61–74.

Kushner, R.B. and Hovel, K.A. (2006). Effects of native predators and eelgrass habitat structure on the introduced Asian mussel *Musculista senhousia* (Benson in Cantor) in southern California. *Journal of Experimental Marine Biology and Ecology*, 332(2), 166–177.

Lawler, S.P., Dritz, D., Strange, T. and Holyoak, M. (1999). Effects of introduced mosquitofish and bullfrogs on the threatened California red-legged frog. *Conservation Biology*, 13(3), 613–622.

Levine, J.M., Adler, P.B. and Yelenik, S.G. (2004). A meta-analysis of biotic resistance to exotic plant invasions. *Ecology Letters*, 7(10), 975–989.

Lodge, D.M. (1993). Biological invasions: lessons for ecology. *Trends in Ecology and Evolution*, 8(4), 133–137.

Lodge, D.M., Williams, S., MacIsaac, H.J., *et al.* (2006). Biological invasions: recommendations for US policy and management. *Ecological Applications*, 16(6), 2035–2054.

Louda, S.M., Pemberton, R.W., Johnson, M.T. and Follett, P. (2003). Nontarget effects – the Achilles' heel of biological control? Retrospective analyses to reduce risk associated with bio-control introductions. *Annual Review of Entomology*, 48(1), 365–396.

MacArthur, R.H. and Pianka, E.R. (1966). On optimal use of a patchy environment. *American Naturalist*, 100(916), 603–609.

Mack, R.N., Simberloff, D., Lonsdale, W., Evans, H., Clout, M. and Bazzaz, F.A. (2000). Biotic invasions: causes, epidemiology, global consequences, and control. *Ecological Applications*, 10(3), 689–710.

Maron, J.L. and Vilà, M. (2001). When do herbivores affect plant invasion? Evidence for the natural enemies and biotic resistance hypotheses. *Oikos*, 95(3), 361–373.

McMahon, K., Conboy, A., O'Byrne-White, E., Thomas, R.J. and Marples, N.M. (2014). Dietary wariness influences the response of foraging birds to competitors. *Animal Behavior*, 89, 63–69.

McWilliam, R.A., Minchinton, T.E. and Ayre, D.J. (2013). Despite prolonged association in closed populations, an intertidal predator does not prefer abundant local prey to novel prey. *Biological Journal of the Linnean Society*, 108(4), 812–820.

Mitchell, C.E. and Power, A.G. (2003). Release of invasive plants from fungal and viral pathogens. *Nature*, 421(6923), 625–627.

Müller-Schärer, H., Schaffner, U. and Steinger, T. (2004). Evolution in invasive plants: implications for biological control. *Trends in Ecology and Evolution*, 19(8), 417–422.

Ogutuohwayo, R. (1990). The decline of the native fishes of Lakes Victoria and Kyoga (East-Africa) and the impact of introduced species, especially the Nile Perch, *Lates niloticus*, and the Nile Tilapia, *Oreochromis niloticus*. *Environmental Biology of Fishes*, 27(2), 81–96.

Parker, J.D., Burkepile, D.E. and Hay, M.E. (2006). Opposing effects of native and exotic herbivores on plant invasions. *Science*, 311(5766), 1459–1461.

Pearse, I.S. and Hipp, A.L. (2009). Phylogenetic and trait similarity to a native species predict herbivory on non-native oaks. *Proceedings of the National Academy of Sciences, USA*, 106(43), 18097–18102.

Perry, D.M. (1987). Optimal diet theory: behavior of a starved predatory snail. *Oecologia*, 72(3), 360–365.

Phillips, B.L. and Shine, R. (2004). Adapting to an invasive species: toxic cane toads induced morphological change in Australian snakes. *Proceedings of the National Academy of Sciences, USA*, 101(49), 17150–17155.

Pierce, G.J. and Ollason, J.G. (1987). Eight reasons why optimal foraging theory is a complete waste of time. *Oikos*, 49(1), 111–118.

Pimentel, D., Zuniga, R. and Morrison, D. (2005). Update on the environmental and economic costs associated with alien-invasive species in the United States. *Ecological Economics*, 52(3), 273–288.

Pyke, G.H., Pulliam, H.R. and Charnov, E.L. (1977). Optimal foraging: selective review of theory and tests. *Quarterly Review of Biology*, 52(2), 137–154.

Reusch, T.B.H. (1998). Native predators contribute to invasion resistance to the non-indigenous bivalve *Musculista senhousia* in southern California, USA. *Marine Ecology Progress Series*, 170, 159–168.

Ricciardi, A. and Mottiar, M. (2006). Does Darwin's naturalization hypothesis explain fish invasions? *Biological Invasions*, 8(6), 1403–1407.

Rittschof, D., Kieber, D. and Merrill, C. (1984). Modification of responses of newly hatched snails by exposure to odors during development. *Chemical Senses*, 9(3), 181–192.

Robbins, T.R., Freidenfelds, N.A. and Langkilde, T. (2013). Native predator eats invasive toxic prey: evidence for increased incidence of consumption rather than aversion-learning. *Biological Invasions*, 15(2), 407–415.

Salo, P., Korpomaki, E., Banks, P.B., Nordstrom, M. and Dickman, C.R. (2007). Alien predators are more dangerous than native predators to prey populations. *Proceedings of the Royal Society B: Biological Sciences*, 274(1615), 1237–1243.

Sih, A., Bolnick, D.I., Luttbeg, B., *et al.* (2010). Predator–prey naïveté, antipredator behavior, and the ecology of predator invasions. *Oikos*, 119(4), 610–621.

Sih, A. and Christensen, B. (2001). Optimal diet theory: when does it work, and when and why does it fail? *Animal Behavior*, 61, 379–390.

Simberloff, D. and Stiling, P. (1996). How risky is biological control? *Ecology*, 77(7), 1965–1974.

Simberloff, D., Dayan, T., Jones, C. and Ogura, G. (2000). Character displacement and release in the small Indian mongoose, *Herpestes javanicus*. *Ecology* 81(8): 2086–2099.

Smith, L.D. (2004). Biogeographic differences in claw size and performance in an introduced crab predator *Carcinus maenas*. *Marine Ecology Progress Series*, 276, 209–222.

Stiling, P. and Cornelissen, T. (2005). What makes a successful biocontrol agent? A meta-analysis of biological control agent performance. *Biological Control*, 34(3), 236–246.

Suboski, M.D. (1992). Releaser-induced recognition learning by gastropod mollusks. *Behavioral Processes*, 27(1), 1–26.

Torchin, M.E. and Mitchell, C.E. (2004). Parasites, pathogens, and invasions by plants and animals. *Frontiers in Ecology and the Environment*, 2(4), 183–190.

Torchin, M.E., Lafferty, K.D., Dobson, A.P., McKenzie, V.J. and Kuris, A.M. (2003). Introduced species and their missing parasites. *Nature*, 421(6923), 628–630.

Trussell, G.C. (1996). Phenotypic plasticity in an intertidal snail: the role of a common crab predator. *Evolution*, 50(1), 448–454.

Trussell, G.C. and Nicklin, M.O. (2002). Cue sensitivity, inducible defense, and trade-offs in a marine snail. *Ecology*, 83(6), 1635–1647.

Turner, A.M. (2008). Predator diet and prey behavior: freshwater snails discriminate among closely related prey in a predator's diet. *Animal Behavior*, 76, 1211–1217.

Verhoeven, K.J.F., Biere, A., Harvey, J.A. and van der Putten, W.H. (2009). Plant invaders and their novel natural enemies: who is naive? *Ecology Letters*, 12(2), 107–117.

Webb, J.K., Brown, G.P., Child, T., *et al.* (2008). A native dasyurid predator (common planigale, *Planigale maculata*) rapidly learns to avoid a toxic invader. *Austral Ecology*, 33(7), 821–829.

Wells, E.H. (2013). Evolutionary novelty and naïveté in invertebrate predator-prey interactions in a benthic marine community. PhD dissertation. Davis, CA: University of California.

Weissburg, M., Smee, D.L. and Ferner, M.C. (2014). The sensory ecology of nonconsumptive predator effects. *American Naturalist*, 184(2), 141–157.

Werner, E.E. and Hall, D.J. (1974). Optimal foraging and size selection of prey by bluegill sunfish (*Lepomis macrochirus*). *Ecology*, 55(5), 1042–1052.

Williamson, M. and Fitter, A. (1996). The varying success of invaders. *Ecology*, 77(6), 1661–1666.

Part III

Case Studies

13 Behaviours Mediating Ant Invasions

Jules Silverman and Grzegorz Buczkowski

Introduction

Invasive ants represent a serious worldwide threat to the integrity of ecological communities by causing dramatic reductions in native ant diversity, with cascading impacts at other trophic levels of biological organization. While climatic and physical features of the recipient environment largely delineate the boundaries for invasive species establishment, intrinsic behavioural traits, of both the invader and the native animal taxa encountered by the invader, largely underlie invasion success within these abiotic constraints.

Phylogeny defines a narrow range of behaviours displayed by animals, which are further modulated by population level and environmental constraints. In turn, differences in behaviour can impact an animal's environment. Considering the many approaches taken to understand the nature of ant invasions, we will narrow our discussion to the behavioural determinants underlying ant invasions, but in the context of the genetic, ecological, biogeographical, and evolutionary constraints and opportunities facing the invader.

Ant behaviour and its influence on population growth can be tied to each stage of the invasion process, from resilience to hardships encountered during the transportation stage to establishment and spread into the new environment. Behaviour during transportation and initial establishment has been largely neglected because newly invasive propagules are often unrecognized during transport and upon arrival, sometimes requiring decades of growth to become noticeable (Chapple *et al.*, 2012). Once established, pre-existing traits, behavioural plasticity and trait evolution can favour invasive ant spread (Holway and Suarez, 1999; Phillips and Suarez, 2012).

Prior to Holway *et al.* (2002), most of the attention on invasive ants was devoted to the two most widespread and destructive species, the red imported fire ant, *Solenopsis invicta*, and the Argentine ant, *Linepithema humile*. Since their review, considerably more has been learned about these two ants and the behaviours suspected of underlying their invasion success. More recently, other ecologically destructive invasive ants have received the attention of behavioural biologists and ecologists. These species include: *Wasmannia auropunctata*, *Pheidole megacephala*, *Anoplolepis gracilipes*, *Nylanderia*

Biological Invasions and Animal Behaviour, eds J.S. Weis and D. Sol. Published by Cambridge University Press. © Cambridge University Press 2016.

fulva, *Lasius neglectus*, *Myrmica rubra* and *Tapinoma sessile*. This chapter will synthesize much of the research conducted since Holway *et al*. (2002) focusing on the behavioural aspects that we consider key to invasive ant ecological dominance: human-mediated dispersal and budding, unicoloniality, competitive dominance, omnivory, dispersed central-place foraging and rapid recruitment.

Dispersal Behaviour: Departure from the Native Range and Spread within the Introduced Range

The ecological success of many invasive ant species can be attributed largely to their efficient passive and active dispersal: passive, via human-assisted jump dispersal (Suarez *et al*., 2001) and active, via nuptial flights and colony budding. Phillips and Suarez (2012) have suggested that selection on population-level variation may favour better dispersers: those opportunistic phenotypes colonizing human cargo for intercontinental transport as well as colony fragments better adapted for dispersal at the invasion front.

Long-distance Dispersal

Long-distance anthropogenic jump dispersal results in large dispersal distances (thousands of kilometres) because it is influenced mainly by patterns of global trade and transportation and is thus responsible for the worldwide movement of invasive ant propagules to previously uninfested areas. Jump dispersal is often facilitated by the nesting behaviour and reproductive strategies of invasive ants. Many species do not construct permanent nests, but rather colonize ephemeral sites such as leaf piles, root masses and cavities. These ants are highly mobile and move nests to escape unfavourable abiotic conditions, environments under which these vagile traits were presumably selected (Lebrun *et al*., 2007). Such nesting habits facilitate the movement of invasive species which frequently infest material destined for transport, with plant material being particularly susceptible (Ward *et al*., 2006).

Long-distance dispersal patterns have been studied particularly well in the Argentine ant, which has attained worldwide distribution. Left to its own devices, the Argentine ant is a rather poor long-distance colonizer. However, through close association with humans and the survivability of relatively small propagules (Hee *et al*., 2000) it has been able to efficiently move around the world (Suarez *et al*., 2001; Roura-Pascual *et al*., 2004).

Long-distance dispersal associated with global transportation networks has also been linked to the worldwide dispersal of other invasive ants, including *S. invicta* (Ascunce *et al*., 2011). Interestingly, invasive populations in Australia, California, China, New Zealand and Taiwan have been genetically traced to introduced populations in the south-eastern United States rather than to native populations in South America. This suggests that behavioural traits and population structure associated with the success of fire ants in the southern United States may have pre-adapted these ants for successful transportation and establishment in other areas of the world. Worldwide distribution patterns based on long-distance dispersal of other invasive ant species are poorly understood; however, close association with landscaping materials (e.g. Groden *et al.*, 2005; Wetterer, 2012) is likely responsible for inter- and intra-continental transport.

Nuptial Flights

Alate nuptial flights, commonly employed by ants, can disperse colony founders up to several hundred metres from their natal nests, depending on intrinsic flight capabilities and weather patterns (Hölldobler and Wilson, 1990). With the exception of *S. invicta*, nuptial flights have been rarely observed in invasive ant species. *Solenopsis invicta* colonies produce male and female alates which disperse via nuptial flights, though monogyne (single queen) and polygyne (multiple queen) populations display profound differences in mating and dispersal behaviours (Tschinkel, 2006): inseminated queens from monogyne colonies disperse much further than polygyne queens, which ultimately affects population genetic structure (Shoemaker *et al.*, 2006). Mating in invasive ants is largely internidal and winged males can enter non-natal nests to mate (Passera and Keller, 1994; Espadaler *et al.*, 2004; Hicks, 2012), which may facilitate gene flow across local populations.

Short-distance Dispersal

Common to invasive ant species is short-distance dispersal via colony budding or fission (Holway and Case, 2000; Espadaler *et al.*, 2004; Wetterer, 2007; LeBrun *et al.*, 2013). Here, a subset of the colony's population disperses away on foot to a new location, establishing as an independent colony, which may or may not retain contact with the parent colony. Budding is independent of gyne production and may be initiated at any time.

Dispersal Distances

Long-distance human-mediated dispersal is largely passive, independent of the stowaway's behaviour. Spread by budding, by contrast, depends on intrinsic colony characteristics such as colony size and caste composition, as well as foraging and aggression behaviour (e.g. Buczkowski and Bennett, 2008; Hoffman and Saul, 2010). Both local and long-distance dispersal are limited by various climatic influences (Brightwell *et al.*, 2010; Bertelsmeier and Courchamp, 2014) and biotic factors; specifically interspecific competition (Rowles and O'Dowd, 2007) and food availability (Abbott, 2005; Rowles and Silverman, 2009). Proximity to human infrastructure influences the spread of invasive ants (Roura-Pascual *et al.*, 2011) due particularly to habitat modification, including irrigation practices (Menke and Holway, 2006). Most established invasive ant populations spread at a rate of from tens to hundreds of metres per year (e.g. Suarez *et al.*, 2001), so it is puzzling that *M. rubra*, first introduced into the northeastern United States in the early 1900s, has spread little since and has only recently emerged as a significant pest in urban and natural ecosystems after more than a century of dormancy (Wetterer and Radchenko, 2011). Such lag times are common in biological invasions and populations may exhibit low numbers until critical adaptive changes (genetic or environmental) happen to enable it to spread widely.

Intraspecific Aggression and Nestmate Recognition

Colony Boundaries and Population Structure

A key worker behaviour responsible for maintaining ant colony cohesiveness and genetic autonomy is nest defence against con- and heterospecific invaders (Hölldobler and Wilson, 1990). The colony boundaries enclosing one or relatively few nests of non-invasive ant species are fairly small and generally well-defined: non-nestmates (non-kin) encountered near nests are frequently attacked and killed. However, intraspecific worker aggression in all invasive ant species is absent or greatly diminished across spatial scales generally much larger than the territories occupied by the native ants they displace (Macom and Porter, 1996; Giraud et al., 2002; Le Breton et al., 2004; Abbott et al., 2007; Fournier et al., 2009; Sunamura et al., 2009) and larger than the colony boundaries of these same invasive species within their native range (Porter et al., 1997; Heller, 2004; Pedersen et al., 2006; Orivel et al., 2009).

Worker aggression is directed towards enemies while securing food resources and defending nests containing brood and reproductives, yet with the exception of single-queen (monogyne) colonies of S. invicta housed within a single well-defined nest (monodomy), the colonies of invasive ants contain numerous, ill-defined, largely ephemeral nests (polydomy). The organizational unit of these functionally connected nests has been termed the supercolony, ranging considerably in size (Giraud et al., 2002; Buczkowski et al., 2004; Hoffman, 2014), and populations consisting of one or more supercolonies are unicolonial (Helantera et al., 2009). However, though worker–worker aggression within supercolony boundaries is absent, local non-aggressive self–non-self discrimination (Steiner et al., 2007), subtle genetic distinctions (Ingram and Gordon, 2003) and food-sharing (Heller et al., 2008) likely limit the flow of information across discontinuous expansive supercolonies (Sunamura et al., 2009). Furthermore, by screening males prior to nest entry and mating, workers can restrict gene flow (Sunamura et al., 2011).

Since colony defence is energetically costly, with reproductive and resource acquisition trade-offs, unicolonial ant colonies grow larger than similar mutually aggressive colonies (Holway et al., 1998), with competitive advantages over some native ant species (Holway and Suarez, 2004). Although large field populations of introduced ants are often correlated with reduced native arthropod biodiversity and abundance (Holway, 1999; Guenard and Dunn, 2010; O'Dowd et al., 2003) there is no clear connection between the lack of aggression within supercolonies and invasion success and the advantages that numerical dominance provides in exploiting resources (Gordon and Heller, 2013).

Several models have been proposed for the evolution of unicoloniality. Tsutsui *et al.* (2000) have suggested that genetically and behaviourally homogeneous exotic ant propagules were transported outside of their native range and subsequently expanded, possibly due to Allee effects of mutually beneficial intraspecific interactions between queens and workers (Luque *et al.*, 2013). The massive Argentine ant supercolonies in southern Europe may have arisen from several South American introductions whose alleles coding for nestmate discrimination were subsequently lost either before or after intercolony interactions (Giraud *et al.*, 2002), with formerly mutually aggressive colonies exchanging queens and fusing (Vasquez and Silverman, 2008a, b), with increased fitness (Vasquez *et al.*, 2012). Thus, because of the high costs, abandoning aggression may be adaptive (Steiner *et al.*, 2007). However, very high levels of intraspecific aggression preclude colony fusion (Vasquez and Silverman, 2008a) with adjacent field colonies maintaining distinct boundaries and workers' increasing aggressiveness with continuous contact (Thomas *et al.*, 2005). Though invasive populations within a species may possess certain unique behavioural syndromes (Jandt *et al.*, 2014) favouring invasion success (e.g. unicoloniality and flexible nesting behaviours), many species display the same behaviours in both their native and introduced ranges and may thus be pre-adapted for invasion success (Heller, 2004; Pedersen *et al.*, 2006; Lebrun *et al.*, 2007; Foucad *et al.*, 2009; Cremer *et al.*, 2008; Buczkowski, 2010; Menke *et al.*, 2010). *Myrmica rubra* represents an interesting exception to the possible connection between unicoloniality and invasiveness. Introduced from Europe in the early twentieth century and spreading slowly across portions of the northeastern United States, *M. rubra* maintains a multicolonial population structure with aggression across nests <10 m apart (Garnas *et al.*, 2007). Though multiple introductions of distinct colony fragments may underlie intraspecific aggression in *M. rubra*, multicoloniality may be a derived trait as a result of intranest mating and loss of gene flow between nests (e.g. Drescher *et al.*, 2010).

Nestmate Recognition

Behavioural, genetic and chemical criteria are employed by investigators in determining colony boundaries and the changes in colony structure following introductions. For the most part, these characteristics vary consistently and allow for assigning colony identity, but colony alignments based on behavioural criterion are by far the most robust and meaningful: there is little doubt that a worker attacked and repelled or killed is not a sibling or nestmate. Ant behaviours, not necessarily those unique to invasive species, such as mouth gaping and twitching, one ant subduing and carrying another, dorsal flexion for chemical defence, biting or pulling legs and antennae, and use of the sting (e.g. Suarez *et al.*, 1999; Foucaud *et al.*, 2009; Gruber *et al.*, 2012) are displayed during aggressive worker encounters. However, measuring aggression behaviour is not straightforward since the costs (C) and benefits (B) of aggression are context dependent. Non-nestmate workers collected apart from nests with brood and confined within a neutral arena are less likely to fight when $C > B$, while an intruder worker placed near a nest entrance will likely be attacked by defending workers ($B > C$) (Roulston *et al.*, 2003; Buczkowski and Silverman, 2005). Furthermore, subtle interactions conveying important colony-level behaviours may be overlooked (Steiner *et al.*, 2007; Bjoerkman-Chiswell *et al.*, 2008) and aggression tests not conducted blind are prone to observer bias (van Wilgenburg and Elgar, 2013). Additionally, aggression in certain instances may only be triggered by a sufficiently aggressive stimulus. For example, young *S. invicta* are spared attack by more aggressive workers from neighbouring colonies by assuming a submissive posture (Cassill *et al.*, 2008).

Worker ants make decisions along the accept-vs-attack continuum following direct antenna-to-body contact with con- and heterospecifics. Ants and other social insects possess unique colony-specific chemical signatures on their cuticle, which are largely genetically encoded but can also contain extrinsic information. These signatures are largely a complex mixture of hydrocarbon alkanes, alkenes and methyl-branched alkanes, varying qualitatively and quantitatively within and between species (Howard and Blomquist, 2005; van Wilgenburg *et al.*, 2010). Current nestmate recognition models suggest that following contact, hydrocarbon signatures are compared with an internal template and produce a graded response depending on the closeness of the match and the context of the encounter (van Zweden and d'Ettorre, 2010; Sturgis and Gordon, 2012). Invasive ants from mutually aggressive, genetically distinct (based on neutral microsatellite markers) colonies, display different hydrocarbon signatures (Errard *et al.*, 2005; Ugelvig *et al.*, 2008; Fournier *et al.*, 2009; Vogel *et al.*, 2009; Blight *et al.*, 2012). However, these signatures are somewhat plastic, changing over time (Drescher *et al.*, 2010) and being modified by extrinsic contributions such as diet (Liang and Silverman, 2000), to the point where single colonies disassociate (Silverman and Liang, 2001). Hydrocarbons somewhat unique to the brown-banded cockroach, *Supella longipalpa*, were responsible for inducing aggression in the Argentine ant (Liang *et al.*, 2001; Brandt *et al.*, 2009). Thus, environmental factors can induce aggression behaviour between nestmates and diminish it between non-nestmates (Buczkowski *et al.*, 2005), but not always (Jaquiéry *et al.*, 2005; Thomas *et al.*, 2006), which may be a function of how high aggression was prior to continuous colony interactions. Furthermore, *L. humile* workers and queens from distinct colonies with nests in close proximity can share cuticular hydrocarbons, leading to adoption of non-nestmate queens and colony fusion (Vasquez *et al.*, 2008, 2009). Besides affecting nestmate recognition, diet affects the outcome of intraspecific competitive interactions, with workers on high carbohydrate diets dominating individuals on high protein diets (Grover *et al.*, 2007).

Degrees of intraspecific aggression behaviour can vary across colonies (Abril and Gomez, 2010). The level of genetic diversity within *L. humile* colonies affects nestmate recognition asymmetrically with workers from less diverse colonies being aggressive towards workers from colonies with greater allelic diversity, presumably because lower genetic diversity translates to less behavioural plasticity and accommodation (Tsutsui *et al.*, 2003). Prey-hydrocarbon-induced aggression varies regionally within the introduced range of the Argentine ant. While nests from within an expansive unicolonial population from California all became highly aggressive towards nestmates following exposure to *S. longipalpa*, aggression was only mildly induced across colonies within the southeastern United States. The distinction in regional behaviours was likely due to an aggression threshold being exceeded in the California colonies following exposure to key *S. longipalpa* cuticular hydrocarbons (Buczkowski and Silverman, 2006).

Interspecific Behaviours

Interactions with Native Ant Species

Invasive ants have been quite successful at driving native ants from their established habitats, but are they more aggressive and better competitors than native ant species? Two of the most successful exotic ants (*L. humile* and *S. invicta*) coexist within species-rich ant communities in South American woodlands and savannahs. The release from intense interspecific competition following introduction events and rapid response to disturbance may, in large part, underlie invasion success (Lebrun *et al.*, 2007; Suarez *et al.*, 2008). So, these and other ant invasions are largely events of opportunity where propagules from a small suite of behaviourally endowed species are transported by human activity to locations with suitable climate, avoiding detection and escape from quarantine procedures, and deposited in habitats with a depauperate competitive native ant fauna, plus suitable food.

While workers of several invasive taxa possess stings (e.g. *S. invicta*, *W. auropunctata*, *P. megacephala*, *P. chinensis*, *M. rubra*) or chemical weapons (e.g. *L. humile*, *T. sessile*, *N. fulva*) that can immobilize formicine heterospecifics, they are no better endowed than the native species they displace. In staged one-on-one encounters, an invasive ant worker is no more likely to emerge as victor than a native ant worker (Buczkowski and Bennett, 2008) and lone invasive ant workers may display behavioural plasticity by acting submissive (Carpintero and Reyes-Lopez, 2008; Sagata and Lester, 2009; Blight *et al.*, 2010) to deflect aggression. However, the outcomes from both laboratory and field colony-level competition studies are entirely different. Here, numerical advantages enjoyed by invasive species in both the number of workers per nest and number of nests per colony translate into gains in direct (interference) competition such as killing native ant workers or preventing forager egress from nests and/or indirect (exploitative) interactions where food is located quickly and dominated and retrieved (Human and Gordon, 1996; Holway, 1999; Zee and Holway, 2006; Paris and Espadaler, 2009; Drescher *et al.*, 2011). Prey, such as termites, exploited by both native and invasive ants (Bednar and Silverman, 2011), may be reduced to levels insufficient to support native species (Dejean *et al.*, 2007).

Individual behaviours underlying invasive ant competitive success include high foraging tempo and worker recruitment (Grover *et al.*, 2007) and continuous diel activity (Human and Gordon, 1996; Alder and Silverman, 2005). However, substantial advantages in both worker and nest number are likely key to bringing individuals to nearby resources quickly, which can then be secured (Holway and Case, 2001).

As discussed above, it is debatable whether a causal relationship exists between uni-coloniality and invasion success: any presumed linkage is based on retrospective studies, conducted after the invasions are fairly advanced, and thus detectable, with the native fauna largely decimated. Ant invasions begin small, with incipient colonies probably numbering at most in the hundreds or thousands, not millions of workers, and thus these invasive propagules do not enjoy a numerical advantage where they are deposited among native ants. Due to both ethical and practical considerations the fate of invasive ant propagules are not tracked over time: once detected they should be eradicated. So studies examining how newly introduced invasive ant propagules interact with the native ants and potential nutrient-bearing mutualists are limited to the laboratory and confined field conditions where the escape of reproductives is minimized (Brightwell and Silverman, 2007; Sagata and Lester, 2009; Shik and Silverman, 2013).

In non-disturbed habitats invasive ants encounter an array of native species which are able to escape detection (e.g. niche non-overlap, Holway, 1999) or actively defend territories and eliminate the invader. The winter ant, *Prenolepis imparis*, armed with a chemical defence (Sorrells *et al.*, 2011) coexists with or outcompetes the exotic *L. humile* in areas away from human disturbance; however, where human activities (e.g. irrigation) promote *L. humile* spread *P. imparis* is largely absent (Gordon and Heller, 2013). Other local ants such as *Tapinoma nigerrimum* and *T. simrothi*, which share characteristics with *L. humile*, such as large polygynous and polydomous colonies with worker mass recruitment to hemipteran exudates, outcompeted *L. humile* in assays with similar worker numbers, however, *L. humile* uses cooperative fighting strategies to overwhelm competitors (Blight *et al.*, 2010). Therefore, these and other native ant species (Walters and Mackay, 2005; Sagata and Lester, 2009) can offer strong resistance and delay or prevent the spread of Argentine ants. However, when environmental conditions favour *L. humile* colony growth, native ants are frequently defenceless when outnumbered (Thomas and Holway, 2005).

Intraspecific differences in invasive ant competitive ability can influence the outcome of interactions with native ants. Argentine ant workers from separate European supercolonies differed in their aggression towards native ants and towards each other with workers from the larger colony dominating in dyad aggression assays (Abril and Gomez, 2010). Abbott *et al.* (2007) identified genetic distinctions between *A. gracilipes* occupying separate Pacific islands and suggested that innate behavioural characteristics of the invader, in contrast to characteristics of the invaded habitat, were responsible for there being higher *A. gracilipes* abundance and 50% fewer other ant species on one of these islands.

LeBreton *et al.* (2007) provided evidence for the 'enemy release hypothesis' in animal invasions by highlighting the importance of enemy recognition and specification in competitive interactions. Ant species of the genus *Pheidole* dominated *W. auropunctata* in regions where both species were native (French Guiana), however, *Pheidole* species were ineffective at killing *W. auropunctata* in places where only species of *Pheidole* were native (New Caledonia). LeBreton *et al.* suggested that naïve *Pheidole* species in New Caledonia lacked the behavioural repertoire required to compete favourably with invasive *W. auropunctata*. Similarly, Dejean *et al.* (2008) reported that invasive *P. megacephala* were unable to raid over 90% of nests of other ants within their native range (Cameroon), whereas in their introduced range (Mexico) only 27% of the local species resisted nest raiding, suggesting that the ants species in the invaded range lacked the behaviours needed to resist displacement.

Interactions Between Invasive Ant Species

Though generally introduced through common ports of entry and occupying similar ecological niches, little attention has been devoted to competitive interactions between invasive ant species within their introduced range. Recently, *S. invicta* colonies were discovered in habitat within California occupied by *L. humile* for over a hundred years (Kabashima *et al.*, 2007). When numerically dominant, *L. humile* defeated *S. invicta* in laboratory colony assays, however, *L. humile* were defeated when confronting an equivalent number of *S. invicta* (Kabashima *et al.*, 2007). Another recently discovered invasive ant, *Nylanderia fulva*, which coexists with *S. invicta* in native South America, is one of the few ants able to compete effectively with *S. invicta* within its introduced range (LeBrun *et al.*, 2013). *N. fulva* is able to neutralize *S. invicta* neurotoxin with formic acid secreted from the acidopore at the tip of the gaster (Lebrun *et al.*, 2014). Though introduced to the southeastern United States from different continents, *Pachycondyla chinensis* and *L. humile* occupy similar habitat in human-maintained landscapes and bordering forest fragments. *P. chinensis* are rapidly displacing *L. humile* by expanding their colonies early in the season (Spicer-Rice and Silverman, 2013a) and displaying submissive behaviours to deflect attacks by *L. humile* (Spicer-Rice and Silverman, 2013b). A competitive hierarchy between invasive ants exploiting floral nectar revealed dominance by *A. gracilipes*, followed by *L. humile* and *Pheidole megacephala* (Lach, 2005). Competition between invasive ant species for limited space and food is predicted to intensify with increased propagule pressure and colony expansion.

Omnivory and Foraging Behaviour

Compared with invasive ants, highly territorial native ant colonies are relatively small. While direct interference competition between native ant conspecifics diverts resource investment from growth to defence, limited access to reliable, high-quality food sources also constrains colony expansion. Ants that are exclusively carnivorous occupying a fixed space have relatively small colonies (Hölldobler and Wilson, 1990), while nomadic predators such as army ants can sustain millions of individuals by exploiting prey over large areas (Gottwald, 1995). Colonies of omnivorous or herbivorous ants can become quite large, the largest among these represented by species which depend on readily renewable food sources in the form of plant material/fungus (Wirth *et al.*, 1997) and plant and insect exudates (Davidson *et al.*, 2003).

Interactions with Mutualist Partners

Invasive ant species are largely omnivorous (Holway *et al.*, 2002; Ness and Bronstein, 2004). The broad diets of invasive ants that completely overlap, and extend beyond, the diets of native ants provide a competitive advantage in resource exploitation (Le Breton *et al.*, 2005; Crowder and Snyder, 2010). With one exception (*Pachycondyla* [*Brachyponera*] *chinensis*), invasive ant species utilize carbohydrate-rich floral and extrafloral nectar and hemipteran honeydew, and in addition scavenge prey (live and dead) to fuel colony growth (Ness and Bronstein, 2004; Lach, 2005; Powell *et al.*, 2009; Helms, 2013). However, unlike prey items that can become rapidly depleted when an ant colony completely saturates an area, plant-derived nutrients are renewable, are of little cost for plants to produce, and their location is readily learned. Thus, foraging trails to these dependable nutrient sources are reinforced over extended periods, improving the efficiency of food collection and distribution. A nearby dependable food resource such as honeydew or nectar may be key to the establishment and spread of invasive ants (Helms and Vinson, 2002; O'Dowd *et al.*, 2003; LeBreton *et al.*, 2005; Brightwell and Silverman, 2011). Argentine ant propagules developed faster, producing more brood, when provided hemipteran honeydew rather than insect prey (Shik and Silverman, 2013; Shik *et al.*, 2014). Also, simple sugars fuelled Argentine ant foraging tempo, aggression (Grover *et al.*, 2007) and colony growth enhancing interference competition (Kay *et al.*, 2010), behaviours that could give an incipient colony a competitive advantage over resident native ants. In contrast, invasive *N. fulva* on low-sugar diets were more aggressive and less likely to be killed by *S. invicta* (Horn *et al.*, 2013). Tillberg *et al.* (2007) reported that behind an invasion-front Argentine ants relied more on plant-derived nutrients than at the advancing edge. This is likely a result of prey becoming depleted and replaced by honeydew-producing hemipterans farmed by ants behind the front. Though largely carnivorous, carbohydrates and not amino acids contributed to increased *S. invicta* worker and brood production (Helms and Vinson, 2008; Wilder *et al.*, 2011).

There is no evidence that the predatory ponerine ant, *P. chinensis*, consumes hemipteran honeydew or extrafloral nectar, but in addition to scavenging dead and living insects on the forest floor, they prey on living subterranean termites (Bednar and Silverman, 2011). In a sense, termites are a renewable resource much like aphids and scale insects: perennial shrubs and trees can support very large numbers of honeydew-excreting insects, while fallen decaying wood can support massive long-lived termite colonies fuelling *P. chinensis* colony growth. Polydomy (Guenard and Dunn, 2010), efficient search behaviour and food retrieval (Guenard and Silverman, 2011) and better utilization of termites by *P. chinensis* than the keystone native ant, *Aphaenogaster rudis* (Bednar *et al.*, 2013), may contribute to the displacement of *A. rudis* by *P. chinensis*.

Because of the nutritional rewards they provide, honeydew-excreting hemipterans are frequently protected by their mutualistic ant partners (Buckley, 1987) and hemipteran population densities typically are higher in ant-invaded areas (Coppler *et al.*, 2007; Daane *et al.*, 2007; McPhee *et al.*, 2012). Red imported fire ants build soil shelters at the base of plants within which their mutualist scale resides (Helms and Vinson, 2002) and they also provide important protection to cotton aphids (Rice and Eubanks, 2013). Furthermore, by attracting *S. invicta* to plant parts with both pest and beneficial insects, aphid–fire ant mutualisms alter agroecosytem community structure (Kaplan and Eubanks, 2005), which may extend to natural communities as well. Little is known about whether on a per capita basis invasive ants are better protectors of mutualist partners than native ants, however, in laboratory studies with equal numbers of ants, cotton aphid numbers increased significantly more when tended by *L. humile* versus *T. sessile* (Powell and Silverman, 2010a). Furthermore, *L. humile* reduced predation on cotton aphids more than *T. sessile* (Powell and Silverman, 2010b).

Foraging and Recruitment Behaviour

By organizing collectively, ants, and other social insects, are particularly efficient at locating food patches and assessing food value, then communicating this information to nestmate workers for retrieval and sharing with non-foraging castes. While not unique to them, invasive ants display certain foraging, recruitment and nutrient transfer behaviours that may contribute substantially to their establishment and spread within new habitat.

In multicolonial monodomous ants, foraging individuals locate food across landscapes of various size and complexity, assess its nutritional value, return to the colony and then recruit nestmates to deliver the resource to the nest (central-place foraging). Except in instances where extremely large colonies dominate nearby high value resources (e.g. *Atta*, *Oecophylla*), this strategy is inherently dangerous. Lengthy foraging distances expose scouting workers to risks of predation, inimical weather and hazardous terrain. Furthermore, there is the risk of losing the resource to competitors between the time that the resource is located and a sufficient number of nestmates are recruited to dominate it. Since nest construction by multicolonial ants frequently requires a substantial investment in resource allocation, nests are spatially fixed and thus must be constructed in a stable environment with a faithful food supply (McGlynn, 2012).

With one exception (monogyne *S. invicta* colonies), all invasive ants are polydomous, constructing impermanent nests and using a dispersed central-place foraging strategy to quickly locate (Holway and Case, 2000) and dominate (Holway and Case, 2001) food distributed within broad colony boundaries. These ants are all highly vagile, moving nests rapidly in response to disturbance and to changes in food availability: Argentine ants vacate nests in food-depleted locations and establish nests near a new food source (Silverman and Nsimba, 2000). Since nest-to-food distances can be fairly short in polydomous ants the foraging distance of a given worker may be minimal, yet the distance and rate that nutrients move between nests across the entire colony could be substantial; e.g. 32 m after 4 days in *A. gracilipes* (Hoffman, 2014); *c.* 50 m after several days in *L. humile* (Markin, 1968; Vega and Rust, 2003). Resources are shared along trails between closely linked genetically similar *L. humile* nests rather than across the long distances within supercolony boundaries (Heller *et al.*, 2008). Argentine ant collective searching networks adjust to local density (Gordon and Heller, 2013), and recruitment to food occurs from nearby existing trails rather than from more distant nests (Flanagan *et al.*, 2013). This flexible foraging behaviour facilitates the early arrival at food resources yielding a competitive advantage over native species (Human and Gordon, 1999) and extends forager activity well beyond the network of interconnected nests, leading to the expansion of the colony boundary when new resources are identified and exploited (Rowles and Silverman, 2009). Food distribution within other invasive ant colonies is somewhat restricted. *Tapinoma sessile* workers utilize the trail being provisioned with the food, but not at unconnected locations (Buczkowski and Bennett, 2006) and workers across nests of polygyne *S. invicta* only share resources with neighbouring nests (Weeks *et al.*, 2004).

Of those invasive ants investigated, diel foraging activity is continuous compared with the discrete foraging behaviour patterns of native ants (Human and Gordon, 1996; Vogt *et al.*, 2004; Abbott, 2005; Alder and Silverman, 2005). This around-the-clock activity by invasive ants may prevent native ants from accessing food resources.

Mass recruitment to food, nests and defensive positions by chemical trail following is fairly ubiquitous across the Formicidae, particularly the more highly evolved ant taxa and uniformly in species with large colonies (Hölldobler and Wilson, 1990). All but one invasive ant species (*P. chinensis* – discussed below) recruit via trail pheromones, yet, it is too early to conclude that mass recruitment behaviour by itself is a key trait underlying invasion success, although mass recruitment may be more efficient in invasive ants, allowing them to establish at the propagule stage and dominate resources once they become established. Sophisticated recruitment behaviour in *Anoplolepis gracilipes* and *Paratrechina longicornis* using a modular system of multiple chemicals improves communication and foraging efficiency and flexibility (Witte *et al.*, 2007; a l'Allemand and Witt, 2010). *Paratrechina longicornis* uses fairly persistent pheromones to recruit stable food sources, plus short-term pheromones to recruit nest mates to assist in the exploitation of large food items. They also employ a somewhat unique 'escort' behaviour where accompanying workers facilitate the handling of live prey by transporting workers (Czaczkes *et al.*, 2013). *Pheidole megacephala* uses two different chemical signals to recruit nestmates: a long-lasting pheromone during exploration that elicits weak recruitment and a less persistent pheromone stimulating much stronger recruitment during food exploitation (Dussutour *et al.*, 2009). The use of different pheromones has been found in Argentine ants (Deneubourg *et al.*, 1990) and fire ants (Vander Meer *et al.*, 1981, 1990). It is not clear whether this trail pheromone complexity is unique to invasive ants.

The currently lone invasive ponerine ant, *P. chinensis*, uses a recruitment strategy resembling other more primitive ponerines of relatively small colony size rather than that of the suite of mass recruiting invasive ants. Upon locating a nearby food resource a *P. chinensis* forager returns to its nest, then antennates other workers near the nest entrance. A contacted nestmate assumes a contracted quiescent posture, is picked up and carried outside the nest by the returning worker, and deposited on the food, which it retrieves (Guenard and Silverman, 2011). No trail pheromones have been detected in this ant. This rather slow recruitment process of tandem carrying does not appear to be consistent with behaviours underlying invasion success. However, invasive *P. chinensis* are polydomous, consequently, food is never very far from a nest entrance and few native ant competitors have access to these same resources (Guenard and Dunn, 2010; Bednar and Silverman, 2011; Bednar *et al.*, 2013).

Conclusions and Future Directions

Despite decades of efforts to prevent, contain and eradicate invasive pests, there is no sign that the rate of invasions is slowing down. Indeed, it is expected that biological invasions will become more pervasive and few ecosystems will be spared the impact

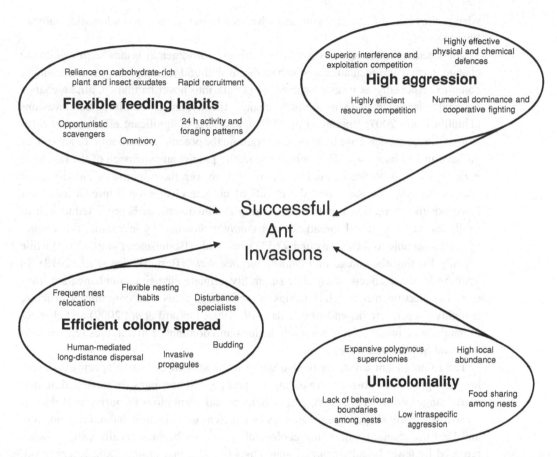

Figure 13.1 Behavioural characteristics promoting the global spread of invasive ants. A variety of factors can affect the prevalence of invasive ants. In this model, unicoloniality, high aggression, flexible feeding habits and efficient colony spread act together to shape the establishment, spread and ecological dominance of invasive ants.

of alien invaders (Butchart *et al.*, 2010; Simberloff *et al.*, 2013). Ants are a particular concern for conservation (Holway *et al.*, 2002; Rabitsch, 2011) because they are easily transported by accident and have a tremendous impact on local biodiversity (Lach and Hooper-Bui, 2010), economics (Harris and Barker, 2007) and ecosystem health (Holway *et al.*, 2002). Invasive ants share a suite of behavioural characteristics that facilitate their introduction, establishment and subsequent spread. These include rapid habitat exploration, reduced intraspecific aggression, high vagility, colony reproduction by budding, hemipteran tending and defence, and omnivory (Holway *et al.*, 2002) (Figure 13.1). These behaviours interact with a wide range of biotic, abiotic and human-mediated factors, and play an important role in invasion success (Roura-Pascual *et al.*, 2010). The accelerating rate of human-induced environmental changes such as urbanization, habitat fragmentation and climate change are leading to greater risks of alien incursions and promote the spread and establishment of biological invaders. Given these changes, what is the future of ant invasions and what trends and behavioural adaptations are we likely to see?

The interaction between invasions and other environmental issues such as climate change, habitat fragmentation and urbanization will undoubtedly become increasingly complex, especially as invasive species penetrate into novel habitats. Climate change especially is expected to profoundly change the dynamics of biological invasions (Thullier *et al.*, 2007; Bradley *et al.*, 2010), and have a significant effect on ant invasions. Climate models have been used to examine the potential distribution of ants under global climate change and these models generally predict an expansion of the ranges of invasive ants (e.g. Roura-Pascual *et al.*, 2004). However, the response of individual ant species may vary considerably; the effects of climate change on future distributions being positive, neutral or negative. For example, climate change is predicted to significantly increase the global spread of *Pachycondyla chinensis* by increasing the amount of habitat suitable to their invasion by 65% worldwide (Bertelsmeier *et al.*, 2013) while actually limiting the spread of *Pheidole megacephala* (Bertelsmeier *et al.*, 2012). In addition to direct effects of climatic suitability, climate change can enhance ant invasions by affecting species relationships in the invaded areas. Ant community structure is highly temperature dependent (Cerda *et al.*, 1997; Lessard *et al.*, 2009) and changes in temperature might shift species interactions and dominance hierarchies (e.g. Spicer-Rice and Silverman, 2013a).

The future of ant invasions is also going to be shaped by native species undergoing demographic explosions, expanding their range, and becoming invasive within their native range (Valery *et al.*, 2008, 2009). Behavioural adaptations favouring unicoloniality will be largely shaped by human activities such as urbanization, habitat degradation, and land use changes with many ecological specialists become locally extinct, being replaced by fewer broadly adapted generalists (e.g. Cremer *et al.*, 2008; Buczkowski, 2010; Menke *et al.*, 2010).

The future of ant invasions is also going to be affected by changes in the taxonomic and biogeographic patterns of invaders. While the majority of invasive ant species established within the last century have originated in tropical and subtropical climates (McGlynn, 1999), the past several decades have seen an increase in invasions by temperate climate ants such as *Lasius neglectus*, *Myrmica rubra*, *Pachycondyla chinensis*, *Tapinoma sessile* and *Tetramorium tsushimae* (Steiner *et al.*, 2006; Guenard and Dunn, 2010; Buczkowski and Krushelnycky, 2012). Climate change will likely modify the environmental impact of these species by enhancing their competitive and predatory effects on native species.

Finally, as invasive species become more widely distributed their ranges may begin to overlap resulting in competitive interactions among multiple invaders. Established populations of widespread invasive ants may become displaced by newly introduced species resulting in dominance shifts, local expatriations, and perhaps even the possibility that invasive species may be used as a tool to combat other invasives (e.g. Spicer-Rice and Silverman, 2013a; LeBrun *et al.*, 2013).

In summary, ant invasions will continue to pose a significant problem for natural and managed ecosystems, but new invasions and the ever-changing nature of current invasions will offer new challenges and opportunities for research on invasive ant biology, prevention and management.

References

a l'Allemand, S.L. and Witte, V. (2010). A sophisticated, modular communication contributes to ecological dominance in the invasive ant *Anoplolepis gracilipes*. *Biological Invasions*, 12, 3551–3561.

Abbott, K.L. (2005). Supercolonies of the invasive yellow crazy ant, *Anoplolepis gracilipes* on an oceanic island: forager activity patterns, density and biomass. *Insectes Sociaux*, 52, 266–273.

Abbott, K.L., Greaves, S.N.J., Ritchie, P.A. and Lester, P.J. (2007). Behaviourally and genetically distinct populations of an invasive ant provide insight into invasion history and impacts on a tropical ant community. *Biological Invasions*, 9, 453–463.

Abril, S. and Gomez, C. (2010). Aggressive behavior of the two European Argentine ant supercolonies (Hymenoptera: Formicidae) towards displaced native ant species of the northeastern Iberian Peninsula. *Myrmecological News*, 14, 99–106.

Alder, P.M. and Silverman, J. (2005). Effects of interspecific competition between two urban ant species, *Linepithema humile* and *Monomorium minimum*, on toxic bait performance. *Journal of Economic Entomology*, 98, 493–501.

Ascunce, M.S., Yang, C-C., Oakey, J., *et al.* (2011). Global invasion history of the fire ant, *Solenopsis invicta*. *Science*, 331, 1066–1068.

Bednar, D.M. and Silverman, J. (2011). Use of termites, *Reticulitermes virginicus*, as a springboard in the invasive success of a predatory ant, *Pachycondyla (=Brachyponera) chinensis*. *Insectes Sociaux*, 58, 459–467.

Bednar, D.M., Shik J.Z. and Silverman, J. (2013). Prey handling performance facilitates competitive dominance of an invasive over native keystone ant. *Behavioral Ecology*, 24, 1312–1319.

Bertelsmeier, C. and Courchamp, F. (2014). Future ant invasions in France. *Environmental Conservation*, 41, 217–228.

Bertelsmeier, C., Luque, G.M. and Courchamp, F. (2012). Global warming may freeze the invasion of big-headed ants. *Biological Invasions*, 15, 1561–1572.

Bertelsmeier, C., Guenard, B. and Courchamp, F. (2013). Climate change may boost the invasion of the Asian needle ant. *PLoS ONE*, 8(10), e75438. doi:10.1371/journal.pone.0075438.

Bjoerkman-Chiswell, B.T., van Wilgenburg, E., Thomas, M.L., Swearer S.E. and Elgar, M.A. (2008). Absence of aggression but not nestmate recognition in an Australian population of the Argentine ant *Linepithema humile*. *Insectes Sociaux*, 55, 207–212.

Blight, O., Provost, E., Renucci, M., Tirard, A. and Orgeas, J. (2010). A native ant armed to limit the spread of the Argentine ant. *Biological Invasions*, 12, 3785–3793.

Blight, O., Berville, L., Vogel, V., *et al.* (2012). Variation in the level of aggression, chemical and genetic distance among three supercolonies of the Argentine ant in Europe. *Molecular Ecology*, 21, 4106–4121.

Bradley, B.A., Blumenthal, D.M., Wilcove, D.S. and Ziska, L.H. (2010). Predicting plant invasions in an era of global change. *Trends in Ecology and Evolution*, 25, 310–318.

Brandt, M., van Wilgenburg, E., Sulc, R., Shea, K.J. and Tsutsui, N.D. (2009). The scent of supercolonies: the discovery, synthesis and behavioural verification of ant colony recognition cues. *BMC Biology*, 7, 71.

Brightwell R.J. and Silverman. J. (2007). Argentine ant foraging activity and interspecific competition in complete vs. queenless and broodless colonies. *Insectes Sociaux*, 54, 329–333.

Brightwell R.J. and Silverman, J. (2011). The Argentine ant persists through unfavorable winters via a mutualism facilitated by a native tree. *Environmental Entomology*, 40, 1019–1026.

Brightwell, R.J., Labadie, P.L. and Silverman, J. (2010). Northward expansion of the invasive Argentine ant, *Linepithema humile* (Hymenoptera: Formicidae) in the eastern U.S. is constrained by winter soil temperatures. *Environmental Entomology*, 39, 1659–1665.

Buckley, R.C. (1987). Interactions involving plants, homoptera, and ants. *Annual Review of Ecology and Systematics*, 18, 111–135.

Buczkowski, G. (2010). Extreme life history plasticity and the evolution of invasive characteristics in a native ant. *Biological Invasions*, 12, 3343–3349.

Buczkowski, G. and Bennett, G.W. (2006). Dispersed central-place foraging in the polydomous odorous house ant, *Tapinoma sessile* as revealed by a protein marker. *Insectes Sociaux*, 53, 282–290.

Buczkowski, G. and Bennett, G.W. (2008). Aggressive interactions between the introduced Argentine ant, *Linepithema humile* and the native odorous house ant, *Tapinoma sessile*. *Biological Invasions*, 10, 1001–1011.

Buczkowski, G. and Kruskelnycky, P. (2012). The odorous house ant, *Tapinoma sessile* (Hymenoptera: Formicidae), as a new temperate-origin invader. *Myrmecological News*, 16, 61–66.

Buczkowski, G. and Silverman, J. (2005). Context-dependent nestmate discrimination and the effect of action thresholds on exogenous cue recognition in the Argentine ant. *Animal Behaviour*, 69, 741–749.

Buczkowski, G. and Silverman, J. (2006). Geographical variation in Argentine ant aggression behavior mediated by environmentally derived nestmate recognition cues. *Animal Behaviour*, 71, 327–335.

Buczkowski, G., Vargo, E. and Silverman, J. (2004). The diminutive supercolony: the Argentine ants of the southeastern United States. *Molecular Ecology*, 13, 2235–2242.

Buczkowski G., Kumar, R., Suib, S.L. and Silverman, J. (2005). Diet-related modification of cuticular hydrocarbon profiles of the Argentine ant, *Linepithema humile*, diminishes intercolony aggression. *Journal of Chemical Ecology*, 31, 829–843.

Butchart, S.H.M., Walpole, M., Collen, B., *et al.* (2010). Global biodiversity: indicators of recent declines. *Science*, 328, 1164–1168.

Carpintero, S. and Reyes-Lopez, J. (2008). The role of competitive dominance in the invasive ability of the Argentine ant (*Linepithema humile*). *Biological Invasions*, 10, 25–35.

Cassill, D.L., Vo, K. and Becker, B. (2008). Young fire ant workers feign death and survive aggressive neighbors. *Naturwissenschaften*, 95, 617–624.

Cerda, X., Retana, J. and Cros, S. (1997). Thermal disruption of transitive hierarchies in Mediterranean ant communities. *Journal of Animal Ecology*, 66, 363–374.

Chapple, D.G., Simmonds, S.M. and Wong, B.M. (2012). Can behavioral and personality traits influence the success of unintentional species introductions? *Trends in Ecology and Evolution*, 27, 57–64.

Coppler, L.B., Murphy, J.F. and Eubanks, M.D. (2007). Red imported fire ants (Hymenoptera: Formicidae) increase the abundance of aphids in tomato. *Florida Entomologist*, 90, 419–425.

Cremer, S., Ugelvig, L.V., Drijfhout, F.P., *et al.* (2008). The evolution of invasiveness in garden ants. *PLoS ONE*, 3(12), e3838.

Crowder, D.W. and Snyder, W.E. (2010). Eating their way to the top? Mechanisms underlying the success of invasive insect generalist predators. *Biological Invasions*, 12, 2857–2876.

Czaczkes, T.J., Vollet-Neto, A. and Ratnieks, F.L.W. (2013). Prey escorting behavior and possible convergent evolution of foraging recruitment mechanisms in an invasive ant. *Behavioral Ecology*, 24, 1177–1184.

Daane, K.M., Sime, K.R., Fallon, J. and Cooper, M.L. (2007). Impacts of Argentine ants on mealybugs and their natural enemies in California's coastal vineyards. *Ecological Entomology*, 32, 583–596.

Davidson, D.W., Cook S.C., Snelling, R.R. and Chua T.H. (2003). Explaining the abundance of ants in lowland tropical rainforest canopies. *Science*, 300, 969–972.

Dejean, A., Kenne, M. and Moreau, C.S. (2007). Predatory abilities favor the success of the invasive ant *Pheidole megacephala* in an introduced area. *Journal of Applied Entomology*, 131, 625–629.

Dejean, A., Moreau, C.S., Kenne, M. and Leponce, M. (2008). The raiding success of *Pheidole megacephala* on other ants in both its native and introduced ranges. *Comptes Rendus Biologies*, 331, 631–635.

Deneubourg, J.-L., Aron, S., Goss, S. and Pasteels, J.M. (1990). The self-organizing exploratory pattern of the Argentine ant. *Journal of Insect Behavior*, 3, 159–168.

Drescher, J., Bluthgen, N., Schmitt, T., Buhler, J. and Feldhaar, H. (2010). Societies drifting apart? behavioural, genetic and chemical differentiation between supercolonies in the yellow crazy ant *Anoplolepis gracilipes*. *PLoS ONE*, 5(10), e13581. doi:10.1371/journal.pone.0013581

Drescher, J, Feldhaar, H. and Bluthgen, N. (2011). Interspecific aggression and resource monopolization of the invasive ant *Anoplolepis gracilipes* in Malaysian Borneo. *Biotropica*, 43, 93–99.

Dussutour, A., Nicolis, S.C., Shephard, G., Beekman, M. and Sumpter, D.J.T. (2009). The role of multiple pheromones in food recruitment by ants. *The Journal of Experimental Biology*, 212, 2337–2348.

Errard, C., Delabie, J, Jourdan, H. and Hefetz, A. (2005). Intercontinental chemical variation in the invasive ant *Wasmannia auropunctata* (Roger) (Hymenoptera Formicidae): a key to the invasive success of a tramp species. *Naturwissenschaften*, 92, 319–323.

Espadaler, X., Rey, S. and Bernal, V. (2004). Queen number in a supercolony of the invasive garden ant, *Lasius neglectus*. *Insectes Sociaux*, 51, 232–238.

Flanagan, T.P., Pinter-Wollman, N.M., Moses, M.E. and Gordon, D.M. (2013). Fast and flexible: Argentine ants recruit from nearby trails. *PLoS ONE*, 8, e70888.

Foucaud, J., Orivel, J., Fournier, D., *et al.* (2009). Reproductive system, social organization, human disturbance and ecological dominance in native populations of the little fire ant, *Wasmannia auropunctata*. *Molecular Ecology*, 18, 5059–5073.

Fournier, D., de Biseau, J.-C. and Aron, S. (2009). Genetics, behaviour and chemical recognition of the invading ant *Pheidole megacephala*. *Molecular Ecology*, 18, 186–199.

Garnas, J.R., Drummond, F.A. and Groden, E. (2007). Intercolony aggression within and among local populations of the invasive ant, *Myrmica rubra* (Hymenoptera: Formicidae), in coastal Maine. *Environmental Entomology*, 36, 105–113.

Giraud, T., Pedersen, J.S. and Keller, L. (2002). Evolution of supercolonies: the Argentine ants of southern Europe. *Proceedings of the National Academy of Sciences of the Unites States of America*, 99, 6075–6079.

Gordon, D.M. and Heller, N.E. (2013). The invasive Argentine ant *Linepithema humile* (Hymenoptera: Formicidae) in Northern California reserves: from foraging behavior to local spread. *Myrmecological News*, 19, 103–110.

Gottwald, W.H. (1995). *Army Ants: The Biology of Social Predation*. Ithaca, NY: Cornell University Press.

Groden, E., Drummond, F.A., Garnas, J. and Franceour, A. (2005). Distribution of an invasive ant, *Myrmica rubra* (Hymenoptera: Formicidae), in Maine. *Journal of Economic Entomology*, 98, 1774–1784.

Grover, C.D., Kay, A.D., Monson, J.A., Marsh, T.C. and Holway, D.A. (2007). Linking nutrition and behavioral dominance: carbohydrate scarcity limits aggression and activity in Argentine ants. *Proceedings of the Royal Society: Biological Sciences B*, 274, 2951–2957.

Gruber, M.A.M., Hoffmann, B.D., Ritchie, P.A. and Lester, P.J. (2012). Recent behavioural and population genetic divergence of an invasive ant in a novel environment. *Diversity and Distributions*, 18, 323–333.

Guenard, B. and Dunn, R.R. (2010). A new (old), invasive ant in the hardwood forests of eastern North America and its potentially widespread impacts. *PLoS ONE*, 5, e11614.

Guenard, B. and Silverman, J. (2011). Tandem carrying, a new foraging strategy in ants: description, function and adaptive significance relative to other described foraging strategies. *Naturwissenschaften*, 98, 651–659.

Harris, R. and Barker, G. (2007). Relative risk of invasive ants (Hymenoptera: Formicidae) establishing in New Zealand invasive social insect. *New Zealand Journal of Zoology*, 34, 161–178.

Hee, J.J., Holway, D.A., Suarez, A.V. and Case, T.J. (2000). Role of propagule size in the success of incipient colonies of the invasive Argentine ant. *Conservation Biology*, 14, 559–563.

Helantera, H., Strassmann, J.E., Carrillo, J. and Queller, D.C. (2009). Unicolonial ants: where do they come from, what are they and where are they going? *Trends in Ecology and Evolution*, 24, 341–349.

Heller, N.E. (2004). Colony structure in native and introduced populations of the invasive Argentine ant, *Linepithema humile*. *Insectes Sociaux*, 51, 378–386.

Heller, N.E., Ingram, K.K. and Gordon, D.M. (2008). Nest connectivity and colony structure in unicolonial Argentine ants. *Insectes Sociaux*, 55, 397–403.

Helms, K.R. (2013). Mutualisms between ants (Hymenoptera: Formicidae) and honeydew-producing insects: are they important in ant invasions? *Myrmecological News*, 18, 61–71.

Helms, K.R. and Vinson, S.B. (2002). Widespread association of the invasive ant *Solenopsis invicta* with an invasive mealybug. *Ecology*, 83, 2425–2438.

Helms, K.R. and Vinson, S.B. (2008). Plant resources and colony growth in an invasive ant: the importance of honeydew-producing Hemiptera in carbohydrate transfer across trophic levels. *Environmental Entomology*, 37, 487–493.

Hicks B.J. (2012). How does *Myrmica rubra* (Hymenoptera: Formicidae) disperse in its native range? Record of male-only swarming flights from Newfoundland. *Myrmecological News*, 16, 31–34.

Hoffmann, B.D. (2014). Quantification of supercolonial traits in the yellow crazy ant, *Anoplolepis gracilipes*. *Journal of Insect Science*, 14, 1–21.

Hoffmann, B.D. and Saul, W.C. (2010). Yellow crazy ant (*Anoplolepis gracilipes*) invasions within undisturbed mainland Australian habitats: no support for biotic resistance hypothesis. *Biological Invasions*, 13, 3093–3108.

Hölldobler, B. and Wilson, E. (1990). *The Ants*. Cambridge, MA: Belknap Press.

Holway, D.A. (1999). Competitive mechanisms underlying the displacement of native ants by the invasive Argentine ant. *Ecology*, 80, 238–251.

Holway, D.A. and Case, T.J. (2000). Mechanisms of dispersed central-place foraging in polydomous colonies of the Argentine ant. *Animal Behaviour*, 59, 433–441.

Holway, D.A. and Case, T.J. (2001). Effects of colony-level variation on competitive ability in the invasive Argentine ant. *Animal Behaviour*, 61, 1181–1192.

Holway, D.A. and Suarez, A.V. (1999). Animal behavior: an essential component of invasion biology. *Trends in Ecology and Evolution*, 14, 328–330.

Holway, D.A. and Suarez, A.V. (2004). Colony-structure variation and interspecific competitive ability in the invasive Argentine ant. *Oecologia*, 138, 216–222.

Holway, D.A., Suarez, A.V. and Case, T.J. (1998). Loss of intraspecific aggression in the success of a widespread invasive social insect. *Science*, 282, 949–952.

Holway, D.A., Lach, L., Suarez, A.V., Tsutsui, N.D. and Case, T.J. (2002). Causes and consequences of ant invasions. *Annual Review of Ecology Evolution and Systematics*, 33, 181–233.

Horn, K.C., Eubanks, M.D. and Siemann, E. (2013). The effect of diet and opponent size on aggressive interactions involving Caribbean crazy ants (*Nylanderia fulva*). *PLoS ONE*, 8(6), e66912.

Howard, R.W. and Blomquist, G.J. (2005). Ecological, behavioral, and biochemical aspects of insect hydrocarbons. *Annual Review of Entomology*, 50, 371–393.

Human, K.G. and Gordon, D.M. (1996). Exploitative and interference competition between the Argentine ant and native ant species. *Oecologia*, 105, 405–412.

Human, K.G. and Gordon, D.M. (1997). Effects of Argentine ants on invertebrate diversity in northern California. *Conservation Biology*, 11, 1242–1248.

Human, K.G. and Gordon, D.M. (1999). Behavioral interactions of the invasive Argentine ant with native ant species. *Insectes Sociaux*, 46, 159–163.

Ingram, K.K. and Gordon, D.M. (2003). Genetic analysis of dispersal dynamics in an invading population of Argentine ants. *Ecology*, 84, 2832–2842.

Jaquiéry, J., Vogel, V. and Keller, L. (2005). Multilevel genetic analyses of two supercolonies of the Argentine ant, *Linepithema humile*. *Molecular Ecology*, 14, 589–598.

Jandt, J.M., Bengston, S., Pinter-Wollman, N., *et al.* (2014). Behavioural syndromes and social insects: personality at multiple levels. *Biological Reviews of the Cambridge Philosophical Society*, 89, 48–67.

Kabashima, J.N., Greenberg, L., Rust, M.K. and Paine, T.D. (2007). Aggressive interactions between *Solenopsis invicta* and *Linepithema humile* (Hymenoptera: Formicidae) under laboratory conditions. *Journal of Economic Entomology*, 100, 148–154.

Kaplan, I. and Eubanks, M.D. (2005). Aphids alter the community-wide impact of fire ants. *Ecology*, 86, 1640–1649.

Kay, A.D., Zumbusch, T.B., Heinen, J.L., Marsh, T.C. and Holway, D.A. (2010). Nutrition and interference competition have interactive effects on the behavior and performance of Argentine ants. *Ecology*, 91, 57e64.

Lach, L. (2005). Interference and exploitation competition of three nectar-thieving invasive ant species. *Insectes Sociaux*, 52, 257–262.

Lach, L. and Hooper-Bui, L.M. (2010). Consequences of ant invasions. In *Ant Ecology*, ed. Lach, L., Parr, C.L. and Abbott, K.L. Oxford, UK: Oxford University Press, pp. 261–286.

Le Breton, J., Delabie, J.H.C., Chazeau, J., Dejean, A. and Jourdan, H. (2004). Experimental evidence of large scale unicoloniality in the tramp ant *Wasmannia auropunctata* (Roger). *Journal of Insect Behaviour*, 17, 263–271.

Le Breton, J., Jourdan, H., Chazeau, J., Orivel, J. and Dejean, A. (2005). Niche opportunity and ant invasion: the case of *Wasmannia auropunctata* in a New Caledonian rain forest. *Journal of Tropical Ecology*, 21, 93–98.

LeBreton, J., Orivel J., Chazeau, J. and Dejean, A. (2007). Unadapted behaviour of native, dominant ant species during the colonization of an aggressive, invasive ant. *Ecological Research*, 22, 107–114.

LeBrun, E.G., Tillberg, C.V., Suarez, A.V., *et al.* (2007). An experimental study of competition between fire and Argentine ants in their native range. *Ecology*, 88, 63–75.

LeBrun, E.G., Abbott, J. and Gilbert, L.E. (2013). Imported crazy ant displaces imported fire ant, reduces and homogenizes grassland ant and arthropod assemblages. *Biological Invasions*, 15, 2429–2442.

LeBrun, E.G., Jones, N.T.J. and Gilbert, L.E. (2014). Chemical warfare among invaders: a detox-ification interaction facilitates an ant invasion. *Science*, 343, 1014–1017.

Lessard, J.P., Fordyce, J.A., Gotelli, N.J. and Sanders, N.J. (2009). Invasive ants alter the phylo-genetic structure of ant communities. *Ecology*, 90, 2664–2669.

Liang, D. and Silverman, J. (2000). "You are what you eat": diet modifies cuticular hydrocarbons and nestmate recognition in the Argentine ant, *Linepithema humile*. *Naturwissenschaften*, 87, 412–416.

Liang, D., Blomquist, G. and Silverman, J. (2001). Hydrocarbon-released nestmate aggression in the Argentine ant, *Linepithema humile*, following encounters with insect prey. *Comparative Biochemistry and Physiology Part B*, 129, 871–882.

Luque, G.M., Giraud, T. and Courchamp, F. (2013). Allee effects in ants. *Journal of Animal Ecology*, 82, 956–965.

Macom, T.E. and Porter, S.D. (1996). Comparison of polygyne and monogyne red imported fire ant (Hymenoptera: Formicidae) population densities. *Annals of the Entomological Society of America*, 89, 535–543.

Markin, G.P. (1968). Nest relationships of the Argentine ant, *Iridomyrmex humilis* (Hymenoptera: Formicidae). *Journal of Economic Entomology*, 41, 511–516.

McGlynn, T.P. (1999). The worldwide transfer of ants: geographic distribution and ecological invasions. *Journal of Biogeography*, 26, 535–548.

McGlynn, T.P. (2012). The ecology of nest movement in social insects. *Annual Review of Ento-mology*, 57, 291–308.

McPhee, K., Garnas, J., Drummond, F. and Groden, E. (2012). Homopterans and an invasive red ant, *Myrmica rubra* (L.), in Maine. *Environmental Entomology*, 41, 59–71.

Menke, S.B. and Holway, D.A. (2006). Abiotic factors control invasion by Argentine ants at the community scale. *Journal of Animal Ecology*, 75, 368–376.

Menke, S.B., Booth, W., Dunn, R.R., *et al.* (2010). Is it easy to be urban? Convergent success in urban habitats among lineages of a widespread native ant. *PLoS ONE*, 5, e9194.

Ness, J.H. and Bronstein, I.L. (2004). The effects of invasive ants on prospective ant mutualists. *Biological Invasions*, 6, 445–461.

O'Dowd, D.J., Green, P.T. and Lake, P.S. (2003). Invasional 'meltdown' on an oceanic island. *Ecology Letters*, 6, 812–817.

Orivel, J., Grangier, J., Foucaud, J., *et al.* (2009). Ecologically heterogeneous populations of the invasive ant *Wasmannia auropunctata* within its native and introduced ranges. *Ecological Ento-mology*, 34, 504–512.

Paris, C.I. and Espadaler, X. (2009). Honeydew collection by the invasive garden ant *Lasius neglectus* versus the native ant *L. grandis*. *Arthropod-Plant Interactions*, 3, 75–85.

Passera, L. and Keller, L. (1994). Mate availability and male dispersal in the Argentine and *Linep-ithema humile* (Mayr) (=*Iridomyrmex humilis*). *Animal Behaviour*, 48, 361–369.

Pedersen, J.S., Krieger, M.J.B., Vogel, V., Giraud, T. and Keller, L. (2006). Native supercolonies of unrelated individuals in the invasive Argentine ant. *Evolution*, 60, 782–791.

Phillips, B.L. and Suarez, A.V. (2012). The role of behavioural variation in invasion of new areas. In *Behavioural Responses to a Changing World: Mechanisms and Consequences*, ed. Candolin, U. and Wong, B.B.M. Oxford: Oxford University Press, pp. 190–200.

Porter, S.D., Williams, D.F., Patterson, R.S. and Fowler, H.G. (1997). Intercontinental differences in the abundance of *Solenopsis* fire ants (Hymenoptera: Formicidae): Escape from natural enemies? *Environmental Entomology*, 26, 373–384.

Powell, B.E. and Silverman, J. (2010a). Population growth of *Aphis gossypii* and *Myzus persicae* (Hemiptera: Aphididae) in the presence of *Linepithema humile* and *Tapinoma sessile* (Hymenoptera: Formicidae). *Environmental Entomology*, 39, 1492–1499.

Powell, B.E. and Silverman, J. (2010b). Impact of *Linepithema humile* and *Tapinoma sessile* (Hymenoptera: Formicidae) on three natural enemies of *Aphis gossypii* (Hemiptera: Aphididae). *Biological Control*, 54, 285–291.

Powell, B.E., Brightwell, R.J. and Silverman, J. (2009). Effect of an invasive and native ant on a field population of the black citrus aphid (Hemiptera: Aphididae). *Environmental Entomology*, 38, 1618–1625.

Rabitsch, W. (2011). The hitchhiker's guide to alien ant invasions. *BioControl*, 56, 551–572.

Rice, K.B. and Eubanks, M.D. (2013). No enemies needed: cotton aphids (Hemiptera: Aphididae) directly benefit from red imported fire ant (Hymenoptera: Formicidae) tending. *Florida Entomologist*, 96, 929–932.

Roulston, T.H., Buczkowski, G., Silverman, J. (2003). Nestmate discrimination in ants: effect of bioassay on aggressive behavior. *Insectes Sociaux*, 50, 151–159.

Roura-Pascual, N., Suarez, A.V., Gomez, *et al.* (2004). Geographical potential of Argentine ants (*Linepithema humile* Mayr) in the face of global climate change. *Proceedings of the Royal Society London B: Biological Sciences*, 271, 2527–2534.

Roura-Pascual, N., Hui, C., Takayoshi I., *et al.* (2011). Relative roles of climatic suitability and anthropogenic influence in determining the pattern of spread in a global invader. *Proceedings of the National Academy of Sciences, USA*, 108, 220–225.

Rowles, A.D. and O'Dowd, D.J. (2007). Interference competition by Argentine ants displaces native ants: implications for biotic resistance to invasion. *Biological Invasions*, 9, 73–85.

Rowles, A.D. and Silverman, J. (2009). Carbohydrate supply limits invasion of natural communities by Argentine ants. *Oecologia*, 161, 161–171.

Sagata, K. and Lester, P.J. (2009). Behavioural plasticity associated with propagule size, resources, and the invasion success of the Argentine ant *Linepithema humile*. *Journal of Applied Ecology*, 46, 19–27.

Shik, J.Z. and Silverman, J. (2013). Towards a nutritional ecology of invasive establishment: aphid mutualists provide better fuel for incipient Argentine ant colonies than insect prey. *Biological Invasions*, 15, 829–836.

Shik, J.Z., Kay, A.D. and Silverman, J. (2014). Aphid honeydew provides a nutritionally balanced resource for incipient Argentine ant mutualists. *Animal Behaviour*, 95, 33–39.

Shoemaker, D.D., DeHeer, C.J., Krieger, M.J.B. and Ross, K.G. (2006). Population genetics of the invasive fire ant *Solenopsis invicta* (Hymenoptera: Formicidae) in the United States. *Annals of the Entomological Society of America*, 99, 1213–1233.

Silverman, J. and Liang, D. (2001). Colony disassociation following diet partitioning in a unicolonial ant. *Naturwissenshaften*, 88, 73–77.

Silverman, J. and Nsimba, B. (2000). Soil-free collection of Argentine ants based on food-directed brood and queen movement. *Florida Entomologist*, 83, 10–16.

Simberloff, D., Martin, J.-L., Genovesi, P., *et al.* (2013). Impacts of biological invasions: what's what and the way forward. *Trends in Ecology and Evolution*, 28, 58–66.

Sorrells, T.R., Kuritzky, L.Y., Kauhanen, P.G., *et al.* (2011). Chemical defense by the native winter ant (*Prenolepis imparis*) against the invasive Argentine ant (*Linepithema humile*). *PLoS ONE*, 6(4), e18717.

Spicer-Rice, E. and Silverman, J. (2013a). Propagule pressure and climate contribute to the displacement of *Linepithema humile* by *Pachycondyla chinensis*. *PLoS ONE*, 8(2), e56281. doi:10.1371/journal.pone.0056281.

Spicer-Rice, E. and Silverman, J. (2013b). Submissive behaviour and habituation facilitate entry into habitat occupied by an invasive ant. *Animal Behaviour*, 86, 497–506.

Steiner, F.M., Schlick-Steiner, B.C., Trager, J., *et al.* (2006). *Tetramorium tsushimae*, a new invasive ant in North America. *Biological Invasions*, 8, 117–123.

Steiner, F.M., Schlick-Steiner B.C., Moder K., *et al.* (2007). Abandoning aggression but maintaining self-nonself discrimination as a first stage in ant supercolony formation. *Current Biology*, 17, 1903–1907.

Sturgis, S.J. and Gordon, D.M. (2012). Nestmate recognition in ants (Hymenoptera: Formicidae): a review. *Myrmecological News*, 16, 101–110.

Suarez, A.V., Tsutsui, N.D., Holway D.A. and Case T.J. (1999). Behavioral and genetic differentiation between native and introduced populations of the Argentine ant. *Biological Invasions*, 1, 43–53.

Suarez, A.V., Holway, D.A. and Case, T.J. (2001). Patterns of spread in biological invasions dominated by long-distance jump dispersal: insights from Argentine ants. *Proceedings of the National Academy of Sciences, USA*, 98, 1095–1100.

Suarez, A.V., Holway, D.A. and Tsutsui, N.D. (2008). Genetics and behavior of a colonizing species: the invasive Argentine ant. *The American Naturalist*, 172, Suppl. 1, S72–84.

Sunamura, E., Espadaler, X., Sakamoto, H., *et al.* (2009). Intercontinental union of Argentine ants: behavioral relationships among introduced populations in Europe, North America, and Asia. *Insectes Sociaux*, 56, 143–147.

Sunamura, E., Hoshizaki, S., Sakamoto, H., *et al.* (2011). Workers select mates for queens: a possible mechanism of gene flow restriction between supercolonies of the invasive Argentine ant. *Naturwissenschaften*, 98, 361–368.

Thomas, M.L. and Holway, D.A. (2005). Condition-specific competition between invasive Argentine ants and Australian *Iridomyrmex*. *Journal of Animal Ecology*, 74, 532–542.

Thomas, M.L., Tsutsui, N.D. and Holway, D.A. (2005). Intraspecific competition influences the symmetry and intensity of aggression in the Argentine ant. *Behavioral Ecology*, 16, 472–481.

Thomas, M.L., Payne-Makrisâ C. M., Suarez A.V., Tsutsui N.D. and Holway, D.A. (2006). When supercolonies collide: territorial aggression in an invasive and unicolonial social insect. *Molecular Ecology*, 15, 4303–4315.

Thullier, W., Richardson, D.M. and Midgley, G.F. (2007). Will climate change promote alien plant invasions? In *Biological Invasions*, ed. Nentwig, W. Berlin: Springer, pp. 197–211.

Tillberg, C.V., Holway, D.A., LeBrun, E.G. and Suarez, A.V. (2007). Trophic ecology of invasive Argentine ants in their native and introduced ranges. *Proceedings of the National Academy of Sciences, USA*, 104, 20856–20861.

Tschinkel, W.R. (2006). *The Fire Ants*. Cambridge, MA: The Belknap Press of Harvard University Press.

Tsutsui, N.D., Suarez, A.V., Holway, D.A. and Case T.J. (2000). Reduced genetic variation and the success of an invasive species. *Proceedings of the National Academy of Sciences, USA*, 97, 5948–5953.

Tsutsui, N.D., Suarez, A.V. and Grosberg, R.K. (2003). Genetic diversity, asymmetrical aggression, and recognition in a widespread invasive species. *Proceedings of the National Academy of Sciences, USA*, 100, 1078–1083.

Ugelvig, L.V., Drijfhout, F.P., Kronauer, D.J.C, *et al.* (2008). The introduction history of invasive garden ants in Europe: integrating genetic, chemical and behavioural approaches. *BMC Biology*, 6, 11.

Valery, L., Fritz, H., Lefeuvre, J-C. and Simberloff, D. (2008). In search of a real definition of the biological invasion phenomenon itself. *Biological Invasions*, 10, 1345–1351.

Valery, L., Fritz, H., Lefeuvre, J-C. and Simberloff, D. (2009). Invasive species can also be native. *Trends in Ecology and Evolution*, 24, 585.

Vander Meer, R.K., Williams, F.D. and Lofgren, C.S. (1981). Hydrocarbon components of the trail pheromone of the red imported fire ant *Solenopsis invicta*. *Tetrahedron Letters*, 22, 1651–1654.

Vander Meer, R.K., Lofgren, C.S. and Alvarez, F.M. (1990). The orientation inducer pheromone of the fire ant *Solenopsis invicta*. *Physiological Entomology*, 15, 483–488.

van Wilgenburg, E. and Elgar, M.A. (2013). Confirmation bias in studies of nestmate recognition: a cautionary note for research into the behavior of animals. *PLoS One*, 8, e53548.

van Wilgenburg, E., Sulc, R., Shea, K.J. and Tssutsui, N.D. (2010). Deciphering the chemical basis of nestmate recognition. *Journal of Chemical Ecology*, 36, 751–758.

van Zweden, J.S. and d'Ettorre, P. (2010). Nestmate recognition in social insects and the role of hydrocarbons. In *Insect Hydrocarbons: Biology, Biochemistry and Chemical Ecology*, ed. Blomquist, G.J. and Bagneres, A.-G. Cambridge, UK: Cambridge University Press.

Vasquez, G.M. and Silverman, J. (2008a). Intraspecific aggression and colony fusion in the Argentine ant. *Animal Behaviour*, 75, 583–593.

Vasquez, G.M and Silverman, J. (2008b). Non-nestmate conspecific acceptance and the complexity of nestmate discrimination in the Argentine ant. *Behavioral Ecology and Sociobiology*, 62, 537–548.

Vasquez, G.M., Schal, C. and Silverman, J. (2008). Cuticular hydrocarbons as queen adoption cues in the invasive Argentine ant. *Journal of Experimental Biology*, 211, 1249–1256.

Vasquez, G.M., Schal, C. and Silverman, J. (2009). Colony fusion in Argentine ants is guided by worker and queen cuticular hydrocarbon profile similarity. *Journal of Chemical Ecology*, 35, 922–932.

Vasquez, G.M., Vargo, E.L. and Silverman, J. (2012). Fusion between southeastern US Argentine ant colonies and its effect on colony size and productivity. *Annals of the Entomological Society of America*, 105, 268–274.

Vega, S.Y. and Rust, M.K. (2003). Determining the foraging range and origin of resurgence after treatment of Argentine ant (Hymenoptera: Formicidae) in urban areas. *Journal of Economic Entomology*, 96, 844–849.

Vogel, V., Pedersen, J.S., D'Ettorre, P., Lehmann, L. and Keller, L. (2009). Dynamics and genetic structure of Argentine ant supercolonies in their native range. *Evolution*, 63, 1627–1639.

Vogt, J.T., Reed, J.T. and Brown, R.L. (2004). Temporal foraging activity of selected ant species in Northern Mississippi during summer months. *Journal of Entomological Science*, 39, 444–452.

Walters, A.C. and Mackay, D.A. (2005). Importance of large colony size for successful invasion by Argentine ants (Hymenoptera: Formicidae): evidence for biotic resistance by native ants. *Austral Ecology*, 30, 395–406.

Ward, P.S., Beggs, J.R., Clout, M.N., Harris, R.J. and O'Connor, S. (2006). The diversity and origin of exotic ant arriving in New Zealand via human-mediated dispersal. *Diversity and Distributions*, 12, 601–609.

Weeks, R.D. Jr., Wilson, L.T., Vinson, S.B. and James, W.D. (2004). Flow of carbohydrates, lipids and protein among colonies of polygyne red imported fire ants, *Solenopsis invicta* (Hymenoptera: Formicidae). *Annals of the Entomological Society of America*, 97, 105–110.

Wetterer, J.K. (2007). Biology and impacts of Pacific Island invasive species. 3. The African big-headed ant, *Pheidole megacephala* (Hymenoptera: Formicidae). *Pacific Science*, 61, 437–456.

Wetterer, J.K. (2012). Worldwide spread of the African big-headed ant, *Pheidole megacephala* (Hymenoptera: Formicidae). *Myrmecological News*, 17, 51–62.

Wetterer, J.K. and Radchenko, A.G. (2011). Worldwide spread of the ruby ant, *Myrmica rubra* (Hymenopetra: Formicidae). *Myrmecological News*, 14, 87–96.

Wilder, S.M., Holway, D.A., Suarez, A.V. and Eubanks, M.D. (2011). Macronutrient content of plant-based food affects growth of a carnivorous arthropod. *Ecology*, 92, 325–332.

Wirth, R., Betschlag, W., Ryel, R.J. and Holldobler, B. (1997). Annual foraging of the leaf-cutting ant *Atta colombica* in a semideciduous rain forest in Panama. *Journal of Tropical Ecology*, 13, 741–757.

Witte, V., Attygalle, A.B. and Meinwald, J. (2007). Complex chemical communication in the crazy ant *Paratrechina longicornis* Latreille (Hymenoptera: Formicidae). *Chemoecology*, 17, 57–62.

Zee, J. and Holway, D.A. (2006). Nest raiding by the invasive Argentine ant on colonies of the harvester ant, *Pogonomyrmex subnitidus*. *Insectes Sociaux*, 53, 161–167.

14 Invasions by Mosquitoes: The Roles of Behaviour Across the Life Cycle

Steven A. Juliano and L. Philip Lounibos

Because of their prominent role as vectors of disease, it is arguable that mosquitoes have had a greater impact on human health, well-being, activities and history than any other family of organisms. What is less appreciated is that many of the truly prominent impacts of mosquitoes on human well-being are the result of a relatively small number of invasive mosquito species that have proved to be highly successful in multiple parts of the world, produced health impacts on humans (Lounibos, 2002; Juliano and Lounibos, 2005), assemblages of native birds (Van Riper *et al.*, 1986; LaPointe *et al.*, 2012), and on distributions of resident mosquitoes (O'Meara *et al.*, 1995; Andreadis and Wolfe, 2010; Kaplan *et al.*, 2010). Investigations of invasive and other non-native mosquitoes document the multiple roles that behaviour has played in mosquito invasions (Figure 14.1). In this chapter, we review three specific roles behaviour has played in mosquito invasions, focusing on the best-studied cases, and drawing attention to questions concerning behaviour that should be addressed in other, less-studied cases. The three roles for behaviour are: (i) behavioural adaptation to anthropogenic environments; (ii) behavioural interactions with native predators; and (iii) mating behaviour and interactions with closely related species

Definitions and Concepts

We follow recent authors (Davis, 2006; Lockwood *et al.*, 2007) in distinguishing between non-native species, which have been introduced via human activities outside their natural range, and invasive species, which have had demonstrable ecological, economic, or health-related impacts. Although the impacts are to some degree subjectively defined, they are strongly associated with objectively defined outcomes of biological processes: the increases in abundance and distribution of a non-native species in its introduced range. Invasion is one end point of a multi-step process (Figure 14.1) (Lockwood *et al.*, 2007). To become invasive, an organism must be transported and introduced into a new area. Then it must become established as a population maintained by its own reproduction (as opposed to persisting solely due to repeated introductions). Such an established population will necessarily be both local and rare unless its

Biological Invasions and Animal Behaviour, eds J.S. Weis and D. Sol. Published by Cambridge University Press. © Cambridge University Press 2016.

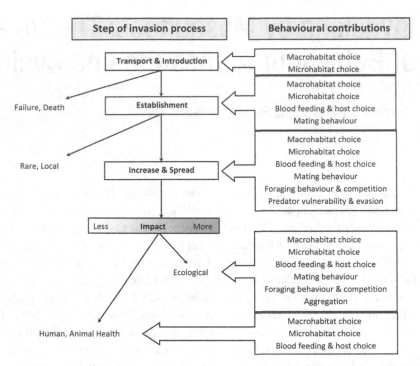

Figure 14.1 Stages in the process of invasion and the potential contributions of behaviour to mosquito success and impact at each stage. 'Macrohabitat' indicates features associated with the vegetation, land use and landscape encountered and chosen by adult mosquitoes. 'Microhabitat' indicates features of the aquatic habitat used by larvae and chosen by ovipositing females.

population dynamics and mobility enable it to increase and spread beyond its initial low abundance and local area. Upon increasing and spreading, the population may have impacts, including effects on resident species, ecosystem processes, economics and human or animal health. We focus on cases wherein behaviours of the adults and larvae have influenced the process of invasion for mosquitoes (Figure 14.1), asking for each case, how behaviour of the invader, or of interacting species, determined the likelihood of success and the extent of impact of the invader.

Invasive and non-native mosquitoes. The global invasive species database (Lowe *et al.*, 2000) lists five mosquitoes: *Aedes aegypti*, *Aedes albopictus*, *Aedes japonicus*, *Anopheles quadrimaculatus* and *Culex quinquefasciatus*. To this list, Juliano and Louni-bos (2005) added: *Aedes atropalpus*, *Aedes notoscriptus*, *Culex pipiens*, *Anopheles gambiae* and *Anopheles darlingi*. Juliano and Lounibos (2005, their Table 1) also list 22 non-native, but non-invasive mosquitoes, which have failed to spread or to become abundant in their non-native range. Juliano and Lounibos (2005) showed that oviposi-tion and larval development in small, water-filled human-made and natural (e.g. tree holes) containers was common among invasive and non-native mosquitoes (>50%), but was not significantly more frequent among invasive versus non-native, non-invasive mosquitoes. The prevalence of these traits among introduced mosquitoes appears to

depend upon the contribution of oviposition behaviour to the transport and colonization steps of an invasion (Figure 14.1) because transport of small containers (e.g. tyres,
water-holding plants) is common. More strikingly, occupying human-dominated macro-
habitats (urban, suburban, domestic) was significantly more frequent among invasive
mosquitoes than among non-native, non-invasive mosquitoes (Juliano and Lounibos,
2005). Association of invasive status with human disturbance occurs in other groups
(e.g. Moyle and Light, 1996; Mack, 2000; Jeschke and Strayer, 2006; Hufbauer *et al.*,
2012). Two biological hypotheses may account for this association of occupying human-
dominated habitats with invasiveness (Juliano and Lounibos, 2005; Hufbauer *et al.*,
2012): exploitation of such habitats may favour general traits (e.g. high population
growth rate) that contribute to the increase and spread or impact stages (Figure 14.1);
or exploitation of human-dominated habitats in the native range may pre-adapt species
to succeed specifically in human-dominated landscapes that are increasingly common
worldwide and ecologically similar everywhere. This latter hypothesis was described
comprehensively by Hufbauer *et al.* (2012) as anthropogenically induced adaptation to
invade (AIAI).We examine the role of behaviour in the AIAI hypothesis with particular reference to *Aedes aegypti*, which we believe provides one of the best examples of
how behaviour contributes to the AIAI route to invasiveness. Other invasive mosquitoes
show some association with human-dominated habitats, including indoor biting and
resting and predominance of human blood meals (e.g. members of the *Anopheles gambiae* complex – Gillies, 1964; Dekker *et al.*, 2002; Takken and Verhulst, 2013), and use
of human-made containers (e.g. *A. albopictus, A. atropalpus, A. japonicus, C. pipiens,
C. quinquefasciatus* – reviewed by Juliano and Lounibos, 2005), but similar behavioural
studies testing for active behavioural choice and a heritable basis of such choice are rare
or absent for most other invasive mosquitoes. Thus, we discuss other species briefly,
indicating where future behavioural studies may contribute to a better general understanding of the AIAI mechanism among mosquitoes.

AIAI and Invasive *Aedes aegypti*

Putative adaptation to human-dominated habitats is likely a result of a suite of
behavioural traits related to macro- and microhabitat choice. These traits include feeding, resting and ovipositing indoors (collectively 'domesticity'; Powell and Tabachnick,
2013; or 'endophily'; Lounibos, 2002). Acceptance of, or preference for, human-made
containers for oviposition, and larval development in human-made containers (indoors
or not), is another behavioural contributor to use of anthropogenic habitats. It is likely
that accepting such human-made containers contributed to invasions by *A. aegypti, A.
albopictus, A. atropalpus, A. japonicus, C. pipiens* and *C. quinquefasciatus*, all of which
use human-made containers as oviposition sites, and make use of domestic or peridomestic habitats to varying degrees (reviewed by Juliano and Lounibos, 2005). Finally,
preference for human blood over that of other vertebrates also would give a mosquito
an advantage in a human-dominated landscape. Domesticity, oviposition in human-
made containers and blood-feeding primarily from humans are likely derived traits for

mosquitoes. The ancestral condition is most likely living outdoors in vegetated habitats, oviposition in phytotelmata (=containers in plants) or rock pools, and general feeding on non-human blood sources. It is for *A. aegypti* that we have the best understanding of how behavioural choices favouring anthropogenically modified habitats may have influenced invasion (Brown *et al.*, 2013; Powell and Tabachnick, 2013), and a review of published work provides good evidence that the AIAI hypothesis accounts for the success of *A. aegypti* as a pantropical invader.

Aedes aegypti is a variable species originating in Africa (Powell and Tabachnick, 2013). At least three subspecies have been named: a domestic form *A. aegypti aegypti* that is the successful invasive in tropical and subtropical areas; a non-domestic form *A. aegypti formosus* found typically away from dwellings and only in sub-Saharan Africa; and another domestic form *A. aegypti queenslandensis* formerly found around the Mediterranean, but which now may be extinct (Powell and Tabachnick, 2013). The subspecies differ morphologically, but it is behavioural variation within and between subspecies that is important for the AIAI hypothesis. Ancestrally, *A. aegypti*, as is the case for *A. aegypti formosus* today, is believed to have oviposited in natural containers and likely fed on many different vertebrate hosts, and was not particularly associated with human habitation (Powell and Tabachnick, 2013). Powell and Tabachnick (2013) postulated that drying of North Africa 4000–6000 years ago selected for a move into domestic habitats, and associated behavioural traits. As human dwellings became the principal habitat with reliable water, individuals that chose those habitats and containers and fed primarily on humans would have been favoured. As humans began transoceanic trade during the colonial period, these domestic mosquitoes were more likely to be taken on ships in water vessels, to reside indoors on the ship, and to feed successfully on humans. Heritable variation in these traits still exists in many populations of *A. aegypti* (examples below). The result is varying degrees of behavioural domesticity, with some populations living primarily indoors, others associated with humans and using human-made containers, but living most of their lives outdoors and feeding from other hosts, and some African populations that are non-domestic and not associated with human habitats or human-made containers.

Behavioural Domesticity

Sympatric populations of domestic and non-domestic forms from the Rabai region, Kenya, harbour heritable quantitative variation in the behavioural tendency to enter buildings (Trpis and Haüsermann, 1978). Trpis and Haüsermann (1978) collected *A. aegypti* from three macro-habitats, all within 2 km of one another, near one village. The domestic line (D) was collected from water-filled clay pots inside houses; the peridomestic line (P) was collected from oviposition sites on palm trees in the village and close to houses; and the non-domestic line (N) was collected from water-filled tree holes in a nearby forest. They propagated these lines in the laboratory, creating three pure populations and six reciprocal hybrid populations. They then assayed house-entering behaviour by releasing 900 individuals (1:1 ♀:♂) of each line, each line marked

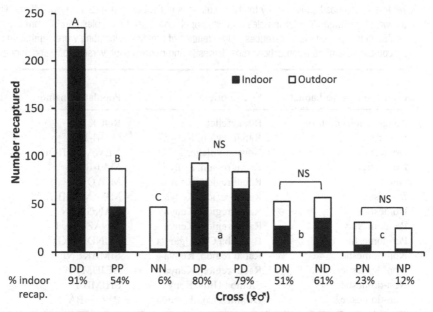

Figure 14.2 Frequencies of recaptures of released populations of *Aedes aegypti* indoors or outdoors in Kenya (Trpis and Haüsermann, 1978). Data from their Table III. D = domestic, P = peridomestic, N = non-domestic. Proportions for pure populations (DD, PP, NN) associated with the same upper case letter are not significantly different. Proportions for reciprocal crosses linked by a bracket labelled NS are not significantly different. Pairs of reciprocal crosses (DP + PD, DN + ND, PN + NP) labelled with the same lower case letter are not significantly different from other such pairs of reciprocal crosses (DN and ND vs. NP and PN: $\chi^2 = 19.48$, df $= 1, p < 0.0001$; DN and ND vs. DP and PD: $\chi^2 = 9.50$, df $= 1, p = 0.0021$; DP and PD vs. NP and PN $\chi^2 = 59.51$, df $= 1, p < 0.0001$).

with a unique colour of fluorescent dust, and recapturing individuals outdoors in the village (peridomestic) and indoors (domestic). Their results (their Table III; presented here as Figure 14.2) were presented without statistical testing of the hypothesis that frequency of recapture within houses is dependent on line. We analysed these data using a categorical model testing for effects of male and female parentage, and interaction, on this behaviour. There were highly significant effects of male ($\chi^2 = 89.8$, df $= 1$, P < 0.0001) and female ($\chi^2 = 72.0$, df $= 1$, P < 0.0001) parentage, but no interaction of parentage ($\chi^2 = 4.3$, df $= 1$, P $= 0.3689$), indicating heritability of this behaviour, but without allelic dominance. The three pure lines all differed significantly, with the D line primarily recaptured indoors, the N line recaptured primarily outdoors, and the P line intermediate (Figure 14.2). Contrasts showed that reciprocal hybrids in all three possible pairings did not differ (χ^2 all <1.8, df $= 1$, all P $\gg 0.10$), indicating no evidence of maternal effects on this behaviour (Figure 14.2). Frequency of recapture indoors for all three hybrid lines was intermediate to the associated parental lines (Figure 14.2), and all hybrid lines differed significantly from one another (Figure 14.2). Hybrids of the D and N lines were very similar to the P line in their frequency of house entry (Figure 14.2) suggesting that the P line was a hybrid of D and N lines (Trpis and Haüsermann, 1978).

Table 14.1 Oviposition frequency for 13 populations of East African *Aedes aegypti* offered vials of plain tap water. Minimum $N = 29$ females for each population. Based on Tables I and II from Leahy *et al.* (1978), who reported that differences in frequencies of females ovipositing were significant ($P < 0.0001$) for comparisons of non-domestic versus domestic and peridomestic versus domestic populations

Collection macro-habitat	Site of origin	Population name	Proportion ovipositing
Long-term laboratory	Rockefeller	ROCK	0.97
Domestic	Rabai region, Kenya	CHIB-IN	0.94
Domestic	Newala village, Tanzania	NEWALA-HOUSE	0.77
Domestic	Rabai region, Kenya	MAJ-IN	0.66
Domestic	Rabai region, Kenya	MOYO	0.65
Domestic	Rabai region, Kenya	MWAMSABU	0.61
Domestic	Rabai region, Kenya	GANGA-IN	0.49
Peridomestic	Rabai region, Kenya	MNAZI-OUT	0.46
Peridomestic	Bwamba Co., Uganda	BUNDIBUGYO	0.03
Non-domestic	Rabai region, Kenya	SIMAKENI	0.49
Non-domestic	Rabai region, Kenya	BEJUMWA	0.22
Non-domestic	Shimba Hills, Kenya	SHIMBA	0.14
Non-domestic	Bwamba Co., Uganda	BWAMBA	0.03

This interpretation is supported by previous mark–release–recapture experiments (Trpis and Haüsermann, 1975).

Although *A. aegypti formosus* is frequently called 'sylvan' (e.g. Trpis and Haüsermann, 1978), in Rabai this subspecies was most abundant at the forest–peridomestic ecotone, and other members of the same subgenus, *Stegomyia*, predominated in peridomestic or forest environments (Lounibos, 1981). Similarly, its invasive counterpart from Asia, *A. albopictus*, is known as the 'forest edge mosquito' in its native range, owing to its association with ecotones (Hawley, 1988). This behavioural choice is also observed in *A. albopictus'* non-native range (Lounibos *et al.*, 2001). Behavioural choice of forest–peridomestic ecotones may be the first step towards a peridomestic or domestic lifestyle.

Oviposition choice by *A. aegypti* also varies among populations in a pattern consistent with the AIAI hypothesis. Leahy *et al.* (1978) showed that populations originating from domestic collections were significantly more likely to oviposit in tap water than populations from peridomestic or non-domestic collections (Table 14.1). Leahy *et al.* (1978) further tested oviposition responses of one domestic and one non-domestic population to containers holding either tap water or coconut husk infusion (containing plant-derived organic compounds). Their data (their Table III, reanalysed and reproduced as Figure 14.3) show that the domestic population had no significant differential response (Figure 14.3), whereas the non-domestic population oviposited significantly more often in coconut husk infusion (Figure 14.3). Two behavioural traits stand out in Figure 14.3: the domestic population oviposits readily and does not discriminate based on the stimulus offered; in contrast, the non-domestic population is more reluctant to oviposit, and avoids oviposition in tap water. Indiscriminate oviposition behaviour is likely to facilitate invasion in multiple ways: first, willingness to oviposit in plain water

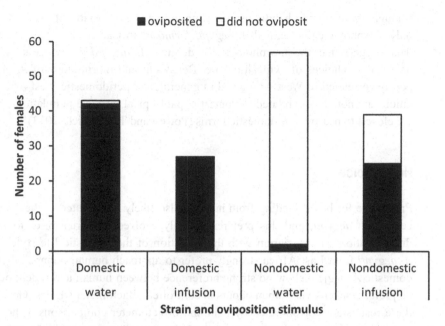

Figure 14.3 Oviposition by female *Aedes aegypti* from domestic and non-domestic populations in water and coconut husk infusion. Oviposition choice did not differ significantly between water and infusion for the domestic strain (likelihood ratio $\chi^2 = 0.98$, df = 1, $p = 0.3213$) but did differ significantly for the non-domestic strain (likelihood ratio $\chi^2 = 45.82$, df = 1, $p < 0.0001$). Data from Leahy *et al.* (1978).

stored in clay jars (Lorimer *et al.*, 1976) enhances the likelihood of laying eggs in water stored by humans as a drinking supply, which we expect would be kept free of extraneous organic inputs. These behaviours would enhance the likelihood of eggs being taken on sea voyages in water storage containers. Second, upon reaching a new location inhabited by humans, such indiscriminate oviposition would increase the likelihood of finding an oviposition site, including stored water (Powell and Tabachnick, 2013). In contrast, a strong preference for oviposition sites with water containing plant organics would predispose individuals to oviposit in some phytotelmata that would be less likely to be transported and perhaps less likely to be found in human-dominated landscapes in a newly colonized area. Such intraspecific variation in oviposition choices may have practical consequences. Failure of a population replacement genetic control trial of domestic *A. aegypti* in Rabai was ascribed in part to behavioural unwillingness of the released translocation homozygote strain to oviposit on clay, which is used for domestic water-holding containers (Lorimer *et al.*, 1976).

West African *A. aegypti* also show variation in choice of oviposition sites. In Senegal, *A. aegypti aegypti* and *A. aegypti formosus* were both collected from human-made containers in urban and rural village environments (Sylla *et al.*, 2009) and natural containers in a sylvan habitat (Sylla *et al.*, 2013). Although no behavioural data were provided, these results suggest considerable variation in habitat and oviposition choices in both putative subspecies (Sylla *et al.*, 2013). Simard *et al.* (2005) reported heavy colonization of human-made and natural containers, and frequent peridomestic (though not domestic) habitat use in Cameroon, and cited Mattingly's (1957) report of only morphologically defined *A. aegypti formosus* in Cameroon. Thus, West African data suggest that the morphologically defined *A. aegypti formosus* show variable behavioural choices of oviposition site. Behaviour and morphology of *A. aegypti* are poorly correlated in West Africa and in general, and peridomestic West African populations are not closely related to domestic, pantropical invasive populations, but rather are closest to nearby non-domestic forms (Powell and Tabachnick, 2013).

Host Choice

Preference for blood-feeding from humans also likely contributed to the invasive success of *A. aegypti*, and this preference likely evolved in Africa prior to worldwide dissemination, in association with the evolution of the domestic life style. Domestic *A. aegypti* from East Africa strongly prefer to approach human odours, whereas non-domestic *A. aegypti* show no strong preference between human and rodent odours, but do prefer chicken over human odours in controlled olfactometer studies (Gouck, 1972); these results are corroborated by field and olfactometer comparisons of host-seeking behaviours of the two subspecies from the Rabai area (Petersen, 1977). Mukwaya (1977) tested olfactory attraction of African strains, one a long-term laboratory strain originating from morphologically domestic and anthropophilic individuals, and the other a morphologically *formosus*-type, non-domestic, non-anthropophilic strain. The domestic strain strongly preferred humans (71% chose human) versus rodents (30%); in contrast, the non-domestic strain had no strong preference (47% human, 53% rodent) (Mukwaya, 1977). Host choice appears to be heritable, as crosses of the two strains yielded F_1 and F_2 individuals with intermediate preferences for humans (Mukwaya, 1977).

   McBride *et al.* (2014) documented this behavioural difference, its heritability and a component of its molecular basis. They showed that multiple domestic populations from Rabai, Kenya, differed dramatically in blood-feeding preferences from outdoor populations from nearby forests. Indoor populations preferred humans and outdoor populations preferred guinea pigs or chickens (McBride *et al.*, 2014). These preferences were associated with morphology, with most indoor populations resembling *A. aegypti aegypti* and all outdoor populations resembling *A. aegypti formosus*. Some of the indoor populations included a mix of morphological types (McBride *et al.*, 2014, their Figure 1). They then showed that behavioural host preference was associated with both differential expression of an odorant receptor protein (OR4) in the antenna and differential allelic sensitivity of OR4 to sulcatone, a compound uniquely abundant on human skin, but rare in other animals (McBride *et al.*, 2014). Thus, McBride *et al.* (2014) showed the molecular and genetic basis of evolution of behavioural host choice, a central trait in putative adaptation of *A. aegypti aegypti* to association with humans and to success as an invader.

   These results take on additional importance for understanding the role of AIAI in *A. aegypti* when compared with phylogeographic analysis of SNPs and nuclear sequences (Brown *et al.*, 2011, 2013; Powell and Tabachnick, 2013). *Aedes aegypti*

aegypti invasive in North and South America, Asia, Pacific Islands and Australia form a monophyletic group that includes the domestic form collected from the Rabai region of eastern Kenya, suggesting that domestication occurred once and gave rise to the worldwide invasive lineage of *A. aegypti aegypti*. With the exception of domestic populations in coastal East Africa, pure domestic *A. aegypti* (i.e. indoors, feeding on humans, and ovipositing in human-made containers) appears to be absent in Africa (Powell and Tabachnick, 2013). Brown *et al.* (2013) presented evidence of hybridization and introgression between domestic and non-domestic forms in West and central Africa, and noted that many African populations that may be assigned morphologically to one of the nominal subspecies of *A. aegypti* show a range of habitat use, host feeding and oviposition choice behaviours (reviewed by Powell and Tabachnick, 2013).

These observations on *A. aegypti* argue against the view that subspecies of *A. aegypti* have discrete suites of distinct behavioural traits, and suggest instead considerable quantitative genetic variation in behaviour in this species (all forms) that serves as the substrate of selection for local adaptation (Powell and Tabachnick, 2013). The range of quantitative genetic variation in house-entering behaviour evident in Trpis and Haüsermann's (1978) data suggest that local selection can shift *A. aegypti* behaviour towards domestic, non-domestic, or intermediate (peridomestic) behaviour patterns. In various parts of its pantropical range, phylogenetically and morphologically 'domestic' *A. aegypti aegypti* have evolved behavioural life styles involving outdoor oviposition, resting and feeding (Chadee *et al.*, 1998; Harrington *et al.*, 2008), choice of non-human hosts (Harrison *et al.*, 1972) and oviposition in natural containers (Wallis and Tabachnick, 1990; Bagny *et al.*, 2009; Bagny Bielhe *et al.*, 2013). We suggest that this quantitative genetic variation in behavioural domesticity enabled ancestral *A. aegypti aegypti* to adapt to human dwellings in Africa, enabled them to be transported via shipping trade around the world, and enabled them to spread and to become invasive in pantropical human-dominated habitats. Further, we suggest that the quantitative genetic variation for domestic behaviour enabled different lineages of invasive *A. aegypti aegypti* to adapt to novel situations in their introduced ranges, and thus to adopt varying degrees of domestic, non-domestic and peridomestic life styles. This scenario closely matches the theory of AIAI (Hufbauer *et al.*, 2012; their Figure 1), including the move by invaders out of human-dominated habitats within the non-native region when ecological opportunity presents itself.

Thus, there is good evidence consistent with pre-adaptation of behavioural traits for macro-habitat choice, microhabitat choice and host choice contributing to the invasive success of *A. aegypti*. Obtaining further empirical data on this process is inhibited because the invasive phase took place centuries to decades ago (Powell and Tabachnick, 2013), and because there has likely been continued evolution of these traits in introduced populations of *A. aegypti*. Circumstantial evidence suggests the AIAI hypothesis may be a factor in other invasions by mosquitoes. A member of the *Anopheles gambiae* complex spread for approximately 10 years as an invasive species in Brazil, prior to its eradication (Soper and Wilson, 1943). Part of both its success as an invader and its health impact likely derive from the behavioural preference for human blood in this species complex (Gillies, 1964; Dekker *et al.*, 2002; Takken and Verhulst, 2013). Gillies (1964) showed that artificial selection for host choice by *A. gambiae* yielded strains that differed in preference for humans within only about three generations. This coupled with a willingness to bite indoors seems likely to have contributed to the temporary success of *A. gambiae*'s invasion. The worldwide invasion success of the members of the *Culex pipiens* species complex is associated with their behavioural domesticity; including use of human-made containers, feeding, resting and sometimes mating indoors, including within the London Underground, and human biting (Fonseca *et al.*, 2004). The AIAI hypothesis also seems likely to explain success of more recently invasive mosquitoes, like *A. albopictus* and *A. japonicus*, both of which use human-made containers, and are common in urban and suburban areas (Juliano and Lounibos, 2005). These invasions are largely <30 years old, and there is still active spread of both species (Benedict *et al.*, 2007; Kaufman and Fonseca, 2014). Empirical data on adaptation to human-dominated habitats in both the native and introduced ranges of these invaders would be valuable. Hufbauer *et al.* (2012; their Table 1) described criteria for testing the AIAI hypothesis including: documenting in the native and introduced ranges use of human-altered environments; adaptation to those environments; heritability of associated traits; and phylogenetic evidence that invasive populations are derived from native populations presumably pre-adapted to human-dominated environments. One of the strengths of the AIAI hypothesis is that it has a chance of being predictive: species adapted behaviourally to human-dominated habitats in their native range may be predicted to have greater potential for invasiveness, and thus deserve greater efforts to prevent invasion.

Invasions by *Aedes albopictus* and Behavioural Interactions with Aquatic Predators

Escape from enemies in their native range is a long-standing hypothesis for the success of invasive species (Lockwood *et al.*, 2007). A corollary is that enemies native to the introduced range may contribute to resistance to invasions (Byers, 2002; DeRivera *et al.*, 2005; Griswold and Lounibos, 2005; Juliano *et al.*, 2010), reducing the likelihood of establishment or increase and spread by a vulnerable invader (Figure 14.1). For aquatic animals like larval mosquitoes, behavioural responses to predators are often central to understanding both impacts of predation and trade-offs between vulnerability to predators and competitive ability (Sih, 1986; Skelly, 1994; Grill and Juliano, 1996; Wellborn, 2002; Brown and Kotler, 2004). In the next section, we review how behavioural vulnerability to predation has influenced invasive success and impact of *A. albopictus* in North America.

Aedes albopictus is a container-dwelling Asian mosquito, with a natural geographic range from Indonesia (~10°S latitude) to Korea (~40°N latitude), and into India (Hawley, 1988; Benedict *et al.*, 2007; Medley, 2010). In its native range its larvae develop in water-filled natural (e.g. tree holes) and human-made (e.g. tyres) containers in sylvan, rural, suburban and urban areas (Hawley, 1988). Its use of human-made containers is likely another example of the contribution of the AIAI mechanism to invasion (see above). It is now the most important invasive mosquito worldwide (Benedict *et al.*,

2007), and is established in western, southeastern and central North America, much of South America, Central America, the Mediterranean basin, west and central Africa, the Caribbean and Pacific islands (Benedict *et al.*, 2007; Carvalho *et al.*, 2014), and Australia's Torres Straits (Ritchie *et al.*, 2006). Its invasions advanced rapidly in the 1980s when it reached the Americas via trade in used tyres (Lounibos, 2002), and it is a serious public health threat for its role in transmission of arboviruses (Benedict *et al.*, 2007). Its invasions have also been implicated in local extinctions or reductions of *A. aegypti* in southern North America and Bermuda, and to a lesser extent, in parts of South America (O'Meara *et al.*, 1995; Juliano, 1998; Braks *et al.*, 2003, 2004; Kaplan *et al.*, 2010). Its competitive ability suggested potential ecological impacts on the native Eastern North American tree hole mosquito *Aedes triseriatus* (Livdahl and Willey, 1991; Aliabadi and Juliano, 2002). These realized and potential impacts depend on the behaviour of larval *A. albopictus* and the species with which it interacts in the aquatic habitat.

Aedes albopictus is a behaviourally flexible generalist at multiple levels: females take blood meals from a variety of vertebrate hosts (Hawley, 1988), and oviposit in many different types of containers in habitats ranging from urban to rural to sylvan (Hawley, 1988). Across this range of habitats *A. albopictus* encounters multiple other mosquito species that are somewhat more restricted in their choices of habitats. In North America, these species include *A. aegypti* in primarily human-made containers in urban, tropical and subtropical habitats, and *A. triseriatus* in both human-made and natural containers in forested and rural habitats. The ecological context in these habitats, coupled with behaviour of larval *Aedes*, leads to pronounced context dependence in the outcome of this invasion and these interactions (Juliano, 2009). Multiple laboratory (Livdahl and Willey, 1991; Aliabadi and Juliano, 2002; Costanzo *et al.*, 2005; Leisnham *et al.*, 2009; Leisnham and Juliano, 2010) and field (Juliano, 1998; Braks *et al.*, 2004; Juliano *et al.*, 2004) investigations have indicated that *A. albopictus* is superior to both *A. aegypti* and *A. triseriatus* in interspecific competition for the typical resources (microorganisms and fine particulate matter) found in human-made containers, though competitive outcomes with *A. triseriatus* may be more equal in tree holes (Livdahl and Willey, 1991; Juliano, 2010). Field evidence clearly shows that invasion by *A. albopictus* is typically associated with declines in *A. aegypti*, sometimes to local extinction (reviewed by Juliano and Lounibos, 2005; Juliano, 2009). In contrast, invasion of habitats occupied by *A. triseriatus* has not led to noticeable declines in this native species (Lounibos *et al.*, 2001), and behaviour of the interacting *Aedes* likely contributes to this difference.

In urban environments in human-made containers, these *Aedes* encounter predators infrequently (Kesavaraju *et al.*, 2008; Leisnham and Juliano, 2010). In contrast, larval predators, particularly the mosquito *Toxorhynchites rutilus* and the midge *Corethrella appendiculata*, are relatively common in natural tree holes in forests of the southeastern United States (Lounibos *et al.*, 1997, 2001; Kesavaraju *et al.*, 2008). *Aedes triseriatus* larvae respond to water-borne cues from *T. rutilus* predation by reducing movement and subsurface foraging (Kesavaraju and Juliano, 2004), which reduces risk of predation by this predator (Juliano and Reminger, 1992; Juliano *et al.*, 1993). Reduced movement and foraging is a common behavioural response to predation cues in aquatic organisms (e.g. Sih, 1986; Skelly, 1994; Grill and Juliano, 1996) and comes at a cost of reduced competitive performance. Selection in the laboratory by exposure of *A. triseriatus* larvae to *T. rutilus* predation can shift the behaviour of subsequent generations from this facultative change in response to predator cues, to constitutive low movement and subsurface foraging, regardless of cues (Juliano and Gravel, 2002). Thus, *A. triseriatus* shows considerable potential to adapt to the predation within a few generations of selection. In contrast, *A. albopictus* showed no significant change in behaviour in response to the same cues, and this interspecific difference in behavioural responses was independent of whether cues arose from predation on *A. albopictus* or on *A. triseriatus* (Kesavaraju and Juliano, 2004). This interspecific difference in behavioural response to predator cues likely contributes to both the relatively limited success of *A. albopictus* in natural tree holes in North America (where *T. rutilus* can be common), and to the limited impact of *A. albopictus* on *A. triseriatus* distribution and abundance.

The pattern of differential responsiveness of *A. triseriatus* and *A. albopictus* to predator cues is similar for *C. appendiculata*, a small, size-selective predator that primarily attacks first and second instar *Aedes* larvae (Kesavaraju *et al.*, 2007; Alto *et al.*, 2009; Juliano *et al.*, 2010). In the field, *A. albopictus* abundance was negatively associated with *C. appendiculata* and *A. triseriatus* abundances, and lower in tree holes than in tyres or cemetery vases, whereas *A. triseriatus* abundance was positively associated with tree holes and with *C. appendiculata* (Kesavaraju *et al.*, 2008). Kesavaraju *et al.* (2007) compared behavioural responses of second instar *A. triseriatus* and *A. albopictus* to predation cues from *C. appendiculata*. As with the responses to *T. rutilus*, there was a significant difference in the responsiveness of the two species, with *A. triseriatus* showing a large increase in time spent resting at the surface in response to cues. *Aedes albopictus* was generally more active, spent more time in subsurface foraging and, in this case, showed a small, significant shift towards more time resting at the surface in response to *C. appendiculata* predatory cues. *Aedes albopictus* was killed by this predator at a significantly greater rate than *A. triseriatus* (Griswold and Lounibos, 2005), and subsurface movement was again the most risky behaviour (Kesavaraju *et al.*, 2007). Despite the small but significant size and stage-dependent behavioural response of *A. albopictus* to *C. appendiculata* (Kesavaraju and Juliano, 2008; Alto *et al.*, 2009), responses of different *A. albopictus* populations to *C. appendiculata* cues did not differ significantly in association with local presence or absence of *C. appendiculata*, nor with origin from tree hole versus tyre versus cemetery vase (Kesavaraju *et al.*, 2008). Laboratory selection via *C. appendiculata* predation yielded no significant evolution of behavioural responses of *A. albopictus* to cues from *C. appendiculata* predation (Kesavaraju and Juliano, 2009). Thus, behaviour of this invader has much less potential to evolve in response to predation than does behaviour of the native *A. triseriatus*. It is unclear why the response to selection on behaviour differs so much for these species. *Aedes albopictus* likely encounters predators (though perhaps not corethrellid midges) in its native range (Hawley, 1988). One hypothesis for the lack of response to selection is that genetic variation for behaviour in North American *A. albopictus* may be limited due to the introduction starting with a limited sample of behavioural genotypes (Kesavaraju and Juliano, 2009).

The importance of this behavioural difference between invader and native can be better understood via the results of a field experiment manipulating abundances of *A. triseriatus* and *C. appendiculata* in tyres and testing for effects on the colonization and abundance of *A. albopictus* (Juliano *et al.*, 2010). Presence or absence of *A. triseriatus* had no significant effect on oviposition by the invader or on survival of larval *A. albopictus*. In contrast, low or high abundance of *C. appendiculata* along with *A. triseriatus* caused a significant reduction of *A. albopictus* larvae, but no significant reduction of egg numbers (Juliano *et al.*, 2010). Thus, *A. albopictus* do not avoid oviposition in containers with predators, but instead, their survival is significantly reduced by increasing abundance of *C. appendiculata*. In this context, *A. triseriatus* individuals may indirectly benefit from the presence of *C. appendiculata*. When *C. appendiculata* larvae were absent, *A. triseriatus* survivorship was better, but because of greater abundance of competing *A. albopictus*, *A. triseriatus* larvae developed more slowly, compared to the high *C. appendiculata* treatment (Juliano *et al.*, 2010). Thus, in this invasion by *A. albopictus*, the presence of a natural enemy may actually reduce the impact of the invader on a native species that can behaviourally cope with that enemy. More generally, success and impact of invasions in habitats with natural predators is likely to depend strongly on behavioural differences between the invader and native species.

Behavioural interactions with predators are less well investigated for other invasive mosquitoes. *Aedes japonicus* shows greater behavioural responses to *T. rutilus* predation cues than does *A. albopictus* (Kesavaraju *et al.*, 2011) suggesting that this invader may fare better than *A. albopictus* in high predation habitats. Like *A. albopictus*, *A. aegypti* is highly vulnerable to predation (Sih, 1986; Grill and Juliano, 1996); but behavioural responses to predation cues are poorly investigated in this species (but see Sih, 1986). *Culex pipiens* larvae respond to cues from multiple predators (Sih, 1986; Beketov and Liess, 2007; Kesavaraju *et al.*, 2011), and exhibit threat-sensitive and victim-specific responses to predation cues (Beketov and Liess, 2007), suggesting that predator evasion behaviour has contributed to *C. pipiens*' invasive success.

Behavioural Interactions of Adults and the Impact of *Aedes albopictus* on *Aedes aegypti*

Transcontinental invasions move species outside their native ranges in which behavioural mechanisms have evolved to prevent interspecific matings (Gröning and Hochkirch, 2008). The result can be hybridization and introgression, and may result in extinction of native species (Lockwood *et al.*, 2007). *Aedes aegypti* native to Africa (Powell and Tabachnick, 2013) and *A. albopictus* native to Asia (Benedict *et al.*, 2007) are two of the most widespread invasive mosquitoes, and their non-native ranges frequently overlap. Laboratory crosses of these species indicate pre- and post-zygotic isolating mechanisms yielding no viable offspring (Leahy and Craig, 1967). However, interspecific mating is facilitated by the behaviours of both species lekking during daylight hours at bloodmeal hosts (Hartberg, 1971; Gubler and Bhattachaya, 1972) and has, by a novel mechanism, had a major impact on rapid displacements of *A. aegypti*.

Satyrization, a form of reproductive competition leading to asymmetric fitness losses because infertile interspecific matings lead to loss of individuals' future reproductive potential, was originally described as a mechanism for maintaining parapatric distributions between closely related species (Ribeiro and Spielman, 1986) and, later, as a potential mechanism for pest control through competitive displacement (Ribeiro, 1988). Satyrization was proposed as one of many possible mechanisms to explain competitive displacements of *A. aegypti* in the southeastern United States by invasive *A. albopictus* in the late 1980s and early 1990s (Nasci *et al.*, 1989). Some experiments with these species were consistent with displacement via satyrization (Nasci *et al.*, 1989). Other laboratory investigations concluded that interspecific matings were too infrequent to cause such an impact (Harper and Paulson, 1994), even though models with low satyrization rates and moderate resource competition predict species extinctions (Ribeiro, 1988; Kishi and Nakazawa, 2013), and empirical estimates for wild-caught females yield low but consistent levels (1.5–3.7%) of interspecific mating between *A. albopictus* and *A. aegypti* from four continents (Tripet *et al.*, 2011; Bargielowski *et al.*, 2015a), which rates have likely been depressed by the evolution of satyrization resistance (see below). Although interspecific matings were bidirectional in nature, fitness losses asymmetrically affected *A. aegypti*, because females of this species, but not *A. albopictus*, were sterilized by male accessory gland products transferred during interspecific copulations (Tripet *et al.*, 2011). Given the potentially large fitness losses attributable to satyrization (Ribeiro, 1988), it was hypothesized that *A. aegypti* in sympatry with *A. albopictus* would evolve behavioural mechanisms to avoid wasteful interspecific matings. Indeed, when compared in a standardized cage environment, female *A. aegypti* allopatric to *A. albopictus* were significantly more susceptible than sympatric counterparts to interspecific mating with *A. albopictus* (Bargielowski *et al.*, 2013). Inseminations of female *A. aegypti* by *A. albopictus* were also significantly more common in cages than the opposite heterospecific insemination, consistent with the unidirectional displacements of *A. aegypti* observed in nature (Bargielowski *et al.*, 2013).

Behavioural resistance to satyrization is selected for extremely rapidly in cage populations of allopatric *A. aegypti*, with resistance increasing detectably after one generation of selection, and cross-insemination dropping from >50% to ~10% after six generations of selection (Bargielowski and Lounibos, 2014). Females from the selected, satyrization-resistant line were significantly slower than females of the unselected line in behavioural acceptance of conspecific mates (Bargielowski and Lounibos, 2014), which may be a by-product of increased 'wariness' of interspecific courtship. Release of selection pressure led to the rapid return of behavioural susceptibility to satyrization (Bargielowski and Lounibos, unpublished data).

Errant mating may also select for changes in male behaviour. When allopatric and sympatric *A. aegypti* males were confined with virgin conspecific and *A. albopictus* females, allopatric males were less successful in avoiding interspecific matings than their sympatric counterparts, suggesting that natural selection to avoid satyrization affects both sexes (Bargielowski *et al.*, 2015b).

Quantifying satyrisation by assessing interspecific sperm in spermathecae underestimates the capacity of male *A. albopictus* to reduce the fitness of virgin female

A. aegypti. Carrasquilla and Lounibos (2015) confined *A. aegypti* females for 3 weeks with *A. albopictus* males, then chose only uninseminated females via *in vivo* determination of insemination status, and housed them with virgin male conspecifics (1♀:2♂). A mean of 68.9% of these females were inseminated by conspecifics, compared to 97.5% of same-aged controls unexposed to *A. albopictus* males. These results indicate that some female *A. aegypti* receive enough male accessory gland products through attempted copulations by *A. albopictus* males to render them sexually unreceptive, even though *A. albopictus* sperm do not reach the female's spermathecae.

The impact of cross-species matings by invasive mosquitoes may depend not only on local conditions but also the interspecific effects of male accessory gland substances, to date documented to have negative asymmetric effects only in the *A. albopictus–A. aegypti* system. If males of the widespread invasive *A. albopictus* attempt to mate in nature and sterilize females of other *Aedes* (*Stegomyia*) species, competitive displacements, or extinctions, could occur. For example, the rapid disappearance of *Aedes* (*Stegomyia*) *guamensis* from Guam following post-World War II establishment of *A. albopictus* on Guam (Ward, 1984) suggests a possible causal relationship between these events. More recent invasions of *A. albopictus* on South Pacific islands with endemic *Stegomyia* species (e.g. Tonga; Guillaumot *et al.*, 2012) could have similar consequences for endemic *Stegomyia* (e.g. *Aedes kesseli* and *Aedes tongae*; Huang and Hitchcock, 1980). Although behavioural divergences of the Scutellaris Group *Aedes* (*Stegomyia*) have been documented (McClain and Rai, 1985), these South Pacific island species are more closely related to *A. albopictus* than *A. aegypti*, which might increase the likelihood of interspecific matings.

Satyrization by other invasive mosquitoes is unstudied. A different aspect of mating behaviour, hybridization within a species complex, may have contributed to the worldwide invasive success and impact of the *Culex pipiens* complex. Hybridization among behaviourally diverse members of this species complex generates variable host preferences that facilitate both invasion success and transmission of arboviruses, such as West Nile, between birds and mammals (Fonseca *et al.*, 2004; Fritz *et al.*, 2015).

Conclusions and Future Directions

Behaviour has diverse and important influences on all phases of mosquito invasions, from transport to impact (Figure 14.1). Some invasive mosquitoes are particularly well-studied because of their obvious roles in disease transmission, and these well-studied species (*A. aegypti, A. albopictus, C. pipiens* complex) have the potential to improve our understanding of mechanisms of invasions in general. Active behavioural evolution is likely to be prominent in the AIAI process described by Hufbauer *et al.* (2012), and the case of *A. aegypti* suggests that behavioural study of other prominent invasive animals in their native and introduced ranges may be useful for testing the AIAI hypothesis, and for establishing its generality. Among the hypotheses for success and impact of invasive species, enemy release and introgressive hybridization (Lockwood *et al.*, 2007) have been prominent, but the behavioural mechanisms evident in the invasion of

A. albopictus suggest alternative, less-investigated, pathways (behavioural vulnerability to native predators, satyrization) by which behaviour of invaders and natives may have strong impacts on the outcomes of invasions. The impact of satyrization in rapid competitive displacements of resident species by invaders warrants further investigations, in *A. albopictus* and in other animal systems. For mosquito biologists, investigating the roles of behaviour in AIAI and other hypotheses for more recent invasive mosquitoes (*A. albopictus, A. japonicus, A. atropalpus*) may enhance understanding of those invasions, and may aid in developing theory and empirical approaches to identifying mosquitoes in their native ranges that are most likely to be the next major invasive vectors of human disease.

References

Aliabadi, B.K. and Juliano, S.A. (2002). Escape from gregarine parasites affects the competitive impact of an invasive mosquito. *Biological Invasions*, 4, 283–297.

Alto, B.E., Kesavaraju, B., Juliano, S.A. and Lounibos, L.P. (2009). Stage-dependent predation on competitors: consequences for the outcome of a mosquito invasion. *Journal of Animal Ecology*, 78, 928–936.

Andreadis, T.G. and Wolfe, R.J. (2010). Evidence for reduction of native mosquitoes with increased expansion of invasive *Ochlerotatus japonicus japonicus* (Diptera: Culicidae) in the northeastern United States. *Journal of Medical Entomology*, 47, 43–52.

Bagny, L., Delatte, H., Quilici, S. and Fontenille, D. (2009). Progressive decrease in *Aedes aegypti* distribution in Reunion Island since the 1900s. *Journal of Medical Entomology*, 46, 1541–1545.

Bagny Beilhe, L., Delatte, H., Juliano, S.A., Fontenille, D. and Quilici, S. (2013). Ecological interactions in *Aedes* species from Reunion Island. *Medical and Veterinary Entomology*, 27, 349–459.

Bargielowski, I. and Lounibos, L.P. (2014). Rapid selection in the dengue vector *Aedes aegypti* in response to satyrization by invasive *Aedes albopictus*. *Evolutionary Ecology*, 28, 193–203.

Bargielowski, I., Lounibos, L.P. and Carrasquilla, M.C. (2013). Evolution of resistance to satyrization: evidence of reproductive character displacement in populations of invasive dengue vectors. *Proceedings of the National Academy of Sciences, USA*, 110, 2888–2892.

Bargielowski, I.E., Lounibos, L.P., Shin, D., *et al.* (2015a). Widespread evidence for interspecific mating between *Aedes aegypti* and *Aedes albopictus* (Diptera: Culicidae) in nature. *Infection, Genetics, and Evolution*, 36, 456–461.

Bargielowski, I., Blosser, E. and Lounibos, L.P. (2015b). The effects of interspecific courtship on the mating success of *Aedes aegypti* and *Aedes albopictus* males. *Annals of the Entomological Society of America*, 108, 513–518.

Beketov, M.A. and Liess, M. (2007). Predation risk perception and food scarcity induce alterations of life-cycle traits of the mosquito *Culex pipiens*. *Ecological Entomology*, 32, 405–410.

Benedict, M.C., Levine, R.S., Hawley, W.A. and Lounibos, L.P. (2007). Spread of the tiger: global risk of invasion by *Aedes albopictus*. *Vector Borne and Zoonotic Diseases*, 7, 76–85.

Braks, M.A.H., Honório, N.A., Lourenço-de-Oliveira, R., Juliano, S.A. and Lounibos, L.P. (2003). Convergent habitat segregation of *Aedes aegypti* and *Aedes albopictus* (Diptera: Culicidae) in southeastern Brazil and Florida, USA. *Journal of Medical Entomology*, 40, 785–794.

Braks, M.A.H., Honório, N.A., Lounibos, L.P., Lourenço-de-Oliveira, R. and Juliano, S.A. (2004). Interspecific competition between two invasive species of container mosquitoes, *Aedes aegypti* and *Aedes albopictus* (Diptera: Culicidae), in Brazil. *Annals of the Entomological Society of America*, 97, 130–139.

Brown, J.E., McBride, C.S., Johnson, P., *et al.* (2011). Worldwide patterns of genetic differentiation imply multiple 'domestications' of *Aedes aegypti*, a major vector of human diseases. *Proceedings of the Royal Society of London B: Biological Sciences*, 278, 2446–2454.

Brown, J.E., Evans, B.R., Zheng, W., *et al.* (2013). Human impacts have shaped historical and recent evolution in *Aedes aegypti*, the dengue and yellow fever mosquito. *Evolution*, 68, 514–525.

Brown, J.S. and Kotler, B.P. (2004). Hazardous duty pay and the foraging cost of predation. *Ecology Letters*, 7, 999–1014.

Byers, J.E. (2002). Physical habitat attribute mediates biotic resistance to non-indigenous species invasion. *Oecologia*, 130, 146–156.

Carrasquilla, M.C. and Lounibos, L.P. (2015). Satyrization without evidence of successful insemination from interspecific mating between invasive mosquitoes. *Biology Letters*, 11, 20150527.

Carvalho, R.G., Lourenço-de-Oliveira, R. and Braga, I.A. (2014). Updating the geographical distribution and frequency of *Aedes albopictus* in Brazil with remarks regarding its range in the Americas. *Memorias do Instituto Oswaldo Cruz*, 109, 787–796.

Chadee, D.D., Ward, R.A. and Novak, R.J. (1998). Natural habitats of *Aedes aegypti* in the Caribbean: a review. *Journal of the American Mosquito Control Association*, 14, 5–11.

Costanzo, K.S., Kesavaraju, B. and Juliano, S.A. (2005). Condition-specific competition in container mosquitoes: the role of non-competing life-history stages. *Ecology*, 86, 3289–3295.

Davis, M.A. (2006). Invasion biology 1958–2004: the pursuit of science and conservation. In *Conceptual Ecology and Invasion Biology: Reciprocal Approaches to Nature*, ed. Cadotte, M.W., McMahon, S.M. and Fukami, T. London: Kluwer Publishers, pp. 35–66.

Dekker, T., Steib, B., Carde, R.T. and Geier, M. (2002). L-lactic acid: a human-signifying host cue for the anthropophilic mosquito *Anopheles gambiae*. *Medical and Veterinary Entomology*, 16, 91–98.

DeRivera, C.E., Ruiz, G.M., Hines, A.H. and Jivo, V.P. (2005). Biotic resistance to invasion: native predator limits abundance and distribution of an introduced crab. *Ecology*, 86, 3364–3376.

Fonseca, D.M., Keyghobadi, N., Malcolm, C.A., *et al.* (2004). Emerging vectors in the *Culex pipiens* complex. *Science* 303, 1535–1538.

Fritz, M.L., Walker, E.D., Miller, J.R., Severson, D.W. and Dworkin, I.D. (2015). Divergent host preferences of above- and below-ground *Culex pipiens* mosquitoes and their hybrid offspring. *Medical and Veterinary Entomology*, 29, 115–123.

Gillies, M.T. (1964). Selection for host preference in *Anopheles gambiae*. *Nature*, 203, 852–854.

Gouck, H.K. (1972). Host preferences of various strains of *Aedes aegypti* and *A. simpsoni* as determined by an olfactometer. *Bulletin of the World Health Organization*, 47, 680–683.

Grill, C.P. and Juliano, S.A. (1996). Predicting species interactions based on behaviour: predation and competition in container-dwelling mosquitoes. *Journal of Animal Ecology*, 65, 63–76.

Griswold, M.W. and Lounibos, L.P. (2005). Does differential predation permit invasive and native mosquito larvae to coexist in Florida? *Ecological Entomology*, 30, 122–127.

Gröning, J. and Hochkirch, A. (2008). Reproductive interference between animal species. *Quarterly Review of Biology*, 83, 257–282.

Gubler, D.J. and Bhattachaya, N.C. (1972). Swarming and mating of *Aedes (S.) albopictus* in nature. *Mosquito News*, 32, 219–223.

Guillaumot, L., Ofanoa, R., Swillen, L., Singh, N., Bossin, H.C. and Schaffner, F. (2012). Distribution of *Aedes albopictus* in southwestern Pacific countries, with a first report from the Kingdom of Tonga. *Parasites and Vectors*, 5, 247.

Harper, J.P. and Paulson, S.L. (1994). Reproductive isolation between Florida strains of *Aedes aegypti* and *Aedes albopictus*. *Journal of the American Mosquito Control Association*, 10, 88–92.

Harrington, L.C., Ponlawat, A., Edman, J.D., Scott, T.W. and Vermeylen, F. (2008). Influence of container size, location, and time of day on oviposition patterns of the dengue vector, *Aedes aegypti*, in Thailand. *Vector Borne and Zoonotic Diseases*, 8, 415–423.

Harrison, B.A., Boonyakanist, P. and Mongkolpanya, K. (1972). Biological observations on *Aedes setoi* Huang in Thailand with notes on rural *Aedes aegypti* (L.) and other *Stegomyia* populations. *Journal of Medical Entomology*, 9, 1–6.

Hartberg, W.K. (1971). Observations on the mating behaviour of *Aedes aegypti* in nature. *Bulletin of the World Health Organization*, 45, 847–850.

Hawley, W.A. (1988). The biology of *Aedes albopictus*. *Journal of the American Mosquito Control Association*, 4 (Suppl.), 1–40.

Huang, Y-M. and Hitchcock, J.C. (1980). Medical entomology studies XII. A revision of the *Aedes scutellaris* Group of Tonga (Diptera: Culicidae). *Contributions of the American Entomological Institute*, 17(3), 1–107.

Hufbauer, R.A., Facon, B., Ravigné, V., et al. (2012). Anthropogenically induced adaptation to invade (AIAI). *Evolutionary Applications*, 5, 89–101.

Jeschke, J.M. and Strayer, D.L. (2006). Determinants of vertebrate invasion success in Europe and North America. *Global Change Biology*, 12, 1608–1619.

Juliano, S.A. (1998). Species introduction and replacement among mosquitoes: interspecific resource competition or apparent competition? *Ecology*, 79, 255–268.

Juliano, S.A. (2009). Species interactions among larval mosquitoes: context dependence across habitat gradients. *Annual Review of Entomology*, 54, 37–56.

Juliano, S.A. (2010). Coexistence, exclusion, or neutrality? A meta-analysis of competition between *Aedes albopictus* and resident mosquitoes. *Israel Journal of Ecology and Evolution*, 56, 325–351.

Juliano, S.A. and Gravel, M.E. (2002). Predation and the evolution of prey behavior: an experiment with tree hole mosquitoes. *Behavioral Ecology*, 13, 301–311.

Juliano, S.A. and Lounibos, L.P. (2005). Ecology of invasive mosquitoes: effects on resident species and on human health. *Ecology Letters*, 8, 558–574.

Juliano, S.A. and Reminger, L. (1992). The relationship between vulnerability to predation and behavior: geographic and ontogenetic differences in larval treehole mosquitoes. *Oikos*, 63, 465–476.

Juliano, S.A., Hechtel, L.J. and Waters, J. (1993). Behavior and risk of predation in larval tree hole mosquitoes: effects of hunger and population history of predation. *Oikos*, 68, 229–241.

Juliano, S. A., O'Meara, G.F., Morrill, J.R. and Cutwa, M.M. (2002). Desiccation and thermal tolerance of eggs and the coexistence of competing mosquitoes. *Oecologia*, 130, 458–469.

Juliano, S.A., Lounibos, L.P. and O'Meara, G.F. (2004). A field test for competitive effects of *Aedes albopictus* on *Aedes aegypti* in South Florida: differences between sites of coexistence and exclusion? *Oecologia*, 139, 583–593.

Juliano, S.A., Lounibos, L.P., Nishimura, N. and Greene, K. (2010). Your worst enemy could be your best friend: predator contributions to invasion resistance and persistence of natives. *Oecologia*, 162, 709–718.

Kaplan, L., Kendell, D., Robertson, D., Livdahl, T. and Khatchikian, C. (2010). *Aedes aegypti* and *Aedes albopictus* in Bermuda: extinction, invasion, invasion and extinction. *Biological Invasions*, 12, 3277–3288.

Kaufman, M.G. and Fonseca, D.M. (2014). Invasion biology of *Aedes japonicus japonicus* (Diptera: Culicidae). *Annual Review of Entomology*, 59, 31–49.

Kesavaraju, B. and Juliano, S.A. (2004). Differential behavioral responses to water-borne cues to predation in two container dwelling mosquitoes. *Annals of the Entomological Society of America*, 97, 194–201.

Kesavaraju, B. and Juliano, S.A. (2008). Behavioral responses of *Aedes albopictus* to a predator are correlated with size-dependent risk of predation. *Annals of the Entomological Society of America*, 101, 1150–1153.

Kesavaraju, B. and Juliano, S.A. (2009). No evolutionary response to four generations of laboratory selection on antipredator behavior of *Aedes albopictus*: potential implications for biotic resistance to invasion. *Journal of Medical Entomology*, 46, 772–781.

Kesavaraju, B., Alto, B.W., Lounibos, L.P. and Juliano, S.A. (2007). Behavioural responses of larval container mosquitoes to a size-selective predator. *Ecological Entomology*, 32, 262–272.

Kesavaraju, B., Damal, K. and Juliano, S.A. (2008). Do natural container habitats impede invader dominance? Predator-mediated coexistence of invasive and native container-dwelling mosquitoes. *Oecologia*, 155, 631–639.

Kesavaraju, B, Kahn, D.F. and Gaugler, R. (2011). Behavioral differences of invasive container-dwelling mosquitoes to a native predator. *Journal of Medical Entomology*, 48, 526–532.

Kishi, S. and Nakazawa, T. (2013). Analysis of species coexistence co-mediated by resource competition and reproductive interference. *Population Ecology*, 55, 305–313.

LaPointe, D.A., Atkinson, C.T. and Samuel, M.D. (2012). Ecology and conservation biology of avian malaria. *Annals of the New York Academy of Sciences*, 1249, 211–226.

Leahy, M.G. and Craig, G.B. (1967). Barriers to hybridization between *Aedes aegypti* and *Aedes albopictus* (Diptera: Culicidae). *Evolution*, 21, 41–58.

Leahy, M.G., VandeHey, R.C. and Booth, K.S. (1978). Differential response to oviposition site by feral and domestic populations of *Aedes aegypti* (L.) (Diptera: Culicidae). *Bulletin of Entomological Research*, 68, 455–463.

Leisnham, P.T. and Juliano, S.A. (2010). Interpopulation differences in competitive effect and response of the mosquito *Aedes aegypti* and resistance to invasion of a superior competitor. *Oecologia*, 164, 221–230.

Leisnham, P.T., Lounibos, L.P., O'Meara, G.F. and Juliano, S.A. (2009). Interpopulation divergence in competitive interactions of the mosquito *Aedes albopictus*. *Ecology*, 90, 2405–2413.

Livdahl, T. and Willey, M.S. (1991). Prospects for an invasion: competition between *Aedes albopictus* and native *Aedes triseriatus*. *Science*, 253, 189–191.

Lockwood, J.L., Hoopes, M.F. and Marchetti, M.P. (2007). *Invasion Ecology*. Malden, MA: Blackwell Publishing.

Lorimer, N., Lounibos, L.P. and Petersen, J.L. (1976). Field trials with a translocation homozygote in *Aedes aegypti* for population replacement. *Journal of Economic Entomology*, 69, 405–409.

Lounibos, L.P. (1981). Habitat segregation among African treehole mosquitoes. *Ecological Entomology*, 6, 129–154.

Lounibos, L.P. (2002). Invasions by insect vectors of human disease. *Annual Review of Entomology*, 47, 233–266.

Lounibos, L.P., Escher, R.L., Nishimura, N. and Juliano, S.A. (1997). Long term dynamics of a predator used for biological control and decoupling from mosquito prey in a subtropical treehole ecosystem. *Oecologia*, 111, 189–200.

Lounibos, L.P., O'Meara, G.F., Escher, R.L., *et al.* (2001). Testing predicted competitive displacement of native *Aedes* by the invasive Asian tiger mosquito *Aedes albopictus* in Florida, USA. *Biological Invasions*, 3, 151–166.

Lowe, S., Browne, M., Boudjelas, S. and De Poorter, M. (2000). *100 of the World's Worst Invasive Alien Species: A selection from the Global Invasive Species Database.* The Invasive Species Specialist Group (ISSG), a specialist group of the Species Survival Commission (SSC) of the World Conservation Union (IUCN), Auckland.

Mack, R.N. (2000). Cultivation fosters plant naturalization by reducing environmental stochasticity. *Biological Invasions*, 2, 111–122.

Mattingly, P.F. (1957). Genetical aspects of the *Aedes aegypti* problem I. Taxonomy and bionomics. *American Journal Tropical Medicine and Parasitology*, 51, 392–408.

McBride, C.S., Baier, F., Omondi, A.B., *et al.* (2014). Evolution of mosquito preference for humans linked to an odorant receptor. *Nature*, 515, 222–227.

McClain, D.K. and Rai, K.S. (1985). Ethological divergence in allopatry and asymmetrical isolation in the South Pacific *Aedes scutellaris* subgroup. *Evolution*, 39, 998–1008.

Medley, K.A. (2010). Niche shifts during the global invasion of the Asian tiger mosquito, *Aedes albopictus* Skuse (Culicidae), revealed by reciprocal distribution models. *Global Ecology and Biogeography*, 19, 122–133.

Moyle, P.B. and Light, T. (1996). Biological invasions of freshwater: empirical rules and assembly theory. *Biological Conservation*, 78, 149–162.

Mukwaya, L.G. (1977). Genetic control of feeding preferences in the mosquitoes *Aedes* (*Stegomyia*) *simpsoni* and *aegypti*. *Physiological Entomology*, 2, 133–145.

Nasci, R.S., Hare, S.G. and Willis, F.S. (1989). Interspecific mating between Louisiana strains of *Aedes albopictus* and *Aedes aegypti* in the field and in laboratory. *Journal of the American Mosquito Control Association*, 5, 416–421.

O'Meara, G.F., Evans Jr., L.F., Gettman, A.D. and Cuda, J.P. (1995). Spread of *Aedes albopictus* and decline of *A. aegypti* (Diptera: Culicidae) in Florida. *Journal of Medical Entomology*, 32, 554–562.

Petersen, J.L. (1977). Behavioral differences in two subspecies of *Aedes aegypti* (L.) (Diptera: Culicidae) in East Africa. PhD dissertation. Notre Dame, IN: University of Notre Dame.

Powell, J.R. and Tabachnick, W.J. (2013). History of domestication and spread of *Aedes aegypti*: a review. *Memorias do Instituto Oswaldo Cruz, Rio de Janeiro*, 108(Suppl. I), 11–17.

Ribeiro, J.M. (1988). Can satyrs control pests and vectors? *Journal of Medical Entomology*, 25, 431–440.

Ribeiro, J.M. and Spielman, A. (1986). The satyr effect: a model predicting parapatry and species extinction. *American Naturalist*, 128, 513–528.

Ritchie. S.A., Moore, P., Carruthers, M., *et al.* (2006). Discovery of a widespread infestation of *Aedes albopictus* in the Torres Strait, Australia. *Journal of the American Mosquito Control Association*, 22, 358–365.

Sih, A. (1986). Antipredator responses and the perception of danger by mosquito larvae. *Ecology*, 67, 434–441.

Simard, F., Nchoutpouen, E., Toto, J.C. and Fontenille, D. (2005). Geographic distribution and breeding site preference of *Aedes albopictus* and *Aedes aegypti* (Diptera: Culicidae) in Cameroon, Central Africa. *Journal of Medical Entomology*, 42, 726–731.

Skelly, D. (1994). Activity level and the susceptibility of anuran larvae to predation. *Animal Behaviour*, 47, 465–468.

Soper, D.L. and Wilson, D.B. (1943). *Anopheles gambiae* in Brazil 1930 to 1940. New York, NY: Rockefeller Foundation.

Sylla, M., Bosio, C., Urdaneta-Marquez, L., Ndiaye, M. and Black, W.C. (2009). Gene flow, subspecies composition, and dengue virus-2 susceptibility among *Aedes aegypti* collections in Senegal. *PLoS Neglected Tropical Diseases*, 3, e408.

Sylla, M., Ndiaye, M. and Black, W.C. (2013). *Aedes* species in treeholes and fruit husks between dry and wet seasons in southeastern Senegal. *Journal of Vector Ecology*, 38, 237–244.

Takken, W. and Verhulst, N.O. (2013). Host preferences of blood-feeding mosquitoes. *Annual Review of Entomology*, 58, 433–453.

Tripet, F., Lounibos, L.P., Robbins, D., Moran, J., Nishimura, N. and Blosser, E.M. (2011). Competitive reduction by satyrization? Evidence for interspecific mating in nature and asymmetric reproductive competition between invasive mosquito vectors. *American Journal of Tropical Medicine and Hygiene*, 85, 265–270.

Trpis, M. and Haüsermann, W. (1975). Demonstration of differential domesticity of *Aedes aegypti* (L.) (Diptera, Culicidae) in Africa by mark–release–recapture. *Bulletin of Entomological Research*, 65, 199–208.

Trpis, M. and Haüsermann, W. (1978). Genetics of house-entering behaviour in East African populations of *Aedes aegypti* (L.) (Diptera, Culicidae) and its relevance to speciation. *Bulletin of Entomological Research*, 68, 521–532.

Van Riper, C., Van Riper, S.G., Goff, M.L. and Laird, M. (1986). The epizootiology and ecological significance of malaria in Hawaiian land birds. *Ecological Monographs*, 56, 327–344.

Wallis, G.P. and Tabachnick, W.J. (1990). Genetic analysis of rock hole and domestic *Aedes aegypti* on the Caribbean island of Anguilla. *Journal of the American Mosquito Control Association*, 6, 625–630.

Ward, R.A. (1984). Mosquito fauna of Guam: case history of an introduced fauna. In *Commerce and the Spread of Pests and Disease Vectors*, ed. Laird, M. New York: Praeger, pp. 143–162.

Wellborn, G.A. (2002). Trade off between competitive ability and antipredator adaptation in a freshwater amphipod species complex. *Ecology*, 83, 129–136.

15 How Behaviour Contributes to the Success of an Invasive Poeciliid Fish: The Trinidadian Guppy (*Poecilia reticulata*) as a Model Species

Amy E. Deacon and Anne E. Magurran

Introduction

Most species introduced to a new habitat fail to establish a self-sustaining population (Williamson and Fitter, 1996). In light of this observation, the documented success of introduced poeciliids worldwide suggests that members of this family of freshwater fish have traits that predispose them to become invasive taxa. FishBase lists 18 poeciliid species as 'established' or 'probably established' outside their native range (Froese and Pauly, 2014); together they are responsible for 10% of freshwater fish species on the Global Invasive Species Database, yet account for just 0.1% of all freshwater fish species. The mosquitofish's (*Gambusia affinis*) presence on the list of 'One Hundred of the World's Worst Invasive Alien Species' (Lowe *et al.*, 2000) highlights the threat that this taxon poses to native freshwater communities.

The negative effects of poeciliid invasions vary across species. *Gambusia holbrooki* and *G. affinis* engage in aggressive interactions, such as fin nipping, leading to elevated mortality in native fish species (Arthington, 1991; Goodsell and Kats, 1999; Mills *et al.*, 2004; Morgan *et al.*, 2004). Other species of poeciliid, including *Phalloceros caudimaculatus, Poecilia latipinna, P. reticulata, Xiphophorus hellerii, X. maculatus* and *X. variatus*, are responsible for the overconsumption of native aquatic invertebrates, and may outcompete other fish by faster reproduction and population growth (Courtenay and Meffe, 1989; Morgan *et al.*, 2004; Deacon *et al.*, 2011; Froese and Pauly, 2014).

Understanding what predisposes some species, such as these poeciliids, to invasion success is vital to the control and management of invasive species worldwide (Sakai *et al.*, 2001). If we can predict which species are most likely to cause a problem on introduction, steps can be taken to reduce the chance of this happening (Kolar and Lodge, 2001). This is particularly important where deliberate introductions are involved. Poeciliids are purposely released for mosquito control reasons (Ghosh *et al.*, 2005; Deacon *et al.*, 2011), and in these scenarios it is important to know whether the

Biological Invasions and Animal Behaviour, eds J.S. Weis and D. Sol. Published by Cambridge University Press. © Cambridge University Press 2016.

ecological costs are likely to outweigh the potential benefits to human health, in which case alternative control measures should be adopted.

It is increasingly clear that behaviour is an important yet often overlooked element of invasion success; in many cases, behavioural traits may be as critical as life history traits in determining whether founder individuals establish. An ideal candidate for investigating the behavioural characters contributing to invasion success is the Trinidadian guppy, *Poecilia reticulata* (Figure 15.1a). There is an extensive body of research on the evolutionary ecology, as well as on the basic biology, life history and ecology of this species (Courtenay and Meffe, 1989; Reznick *et al.*, 2001; Magurran, 2005). Moreover, it has proven a particularly fruitful model for conducting behavioural research, and much of the recent literature relates to its behavioural ecology (e.g. Magurran, 2005).

The guppy is native to, and nearly ubiquitous in, Trinidad, making the island's numerous geographically isolated rivers and streams a 'natural laboratory' where populations exhibit clear behavioural, morphological and life history differences associated with variation in predation regime and other habitat factors (e.g. Reznick and Endler, 1982; Magurran *et al.*, 1995). Field observations and manipulations are carried out with relative ease in the shallow, clear habitats (Magurran and Seghers, 1991) and likewise in outdoor mesocosms when a semi-naturalistic approach is required (e.g. Van Oosterhout *et al.*, 2007; Deacon *et al.*, 2011; Fraser and Lamphere, 2013). Its small size and hardiness, as well as its readiness to reproduce, mean that the guppy is also extremely easily maintained in the laboratory for more controlled experiments (Magurran and Seghers, 1990a).

Invasive Status

The global range of the guppy has expanded dramatically over the past century. Native to northeastern South America and the islands of Trinidad and Tobago, the species is now established in at least 70 countries outside of this range (Deacon *et al.*, 2011), spanning every continent with the exception of Antarctica (see Figure 15.1b).

As mentioned above, the guppy tends to be introduced as a result of mosquito control efforts and as unwanted pet releases (Deacon *et al.*, 2011). However, we know that, in general, most species fail to establish when introduced (Williamson and Fitter, 1996), which raises the question: why does the guppy appear to succeed much more often than most?

An introduced species must succeed at every stage – introduction, establishment and spread – in order to become invasive (see Figure 15.2). In this chapter, we will consider these stages one by one, examining the traits that have been associated with success at each (see Figure 15.1 and Table 15.1).

Stage 1: Transport

Being given the opportunity to invade in the first place is, naturally, an essential component of how likely a species is to become invasive (Rowley *et al.*, 2005). Some species

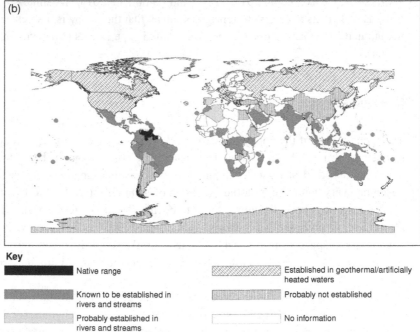

Key

- ▮ Native range
- ▤ Known to be established in rivers and streams
- ▨ Probably established in rivers and streams
- ▨ Established in geothermal/artificially heated waters
- ▥ Probably not established
- ☐ No information

Figure 15.1 (a) A male (above) and female guppy, *Poecilia reticulata* (photo: Sean Earnshaw) and (b) the global distribution of guppies, updated from Deacon *et al.* (2011).

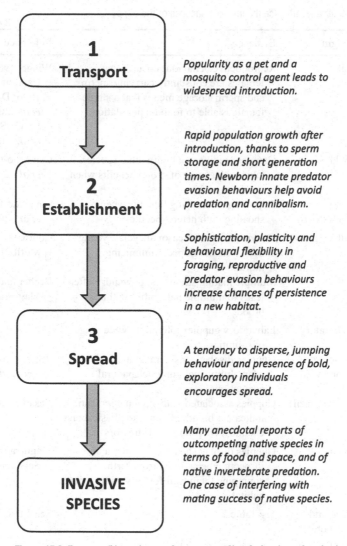

Figure 15.2 Stages of invasion and corresponding behavioural traits in the guppy.

are rarely or never transported outside their native range, while some – usually those of human interest – are frequently introduced, whether deliberately or inadvertently. As popular pets and mosquito control agents, guppies fall very solidly into the latter category. Rixon and colleagues (2005) found that 95% of pet stores surrounding the Laurentian Great Lakes stocked guppies, making them the second most popular aquarium fish of their survey. Furthermore, of the 20 most popular, five were poeciliids. In a 1992 survey, Chapman *et al.* (1997) found that the guppy was the most imported aquarium fish to the United States in terms of volume, accounting for 26% of all imported freshwater fish. The use of the guppy as a mosquito control agent is similarly widespread (Ghosh *et al.*, 2005; Lindholm *et al.*, 2005; Deacon *et al.*, 2011).

Table 15.1 Behavioural traits that may influence the invasive success of the guppy

Trait	Origin*	Evidence	Reference
Reproductive behaviour and sperm storage	All	Reproductive behaviour ensures that females are constantly carrying sperm, and sperm storage means that a single female is able to found a population.	Winge, 1937; Carvalho et al., 1996; Deacon et al., 2011; López-Sepulcre et al., 2013
Heterospecific shoaling	Wild	Guppies readily shoal with other species, taking advantage of group benefits when numbers are low.	Camacho-Cervantes et al., 2014b
	Feral (Mexico)	Guppies gain foraging benefits when shoaling with heterospecifics.	Camacho-Cervantes et al., 2014a
Reproductive behaviour and multiple mating	All	Ovoviviparity. No seasonal cycles, young born well-developed, minimizing mortality.	Courtenay and Meffe, 1989
	All	Multiple mating strategies: polyandry often leads to multiple paternities within a brood.	Becher and Magurran, 2004
Eurytopy	Laboratory	Laboratory guppies tolerant to wide range of salinities.	Chervinski, 1984
	Laboratory and wild	Guppies display considerable thermal adaptability, including behavioural adaptations.	Chung, 2001, Reeve et al., 2014
	Feral (Brazil)	Guppies associated with stream degradation, indicating broad tolerance to physical and chemical environmental stressors.	Casatti et al., 2006
Innate anti-predator behaviours	Wild	Guppies display shoaling, evasion and inspection behaviours from birth, equipping them well for survival in a new habitat.	Magurran and Seghers, 1990a
Behavioural flexibility	Wild and laboratory	*See* Table 2	See Table 15.2
Exploration, boldness and dispersal	Wild	There is considerable gene flow between natural populations.	Crispo et al., 2006
	Laboratory	Some individuals bolder than others. Bolder, more active individuals survive better when faced with predators.	Smith and Blumstein, 2010
	Laboratory	Dispersal probable function of male jumping behaviour	Soares and Bierman, 2013
Anti-predator behaviour modified by selection	Wild	Schooling and predator inspection behaviours are modified by selection in a short period of time.	Magurran et al., 1992

* Key = 'Wild': those guppies either studied in their natural habitat, or caught and observed in the laboratory; 'Laboratory': those guppies bred for several generations in the laboratory; 'Feral': those guppies introduced and established outside their natural range.

However, it is also true that, while necessary, opportunity alone is not sufficient for invasive success; the majority of species fail to thrive even when given the chance. As mentioned earlier, the Poeciliidae are over-represented among invasive species. However, many species from this family have *not* been reported as established outside their native range despite having been adopted by the pet trade (Froese and Pauly, 2014). Therefore it appears that a combination of opportunity and a predisposition to invasiveness determine the likelihood of a particular species being successful at establishing and spreading.

Stage 2: Establishment

Many species that are introduced to new locations will fail to survive for long enough to reproduce, as a result of unfavourable abiotic conditions, lack of suitable food or the presence of a novel predator. However, the guppy is well equipped to deal with these challenges. For example, it has a wide tolerance to abiotic conditions, such as temperature and salinity (Chervinski, 1984; Chung, 2001). This tolerance is not only physiological, but also behavioural; indeed, male guppies have been shown to adjust reproductive behaviour according to changes in temperature and light (Chapman *et al.*, 2009; Reeve *et al.*, 2014). The foraging and predator evasion abilities of the guppy will be discussed in the next section.

Once it has survived initial introduction, the guppy is an exceptional invader, partly thanks to its ability to store sperm for at least 10 months (Winge, 1937; López-Sepulcre *et al.*, 2013). Even a lone pregnant female is capable of establishing a thriving population within a few months, irrespective of her population of origin or whether she has had access to multiple partners (Deacon *et al.*, 2011, 2014). This is helped by the fact that, thanks to the persistent mating behaviour of male guppies, sexually mature females are almost always in possession of sperm. Self-compatibility and asexual reproduction have long been associated with invasiveness, especially in plants (Baker, 1955); sperm storage provides guppies with some of the same benefits while allowing them to retain the advantages of sexual reproduction (Barton and Charlesworth, 1998).

The guppy is ovoviviparous; it gives birth to live young that require no parental care. Indeed, a guppy displays remarkably sophisticated anti-predator behaviours from the moment it is born (Magurran and Seghers, 1990a). As well as possessing an innate shoaling tendency, it will also cautiously approach and inspect predators and has an extremely quick reaction time (Evans and Magurran, 2000; Deacon *et al.*, 2014). The extent to which these behaviours are used varies according to the evolutionary history of the population and the sex of the individual, but each possesses behavioural defences to some degree, increasing its survival chances – even if it finds itself in high predation conditions – and allowing rapid population growth. These defensive behaviours also help mitigate the cannibalistic impact of adult guppies when population density is high, as may be the case in situations where the guppy has been released in ponds and water tanks (Magurran and Seghers, 1990a).

Rapid population growth reduces the likelihood of extinction due to demographic stochasticity. How fast a population grows will depend on how early individuals start producing offspring, how many offspring are produced in each brood and how frequently broods are produced. In guppies, sexual maturity occurs at around 3 months in the wild, at which time a female begins to produce broods of up to 30 young on a monthly basis, enabling rapid population growth (Reznick *et al.*, 2001). By dividing their reproductive output into many separate broods, guppies are effectively 'hedging their bets' to maximize the chance that at least some of their offspring will be born when conditions are favourable (Wilbur and Rudolf, 2006).

The idea of rapid reproduction is closely related to the concept of propagule pressure, as both are measures of how many individuals are present in the early stages of a potential invasion. Propagule pressure describes the total number of individuals of a particular species released into a non-native region (Lockwood *et al.*, 2009). It has frequently been proposed that propagule pressure is the single most critical factor in determining invasive success (Lockwood *et al.*, 2005; Simberloff, 2009). In the case of the guppy, propagule pressure is potentially high, both due to large numbers being introduced at once as part of mosquito control schemes and also, in many situations, due to multiple introductions (Lindholm *et al.*, 2005). The latter may dramatically increase the genetic and phenotypic variation in the population (Kolbe *et al.*, 2004). Furthermore, sperm storage and ovoviviparity expand the gene pool available for the next generation beyond the founding individuals alone.

However, the success of an introduced species is more complex than simply number of individuals; many species fail to establish despite large and multiple introductions (Chapple *et al.*, 2012). Furthermore, establishment does not constitute invasive success; established populations must then spread from the initial place of introduction (see Figure 15.2), and the probability of this occurring can be affected by an overlapping but different suite of factors – many of them behavioural (Holway and Suarez, 1999; Chapple *et al.*, 2012) and it is here that the guppy may have an additional advantage. A new habitat will almost certainly present the guppy with a new suite of predators, competitors and food sources, therefore traits that enable individuals to deal with these challenges will be especially important (see Table 15.1).

Many of the behaviours displayed by guppies are advantageous in an introduction scenario in their own right, but what makes guppies stand out is their remarkable adaptability. This is partly due to their generalist, omnivorous diet (Dussault and Kramer, 1981), and their tolerance of, and ability to adapt to, a wide range of abiotic conditions (Chervinski, 1984; Chung, 2001; Reeve *et al.*, 2014), but can also be seen in the incredible extent of plasticity demonstrated in a wide range of different behaviours (Huizinga *et al.*, 2009).

Behavioural flexibility is a rapidly expanding field in biology. A 'Google Scholar' search revealed that almost five times as many published papers mentioned behavioural flexibility in 2013 (958) than did so in 2003 (202). It has been described by Sol (2003) as 'the general ability of an animal to invent new behaviour or modify its behaviour in an adaptive way'. It can usefully be split into 'activational plasticity', which refers to an animal's innate, direct response to stimuli, and 'developmental plasticity' which is when an animal adjusts their response according to experience, primarily by learning (Snell-Rood, 2013). In this chapter, we largely focus on the latter, as this is the most

Table 15.2 What we know about behavioural flexibility in the guppy

Behaviour	Finding	Relevance to invasive success	Reference
Foraging	The guppy can navigate a maze to reach food, utilize novel feeding apparatus and exploit a novel food source.	Readiness to innovate could be an advantage in a novel environment, both during the establishment and spread stages.	Laland and Reader, 1999
Foraging	Innovated behaviours relating to foraging shown to diffuse through a group by social learning.	Social learning is an effective way for advantageous innovations to spread within an introduced population.	Laland and Williams, 1997
Foraging Anti-predator	Guppies reduce time spent foraging in the presence of predators.	Being able to switch attention between different tasks as appropriate could be an important survival advantage, especially when predators are likely to be novel.	Dugatkin and Godin, 1992
Anti-predator	Guppies display great behavioural flexibility in frequency and duration of anti-predator behaviours.	Allows rapid adaptation to novel predation regime in a new habitat.	Magurran and Seghers, 1990a; Huizinga et al., 2009
Anti-predator	Guppies can adjust their anti-predator behaviour according to the particular predator present.	Plasticity allows rapid adaptation to novel predator species in a new habitat.	Botham et al., 2006; Templeton and Shriner, 2004
Anti-predator	Guppies can build upon their innate behaviour through experience and learning.	Allows adaptation to novel predators and regime through learning.	Kelley and Magurran, 2003b
Anti-predator	Anti-predator responses enhanced in present of experienced conspecifics.	Individuals can learn from the experience of others, increasing likelihood of survival and persistence in new habitat.	Kelley et al., 2003
Anti-predator	Anti-predator behaviour can be induced by the release of chemical cues from conspecifics.	Plasticity allows adaptation to changing environment. Ensures effective and appropriate anti-predator behaviour, even in new habitats.	Brown and Godin, 1999a, b
Reproduction Anti-predator	Males shift mating tactics in favour of sneaky mating when female perceived predation risk is high.	Allows a response to quickly changing conditions, and the adjustment of priorities to optimize reproductive success.	Evans et al., 2002
Reproduction Anti-predator	Strength of female mate choice is weaker under high perceived predation risk.	Allows a response to quickly changing conditions, and the adjustment of priorities to optimize reproductive success.	Godin and Briggs, 1996; Breden and Stoner, 1987; Endler and Houde, 1995
Reproduction	Female mate choice is affected by previous mates.	Flexibility in mate choice helps maximize offspring fitness.	Pitcher et al., 2003

(cont.)

Table 15.2 (cont.)

Behaviour	Finding	Relevance to invasive success	Reference
Reproduction	Adjust mating behaviour depending on receptivity/mated status (males and females).	Demonstrates plasticity of male mating behaviour and allows appropriate prioritization of effort.	Guevara-Fiore et al., 2009; Ojanguren and Magurran, 2004; Eakley and Houde, 2004
Reproduction	Mating behaviour influenced by social environment. Male courtship plastic with respect to density: fewer displays when density high and reduced mating intensity under competition.	Allows adaptation to changing social environment.	Jirotkul, 1999; Magellan et al., 2005; Rodd and Sokolowski, 1995
Reproduction	Males perform more sigmoid displays in high-light environments, and more 'sneaky matings' when light is lower.	Plasticity allows adaptation to changing environment – especially anthropogenic disturbance.	Chapman et al., 2009; Reeve et al., 2014

relevant to surviving in a novel or changing environment (Sol, 2013). Unlike other forms of plasticity, behavioural flexibility allows immediate adjustments in response to novel circumstances. Consequently, it offers individuals a chance of at least short-term survival in order to reproduce in a new habitat, while other adaptations are occurring on a longer time-scale (Sol et al., 2002; Wright et al., 2010; Sol et al., 2013).

Central to behavioural flexibility are individual innovation and social learning, which both involve an individual's behaviour being modified by experience – whether of the environment or of a conspecific (Reader and Laland, 2003). The relative benefits of social learning and innovation depend upon the extent of environmental variability that the individuals are exposed to (Boyd and Richerson, 1988).

In this section we will discuss aspects of guppy behaviour that may contribute to its persistence once introduced and established; those examples that relate to behavioural flexibility specifically are summarized in Table 15.2.

Foraging

As a generalist omnivore, the guppy possesses an innate flexibility in terms of diet, allowing it to capitalize on whatever is available in a new habitat, from insect larvae to algae (Dussault and Kramer, 1981). In their native habitat, different guppy populations display considerable variation in their diet; specifically, guppies from low predation streams tend to have more generalist feeding habits than those coexisting with many piscivorous predators (Bassar et al., 2010; Zandonà et al., 2011). On top of this, it has been demonstrated that, when offered two distinct food types in varying relative densities, a guppy will tend to 'switch' to consume disproportionally more of the most abundant food type (Murdoch et al., 1975). Switching behaviour is by no means unique to the guppy (e.g. Kiørboe et al., 1996; Elliott, 2004), but the species has proved a useful model for demonstrating the phenomenon, not least because it was easy to find two very distinct foods that it readily accepted. Switching behaviour in itself can be viewed as an application of optimal foraging theory and, as such, is a means by which an organism might most efficiently exploit the food resources in a habitat (Ringler, 1985). Possessing the flexibility to switch to the most abundant prey may help introduced populations exploit new or changing food sources as they are encountered in their new habitat.

Developmental behavioural plasticity by means of learning has also been convincingly demonstrated in relation to foraging in the guppy. The tendency to innovate has been linked to motivational state, where a combination of exploration and problem solving are employed readily in the face of novel circumstances (Laland and Reader, 1999). In a series of simple experiments by Laland and Reader (1999), guppies learned to navigate a maze to reach food, to utilize novel feeding apparatus and to exploit a novel food. Innovated behaviours can then diffuse through a population by means of social learning; for example, a guppy can learn the route to a new food source by following conspecifics through a maze (Laland and Williams, 1997). A readiness to employ social learning when it comes to foraging could be a huge advantage in a novel environment, as the discovery of a valuable new food source by one individual can then be shared with others (Laland and Williams, 1997; Laland and Reader, 1999; Reader and Laland, 2000; Reader et al., 2003). Furthermore, the guppy can help mitigate Allee effects during the early stages of an invasion by exploiting local knowledge about food sources from native fish already familiar with the habitat (Camacho-Cervantes et al., 2014a).

Reproduction

As any tropical fish enthusiast will tell you, it is almost impossible *not* to breed guppies. They have a specialized, yet at the same time very adaptable, mating strategy (Thibault and Schultz, 1978). Freshwater habitats are notoriously variable, and many environmental factors have an effect on guppy fecundity, growth and mortality (Magurran, 2005). Therefore, like many freshwater fishes, the guppy seems to have evolved to respond to fluctuating environmental conditions in a way that maximizes reproductive success (Barbosa and Magurran, 2006). This has led to a remarkable plasticity in mating behaviour, which equips them well for novel introduced environments as well as success in their changeable native habitat.

The guppy is sexually dimorphic (see Figure 15.1a), and in general, females tend to have a preference for brightly coloured males. However, they are flexible in how they exert their choice of mate. This occurs at several scales: at the population level, over the lifetime of an individual and in response to recent conditions. First, populations under strong predation pressure have evolved to be less choosy (Breden and Stoner, 1987). Second, early experience can affect female mate choice later in life (Breden et al., 1995; Rosenqvist and Houde, 1997), and finally an individual's choices can be adjusted according to recent conditions. For example, when there is a perceived high predation risk, females will tend to be less choosy, prioritizing predator avoidance (Godin and Briggs, 1996).

Using female preferences as a cue, a male guppy can choose whether to perform his characteristic sigmoid display to convince a female to mate consensually, or to engage in sneaky mating (Evans *et al.*, 2002). All males are capable of engaging in both of these strategies, but individuals and populations do so with varying frequencies, depending on conditions (Endler, 1987; Magurran and Seghers, 1990b; Luyten and Liley, 1991). A sigmoid display can take up to 5 s to perform, leaving the male vulnerable to predation (Luyten and Liley, 1985). Males that switch strategy according to context will reduce their own risk of predation, but equally deployment of sneaking mating when females are preoccupied by predator avoidance may also increase the probability of successful sperm transfer. Indeed, under high predation conditions, males tend to employ sneaky mating strategies more frequently; females have been shown to receive an average of one sneaky mating attempt per minute in such habitats (Magurran and Seghers, 1994).

Multiple mating is ubiquitous in the guppy (Haskins *et al.*, 1961; Neff *et al.*, 2008). In fact, the guppy holds the record for the highest total number of putative sires per brood for a vertebrate species (Avise and Liu, 2011). Multiply-mated females have been shown to produce larger broods (Evans and Magurran, 2000; Barbosa *et al.*, 2012), with greater levels of genetic and phenotypic variability (Barbosa *et al.*, 2010; Gasparini and Pilastro, 2011). All of these multiple mating benefits could be advantageous in an introduction scenario, where founding numbers are often low (Lee, 2002). For example, by giving birth to multiply-sired broods, potential genetic bottlenecks may be minimized, and more variation maintained in the resulting population (Sol *et al.*, 2012; Holman and Kokko, 2013). Thus far, no evidence has been found for this effect in guppies, at least at the establishment stage (Deacon *et al.*, 2014), but it may yet prove important to long-term persistence.

Other advantages of multiple mating include the option for females to employ strategies such as 'trading up', where females progressively choose 'better' males over time, in order to first secure a mating and then shift focus to increasing brood quality (Pitcher *et al.*, 2003). It also encourages post-copulatory mechanisms that may act to minimize inbreeding and improve offspring fitness (Evans and Magurran, 2000; Barbosa *et al.*, 2012).

Anti-Predator Behaviour

Behaviours relating to predator avoidance in the guppy include boldness, predator inspection (Magurran *et al.*, 1992; Godin and Davis, 1995), predator evasion, schooling (Seghers, 1974), foraging (Godin and Smith, 1988; Dugatkin and Godin, 1992; Krause and Godin, 1996), courtship intensity (Godin, 1995) and predator recognition (Kelley and Magurran, 2003b). Aside from the obvious survival advantage that the possession of anti-predator behaviours provides, being able to adapt these to particular circumstances is something that could offer a real advantage when it comes to persisting in a new habitat, as well as when moving beyond the initial point of introduction and spreading more widely.

In its native habitat of Trinidad, suites of anti-predator behaviours vary predictably with habitat and predation regime. For example, when coexisting with ambush-hunting predators such as freshwater prawns (*Macrobrachium* spp.), schooling behaviour can cease to be advantageous, and may even increase detectability (Magurran and Seghers, 1990a). However, other predators, such as the pike cichlid (*Crenicichla frenata*), are 'chasers', and in cases where these are dominant, schooling is a highly advantageous strategy (Seghers, 1974). Experimental introductions and manipulations have demonstrated that these differences can evolve incredibly quickly (Magurran *et al.*, 1992; O'Steen *et al.*, 2002). For example, O'Steen *et al.* (2002) revisited introduced guppy populations in Trinidad from previous translocation experiments. Not only did they detect behavioural differences between high and low predation regime sites, but they also found that these differences persisted in the laboratory-raised F2 generation. This confirmed that the population differences had a genetic basis and therefore that measurable evolution had occurred within 26–36 generations, which is comparable with the time-scale demonstrated for life history evolution in similar circumstances (Reznick *et al.*, 1990). Given that a female guppy can have her first brood at just 12 weeks old (Reznick *et al.*, 2001), measurable differences can evolve in surprisingly short periods of time (Reznick *et al.*, 1997). This has led to the guppy becoming a popular 'classroom' example of the great speed at which natural selection can act on a population under changing conditions.

As well as the shoaling example described above, rapid evolution of behavioural traits has been studied in relation to swimming speed (Ghalambor *et al.*, 2004) and predator evasion (Magurran, 1998). Recently, Kotrschal and colleagues (2013, 2014) demonstrated that strong directional selection for larger brain size in guppies can lead to a rapid shift in this trait in just a few generations. Moreover, they found that this shift was associated with increased cognitive abilities, relevant to many of the behaviours discussed earlier in this chapter – foraging, courtship and predator evasion.

As a guppy gains experience, it is able to build upon its innate behaviours through individual learning. It is extremely likely that all fish are capable of learning (modifying behaviour through experience), but some are more flexible in this than others (Laland *et al.*, 2003). It has long been suggested that species that inhabit unstable environments are likely to require a greater degree of behavioural flexibility, facilitated by learning (Klopfer and MacArthur, 1960; Kieffer and Colgan, 1992). In its native habitat, the guppy is found in a wide variety of communities, including those with many dangerous predators as well as those with very few. As a result, a guppy's reaction to a predator depends very much on its experience. For example, Kelley and Magurran (2003a) found that guppies with previous exposure to predators subsequently reacted more strongly when shown predator models, demonstrating that innate anti-predator defences are moulded by experience.

Further work by Kelley and colleagues (2003) showed that when a guppy with no previous experience of predators was exposed to a predator model in the presence of experienced guppies, its anti-predator reactions were enhanced. This demonstrates that the guppy additionally employs social learning to acquire information concerning novel predators and food sources from conspecifics and thereby allows it to further enhance anti-predator behaviour and survival (Laland and Williams, 1997; Kelley *et al.*, 2003).

Often, a guppy is required to assess the type of predator and relative degree of risk on each encounter, revealing an even greater sophistication to its behavioural plasticity. For example, in a laboratory experiment using wild-caught fish, Templeton and Shriner (2004), found that guppies responded very differently to aerial versus aquatic predation risk. They also found that the nature of their response to aerial predators depended on the type of habitat in which they had evolved; guppies that were from a shallow riffle-dominated stream tended to freeze on the bottom, while those from a vegetation-filled stream were more likely to hide. Therefore, although behaviours are likely to be genetically determined to a large extent, the frequency and duration of behaviour appears to be more flexible (Magurran and Seghers, 1990a). That strategies can be so divergent between geographically close populations, and that behavioural tactics can change dramatically following manipulation of the predation regime, emphasizes the extent of their behavioural flexibility.

For example, Huizinga et al. (2009) demonstrate that despite measurable differences in mean shoaling tendency between high and low predation regime populations, both showed plasticity in their shoaling in response to alarm pheromones. This plasticity allows for a faster phenotypic change than facilitated by selection alone, which is essential when faced with a sudden change of environmental and/or predation regime. It also provides a window of opportunity for guppies to survive, establish and persist before natural selection has had a chance to act to alter the genetic basis for the behaviours (Huizinga et al., 2009). It has even been postulated that this temporal refuge provided by plasticity may allow variation to accumulate in the population, which could ultimately enhance the rate of adaptation by natural selection (Ghalambor et al., 2007).

Most fish shoal at some point during their lives. Gregarious behaviour such as shoaling can promote contact between established and invading species, and can enable small invasive populations to take advantage of the benefits of grouping, including the 'dilution' and 'confusion' anti-predator effects, greater foraging efficiency and increased opportunities for social learning (Pitcher et al., 1982; Pitcher, 1986). Although preferring to shoal with their own species, studies have shown that guppies will readily shoal with heterospecifics (Warburton and Lees, 1996; Camacho-Cervantes et al., 2014b), and other species of poeciliid have demonstrated a similar tendency (Schlupp and Ryan, 1996). This may confer a survival advantage by offering group protection until introduced populations have increased in numbers, and buffering against the Allee effect. Furthermore, a recent study using several species of native Mexican topminnows and feral guppies has shown that guppies gain foraging benefits from shoaling, even when in heterospecific groups (Camacho-Cervantes et al., 2014a).

Behaviour, and the ability to adapt behaviourally to environmental conditions rapidly, is clearly important when it comes to guppies surviving in and adapting to a new habitat. In the next section we will examine how it might contribute to the final stage of invasion: spreading from the initial point of establishment.

Stage 3: Spread

The guppy has been extremely successful at dispersing in its native habitat. In Trinidad, the species is found throughout the island – from lowland plains and swamps to all but the uppermost reaches of the mountain rainforest streams (Phillip, 1998). Moreover, studies have revealed significant gene flow between some of these populations (Crispo *et al.*, 2006).

One particular intraspecific invasion event in Trinidad is well studied. The rivers of the Northern Range can be divided into two distinct drainages – Caroni and Oropuche – with guppy populations that are highly genetically differentiated (Carvalho *et al.*, 1991). In 1957, 200 guppies from the Caroni drainage were deliberately translocated by Caryl Haskins to a previously guppy-free pool above a waterfall barrier in the Oropuche drainage. Over the next 50 years, the descendants of the Caroni stock established, dispersed below the waterfall and admixed with the Oropuche guppy population downstream (Shaw *et al.*, 1992). Subsequent genetic studies have tracked the movement of the introduced genotype over time, revealing that it has now spread many kilometres downstream (Shaw *et al.*, 1992; Becher and Magurran, 2000; Sievers *et al.*, 2012). Sievers and colleagues (2012) demonstrated that this successful spread was at least partly due to the absence of behavioural barriers between the two populations; they interbreed and produce viable offspring. This case study is also a clear example of a general tendency to disperse and spread in this species, and helps explain its remarkably wide distribution in Trinidad.

As mentioned earlier, the streams of the guppy's native habitat are punctuated by numerous waterfalls, riffles and pools. These habitats also differ dramatically between the dry and rainy seasons, which can affect the accessibility of some stretches. In the dry season, waterfalls often dry up, and rivers become a series of isolated pools. To help dispersal in these barrier-filled streams, Soares and Bierman (2013) suggest that the guppy has developed behavioural adaptations: specifically the ability to jump out of water. They studied the jumping behaviour of male guppies and concluded that, rather than for predator avoidance or prey capture, it was most likely a trait developed to aid dispersal. An instinct to jump when trapped in a small body of water could be a great advantage for an introduced fish that is spreading from an initial area of introduction. Croft and colleagues also found that male guppies showed a strong dispersal tendency, and suggested that this might be a strategy to increase the number of females they encounter and thereby their reproductive success (Croft *et al.*, 2003).

Behavioural 'types' linked to dispersal may also affect invasive success. Cote *et al.* (2010) found a correlation between sociability and dispersal behaviour in invasive *Gambusia affinis*, with the more asocial individuals dispersing further. Furthermore, when compared with non-invasive close relatives, invasive species of *Gambusia* displayed higher dispersal tendencies (Rehage and Sih, 2004). There is even evidence that certain individuals have a more innovative 'personality' when it comes to foraging (Laland and Reader, 1999). Indeed, foraging success appears to be greater in mixed shoal compositions – suggesting that a mix of behavioural types in an introduced population may enhance success (Dyer *et al.*, 2009). Smith and Blumstein (2012) found that bolder, more active and exploratory guppies were better at avoiding predators, but suggest that the interaction between predator and guppy behaviour is likely to be important in maintaining behavioural variation.

Recent evidence from *Gambusia* suggests that individuals with greater variability in behavioural traits may be more likely to establish viable populations in an invasion scenario (Cote *et al.*, 2010). It has also been speculated that the dispersal process would naturally select for bold-aggressive individuals, which could in turn confer a stronger tendency to disrupt existing communities (Sih *et al.*, 2004).

Gaps in Our Knowledge

The abundance of traits that can be linked to a potential invasive advantage in the guppy leads us to the question of which of these *actually* contribute to its worldwide distribution, and whether it is a suite of many traits or one or two key traits that enable success. The challenge is to pick these apart and evaluate their relative contributions. Moreover, while some of these traits are easily recorded as present or absent, such as presence of ovovivipary, or easily measured, such as gestation time or brood size, others, such as degree of flexibility in foraging behaviour, are much harder to quantify (Sol *et al.*, 2012).

Like the guppy, the invasive pine tree *Pinus contorta* has also been extensively studied, and is considered a model species for advancing our understanding of invasion biology in plants (Rejmánek and Richardson, 1996; Gundale *et al.*, 2014). We suggest that the guppy could play a parallel role as a model species in our understanding of animal invaders. Approaches used in understanding pine invasions might also be fruitfully used to help explain fish invasions.

For instance, we believe there is great potential here for a phylogenetically controlled comparative analysis of the family Poeciliidae, emulating work on the genus *Pinus* (e.g. Grotkopp *et al.*, 2002), to further identify which of these behavioural and life history traits are most associated with invasive success. As demonstrated by Pollux *et al.* (2014), the Poeciliidae are ideal candidates for this approach; they include species that have been introduced but not become invasive, those that have been introduced and become invasive, and those that have never been introduced outside their native range (Froese and Pauly, 2014). Furthermore, as a group they have been well-studied and so data are already available for many traits in a range of species.

Several attempts have been made to profile invasive fish species, but most of these have focused on life history traits (Vila-Gispert *et al.*, 2005). Nonetheless, they emphasize the importance of trait variability in such analyses (García-Berthou, 2007) as well the importance of controlling for phylogeny (Alcaraz *et al.*, 2005). There is little reason why these methods cannot be extended to include behavioural characters and measures of comparative behavioural flexibility, in the same way that they have been included in comparative studies of avian taxa (Sol *et al.*, 2002).

Impact

As well as influencing the likelihood that introduced guppies will establish and spread, behaviour can also be part of their negative impact as an invasive species.

For example, in Mexico, the guppy's promiscuous mating behaviour is causing problems for the native endangered goodeid, *Skiffia bilineata*. Female *Skiffia* resemble large, fecund female guppies, and as such are extremely attractive to male guppies. Introduced male guppies court these females vigorously and subject them to forced copulations (Valero *et al.*, 2008).

The generalist foraging habits of the guppy can also be a problem in new habitats. In their native habitat there is strong intraguild predation between guppies and their common cohabitant, Hart's rivulus (*Aneblepsoides hartii*, formerly *Rivulus hartii*), with each having a marked influence on the demography of the other (Fraser and Lamphere, 2013). While Hart's rivulus will consume newborn and juvenile guppies, guppies have been witnessed hovering above the almost-hatched eggs of this species, in order to consume them before they bury into the substratum (Douglas Fraser, pers. comm.). This suggests that despite not being strictly predatory, the guppy is still capable of directly impacting heterospecific populations. Other studies have revealed that guppies in their native habitat have a considerable effect on ecosystem structure and function (Bassar *et al.*, 2010; Marshall *et al.*, 2012).

There is already a convincing body of scientific literature regarding the negative effects of invasive *Gambusia holbrooki* and *G. affinis*, which are largely attributed to predation and aggressive interactions (Arthington, 1991; Goodsell and Kats, 1999; Mills *et al.*, 2004; Morgan *et al.*, 2004). Negative ecological impacts have also been associated with poeciliids *Phalloceros caudimaculatus*, *Poecilia latipinna*, *Xiphophorus hellerii*, *X. maculatus* and *X. variatus* (Courtenay and Meffe, 1989; Morgan *et al.*, 2004; Froese and Pauly, 2014). The relative impact of different poeciliid species would benefit greatly from further empirical studies, as reports of negative effects are numerous yet almost entirely anecdotal (Deacon *et al.*, 2011). For example, invasive guppies have been linked to declining populations of cyprinodontids in East Africa, and are similarly thought to have played a role in the decline of the Utah sucker, *Catostomas ardens*, at a thermal spring location in Wyoming (Courtenay and Meffe, 1989). As described earlier, they are also believed to pose a threat to vulnerable goodeids in Mexico (Magurran, 2005; Valero *et al.*, 2008).

Conclusions

The extensive body of research investigating the genetic, behavioural and life history traits of the guppy makes it an ideal species for understanding the factors that contribute to invasion success. It seems likely that the ability of this species to colonize novel environments is at least partly attributable to its evolutionary history in a temporally and spatially variable environment in Trinidad and northern South America. Wild guppies in their native habitat have repeatedly colonized low-predation upstream sites from downstream high-predation areas, and then been washed back down again. This exposure to naturally varying conditions makes it highly advantageous for populations to retain the flexibility to cope with different environments and predator regimes (Huizinga *et al.*, 2009). It probably also helps maintain the genetic variation that underpins rapid evolution. The adaptations that have facilitated survival and persistence under native conditions almost certainly contribute to the guppy's success in habitats worldwide (Thibault and Schultz, 1978; Deacon *et al.*, 2011).

In this chapter, we have tried to show that, alongside other traits favourable to invasion success, the guppy possesses remarkable behavioural flexibility. We have argued that this flexibility plays an important role in enabling newly founded populations to thrive. However, as the guppy is one of the best-studied freshwater fish in the world, an open question is whether other, less-comprehensively investigated, taxa also exhibit equivalent flexibility. The next step would be to conduct comparative analyses in order to test the prediction that increased behavioural flexibility in a species leads to greater invasive success. We propose the Poeciliidae as an excellent candidate group for such work.

A phylogenetically controlled trait analysis could help narrow down the most important characteristics, and a greater understanding of how these relate to invasiveness may contribute to the understanding of invasion biology in general, as well as working towards better predictions of the impacts of introductions in the future. Furthermore, it may enable more effective conservation and management strategies regarding introduced populations of *P. reticulata* and other poeciliid and non-poeciliid invasive species.

Climate change is likely to have far-reaching consequences for the distribution of aquatic invasive species worldwide (Rahel and Olden, 2008; Walther *et al.*, 2009). Increasing and/or more variable temperatures are predicted to impact on invasive species in a number of ways; Hellmann *et al.* (2008) argue that the lower temperature limits of some invaders will be reduced, allowing colonization at higher latitudes and elevations. It is also likely that invasive species will spread to colonize additional habitats. This is because, as well demonstrated in the guppy, invasive species tend to have traits that make them better at adapting to a changing environment – such as broad environmental tolerances, enhanced behavioural flexibility, short generation times and high rates of dispersal (Hellmann *et al.*, 2008).

As we have shown in this chapter, the guppy possesses many of the physiological, life history and behavioural characters associated with extreme adaptability. It is clear that its current range is at least partly dictated by temperature constraints. Furthermore, there are numerous examples in Europe and North America where introduced guppies exist in established, persistent populations near industrial power plants or geothermal vents, where the only barrier preventing spread is temperature (see Figure 15.1). Inevitable escapees and releases from the pet trade mean that the guppy is frequently introduced to places that are outside its environmental tolerance range, but as water temperatures rise, more of these introductions may result in the establishment of self-sustaining populations and subsequent spread (McDowall, 2004).

Invasion biologists tend to be in agreement that identifying potential invaders before they become a problem is far more cost-effective than attempting eradication after they have spread (Leung *et al.*, 2002). This is especially true of aquatic invaders, which are notoriously difficult to eradicate. If particular behavioural traits are found to be characteristic of invasive success, then this could help predict which species are at risk of becoming invasive before they are established or while they are still found only in small numbers (Holway and Suarez, 1999). Indeed, several studies have begun to attempt invasive species profiling for fish, with an emphasis on life history traits (e.g. Marchetti *et al.*, 2004a, b). We argue that behavioural traits should be given greater attention in these analyses, and hope that this chapter will encourage the widespread consideration of behavioural traits in species profiling. To do so will ultimately enhance our understanding of invasive species.

Acknowledgements

The authors gratefully acknowledge an ERC grant (BioTIME 250189) and a NERC grant. We thank Luke Rostant for his assistance in creating Figure 15.1b and Daniel Sol and an anonymous reviewer for their constructive comments.

References

Alcaraz, C., Vila-Gispert, A. and Garcia-Berthou, E. (2005). Profiling invasive fish species: the importance of phylogeny and human use. *Diversity and Distributions*, 11, 289–298.

Arthington, A.H. (1991). Ecological and genetic impacts of introduced and translocated freshwater fishes in Australia. *Canadian Journal of Fisheries and Aquatic Sciences*, 48, 33–43.

Avise, J.C. and Liu, J.-X. (2011). Multiple mating and its relationship to brood size in pregnant fishes versus pregnant mammals and other viviparous vertebrates. *Proceedings of the National Academy of Sciences, USA*, 108, 7091–7095.

Baker, H.G. (1955). Self-compatibility and establishment after 'long-distance' dispersal. *Evolution*, 9, 347–349.

Barbosa, M. and Magurran, A.E. (2006). Female mating decisions: maximizing fitness? *Journal of Fish Biology*, 68, 1636–1661.

Barbosa, M., Dornelas, M. and Magurran, A.E. (2010). Effects of polyandry on male phenotypic diversity. *Journal of Evolutionary Biology*, 23, 2442–2452.

Barbosa, M., Connolly, S.R., Hisano, M., Dornelas, M. and Magurran, A.E. (2012). Fitness consequences of female multiple mating: a direct test of indirect benefits. *BMC Evolutionary Biology*, 12, 1471–2148.

Barton, N.H. and Charlesworth, B. (1998). Why sex and recombination? *Science*, 281, 1986–1990.

Bassar, R.D., Marshall, M.C., López-Sepulcre, A., *et al.* (2010). Local adaptation in Trinidadian guppies alters ecosystem processes. *Proceedings of the National Academy of Sciences, USA*, 107, 3616–3621.

Becher, S.A. and Magurran, A.E. (2000). Gene flow in Trinidadian guppies. *Journal of Fish Biology*, 56, 241–249.

Becher, S.A. and Magurran, A.E. (2004). Multiple mating and reproductive skew in Trinidadian guppies. *Proceedings of the Royal Society B: Biological Sciences*, 271, 1009–1014.

Botham, M.S., Kerfoot, C.J., Louca, V. and Krause, J. (2006). The effects of different predator species on antipredator behavior in the Trinidadian guppy, *Poecilia reticulata*. *Naturwissenschaften*, 93, 431–439.

Boyd, R. and Richerson, P.J. (1988). An evolutionary model of social learning: the effects of spatial and temporal variation. In *Social Learning: Psychological and Biological Perspectives*, ed. Thomas, R. and Zentall, B.G.G. New Jersey: Lawrence Erlbaum Associates, Inc.

Breden, F. and Stoner, G. (1987). Male predation risk determines female preference in the Trinidad guppy. *Nature*, 329, 831–833.

Breden, F., Novinger, D. and Schubert, A. (1995). The effect of experience on mate choice in the Trinidad guppy, *Poecilia reticulata*. *Environmental Biology of Fishes*, 42, 323–328.

Brown, G.E. and Godin, J.-G.J. (1999a). Chemical alarm signals in wild Trinidadian guppies (*Poecilia reticulata*). *Canadian Journal of Zoology*, 77, 562–570.

Brown, G.E. and Godin, J.-G. J. (1999b). Who dares, learns: chemical inspection behaviour and acquired predator recognition in a characin fish. *Animal Behaviour*, 57, 475–481.

Camacho-Cervantes, M., Garcia, C.M., Ojanguren, A.F. and Magurran, A.E. (2014a). Exotic invaders gain foraging benefits by shoaling with native fish. *Open Science*, 1.

Camacho-Cervantes, M., Ojanguren, A., Deacon, A., Ramnarine, I. and Magurran, A. (2014b). Association tendency and preference for heterospecifics in an invasive species. *Behaviour*, 151, 769–780.

Carvalho, G.R., Shaw, P.W., Magurran, A.E. and Seghers, B.H. (1991). Marked genetic-divergence revealed by allozymes among populations of the guppy *Poecilia reticulata* (Poeciliidae), in Trinidad. *Biological Journal of the Linnean Society*, 42, 389–405.

Carvalho, G.R., Shaw, P.W., Hauser, L., Seghers, B.H. and Magurran, A.E. (1996). Artificial introductions, evolutionary change and population differentiation in Trinidadian guppies (*Poecilia reticulata*: Poeciliidae). *Biological Journal of the Linnean Society*, 57, 219–234.

Casatti, L., Langeani, F. and Ferreira, C.P. (2006). Effects of physical habitat degradation on the stream fish assemblage structure in a pasture region. *Environmental Management, V*, 38, 974–982.

Chapman, B.B., Morrell, L.J. and Krause, J. (2009). Plasticity in male courtship behaviour as a function of light intensity in guppies. *Behavioral Ecology and Sociobiology*, 63, 1757–1763.

Chapman, F.A., Fitz-Coy, S.A., Thunberg, E.M. and Adams, C.M. (1997). United States of America trade in ornamental fish. *Journal of the World Aquaculture Society*, 28, 1–10.

Chapple, D.G., Simmonds, S.M. and Wong, B. (2012). Can behavioral and personality traits influence the success of unintentional species introductions? *Trends in Ecology and Evolution*, 27, 57–64.

Chervinski, J. (1984). Salinity tolerance of the guppy, *Poecilia reticulata* Peters. *Journal of Fish Biology*, 24, 449–452.

Chung, K.S. (2001). Critical thermal maxima and acclimation rate of the tropical guppy *Poecilla reticulata*. *Hydrobiologia, V*, 462, 253–257.

Cote, J., Clobert, J., Brodin, T., Fogarty, S. and Sih, A. (2010). Personality-dependent dispersal: characterization, ontogeny and consequences for spatially structured populations. *Philosophical Transactions of the Royal Society B: Biological Sciences*, 365, 4065–4076.

Courtenay, W.R. and Meffe, G.K. (1989). Small fishes in strange places: a review of introduced poeciliids. In *Ecology and Evolution of Livebearing Fishes (Poeciliidae)*, ed. Meffe, G.K. and Snelson F.F. New Jersey: Prentice Hall.

Crispo, E., Bentzen, P., Reznick, D.N., Kinnison, M.T. and Hendry, A.P. (2006). The relative influence of natural selection and geography on gene flow in guppies. *Molecular Ecology*, 15, 49–62.

Croft, D.P., Albanese, B., Arrowsmith, B.J., Botham, M., Webster, M. and Krause, J. (2003). Sex-biased movement in the guppy (*Poecilia reticulata*). *Oecologia*, 137, 62–68.

Deacon, A.E., Ramnarine, I.W. and Magurran, A.E. (2011). How reproductive ecology contributes to the spread of a globally invasive fish. *PLoS ONE*, 6, e24416.

Deacon, A.E., Barbosa, M. and Magurran, A.E. (2014). Forced monogamy in a multiply mating species does not impede colonisation success. *BMC Ecology*, 14, 18.

Dugatkin, L.A. and Godin, J.G.J. (1992). Predator inspection, shoaling and foraging under predation hazard in the Trinidadian guppy, *Poecilia-reticulata*. *Environmental Biology of Fishes*, 34, 265–276.

Dussault, G.V. and Kramer, D.L. (1981). Food and feeding behavior of the guppy, *Poecilia reticulata* (Pisces: Poeciliidae). *Canadian Journal of Zoology*, 59, 684–701.

Dyer, J.R.G., Croft, D.P., Morrell, L.J. and Krause, J. (2009). Shoal composition determines foraging success in the guppy. *Behavioral Ecology*, 20, 165–171.

Eakley, A.L. and Houde, A.E. (2004). Possible role of female discrimination against 'redundant' males in the evolution of colour pattern polymorphism in guppies. *Proceedings of the Royal Society B: Biological Sciences*, 271, S299–S301.

Elliott, J.M. (2004). Prey switching in four species of carnivorous stoneflies. *Freshwater Biology*, 49, 709–720.

Endler, J.A. (1987). Predation, light intensity and courtship behaviour in *Poecilia reticulata* (Pisces: Poeciliidae). *Animal Behaviour*, 35, 1376–1385.

Endler, J.A. and Houde, A.E. (1995). Geographic variation in female preferences for male traits in *Poecilia reticulata*. *Evolution*, 456–468.

Evans, J.P. and Magurran, A.E. (2000). Multiple benefits of multiple mating in guppies. *Proceedings of the National Academy of Sciences, USA*, 97, 10074–10076.

Evans, J.P., Kelley, J.L., Ramnarine, I.W. and Pilastro, A. (2002). Female behaviour mediates male courtship under predation risk in the guppy (*Poecilia reticulata*). *Behavioral Ecology and Sociobiology*, 52, 496–502.

Fraser, D.F. and Lamphere, B.A. (2013). Experimental evaluation of predation as a facilitator of invasion success in a stream fish. *Ecology*, 94, 640–649.

Froese, R. and Pauly, D. (2014). FishBase. Available at: www.fishbase.org,accessed21April2016.

García-Berthou, E. (2007). The characteristics of invasive fishes: what has been learned so far? *Journal of Fish Biology*, 71, 33–55.

Gasparini, C. and Pilastro, A. (2011). Cryptic female preference for genetically unrelated males is mediated by ovarian fluid in the guppy. *Proceedings of the Royal Society B: Biological Sciences*, 278, 2495–2501.

Ghalambor, C.K., Reznick, D.N. and Walker, J.A. (2004). Constraints on adaptive evolution: the functional trade-off between reproduction and fast-start swimming performance in the Trinidadian guppy (*Poecilia reticulata*). *The American Naturalist*, 164, 38–50.

Ghalambor, C.K., Mckay, J.K., Carroll, S.P. and Reznick, D.N. (2007). Adaptive versus non-adaptive phenotypic plasticity and the potential for contemporary adaptation in new environments. *Functional Ecology*, 21, 394–407.

Ghosh, S.K., Tiwari, S.N., Sathyanarayan, T.S., *et al.* (2005). Larvivorous fish in wells target the malaria vector sibling species of the *Anopheles culicifacies* complex in villages in Karnataka, India. *Transactions of the Royal Society of Tropical Medicine and Hygiene*, 99, 101–105.

Godin, J.-G.J. (1995). Predation risk and alternative mating tactics in male Trinidadian guppies (*Poecilia reticulata*). *Oecologia*, 103, 224–229.

Godin, J.-G.J. and Briggs, S.E. (1996). Female mate choice under predation risk in the guppy. *Animal Behaviour*, 51, 117–130.

Godin, J.-G.J. and Davis, S.A. (1995). Who dares, benefits: predator approach behavior in the guppy (*Poecilia-reticulata*) deters predator pursuit. *Proceedings of the Royal Society of London Series B – Biological Sciences*, 259, 193–200.

Godin, J.-G.J. and Smith, S.A. (1988). A fitness cost of foraging in the guppy. *Nature*, 333, 69–71.

Goodsell, J.A. and Kats, L.B. (1999). Effect of introduced mosquitofish on Pacific treefrogs and the role of alternative prey. *Conservation Biology*, 13, 921–924.

Grotkopp, E., Rejmánek, M. and Rost, T.L. (2002). Toward a causal explanation of plant invasiveness: seedling growth and life-history strategies of 29 pine (*Pinus*) species. *The American Naturalist*, 159, 396–419.

Guevara-Fiore, P., Skinner, A. and Watt, P. (2009). Do male guppies distinguish virgin females from recently mated ones? *Animal Behaviour*, 77, 425–431.

Gundale, M.J., Pauchard, A., Langdon, B., *et al.* (2014). Can model species be used to advance the field of invasion ecology? *Biological Invasions*, 16, 591–607.

Haskins, C.P., Haskins, E.F., Mclaughlin, J.J.A. and Hewitt, R.E. (1961). Polymorphism and population structure in *Lebistes reticulatus*, an ecological study. In *Vertebrate Speciation*, ed. Blair, W. F. Austin, TX: University of Texas Press.

Hellmann, J.J., Byers, J.E., Bierwagen, B.G. and Dukes, J.S. (2008). Five potential consequences of climate change for invasive species. *Conservation Biology*, 22, 534–543.

Holman, L. and Kokko, H. (2013). The consequences of polyandry for population viability, extinction risk and conservation. *Philosophical Transactions of the Royal Society B: Biological Sciences*, 368.

Holway, D.A. and Suarez, A.V. (1999). Animal behavior: an essential component of invasion biology. *Trends in Ecology and Evolution*, 14, 328–330.

Huizinga, M., Ghalambor, C. and Reznick, D. (2009). The genetic and environmental basis of adaptive differences in shoaling behaviour among populations of Trinidadian guppies, *Poecilia reticulata*. *Journal of Evolutionary Biology*, 22, 1860–1866.

Jirotkul, M. (1999). Population density influences male–male competition in guppies. *Animal Behaviour*, 58, 1169–1175.

Kelley, J.L. and Magurran, A.E. (2003a). Effects of relaxed predation pressure on visual predator recognition in the guppy. *Behavioral Ecology and Sociobiology*, 54, 225–232.

Kelley, J.L. and Magurran, A.E. (2003b). Learned predator recognition and antipredator responses in fishes. *Fish and Fisheries*, 4, 216–226.

Kelley, J.L., Evans, J.P., Ramnarine, I.W. and Magurran, A.E. (2003). Back to school: can antipredator behaviour in guppies be enhanced through social learning? *Animal Behaviour*, 65, 655–662.

Kieffer, J.D. and Colgan, P.W. (1992). The role of learning in fish behaviour. *Reviews in Fish Biology and Fisheries*, 2, 125–143.

Kiørboe, T., Saiz, E. and Viitasalo, M. (1996). Prey switching behaviour in the planktonic copepod *Acartia tonsa*. *Marine Ecology Progress Series*, 143, 65–75.

Klopfer, P.H. and Macarthur, R.H. (1960). Niche size and faunal diversity. *American Naturalist*, 94(877), 293–300.

Kolar, C.S. and Lodge, D.M. (2001). Progress in invasion biology: predicting invaders. *Trends in Ecology and Evolution*, 16, 199–204.

Kolbe, J.J., Glor, R.E., Schettino, L.R.G., *et al.* (2004). Genetic variation increases during biological invasion by a Cuban lizard. *Nature*, 431, 177–181.

Kotrschal, A., Rogell, B., Bundsen, A., *et al.* (2013). Artificial selection on relative brain size in the guppy reveals costs and benefits of evolving a larger brain. *Current Biology*, 23, 168–171.

Kotrschal, A., Corral-Lopez, A., Amcoff, M. and Kolm, N. (2014). A larger brain confers a benefit in a spatial mate search learning task in male guppies. *Behavioral Ecology*, aru227v1.

Krause, J. and Godin, J.-G.J. (1996). Influence of prey foraging posture on flight behavior and predation risk: predators take advantage of unwary prey. *Behavioral Ecology*, 7, 264–271.

Laland, K.N. and Reader, S.M. (1999). Foraging innovation in the guppy. *Animal Behaviour*, 57, 331–340.

Laland, K.N. and Williams, K. (1997). Shoaling generates social learning of foraging information in guppies. *Animal Behaviour*, 53, 1161–1169.

Laland, K.N., Brown, C. and Krause, J. (2003). Learning in fishes: from three-second memory to culture. *Fish and Fisheries*, 4, 199–202.

Lee, C.E. (2002). Evolutionary genetics of invasive species. *Trends in Ecology and Evolution*, 17, 386–391.

Leung, B., Lodge, D.M., Finnoff, D., *et al.* (2002). An ounce of prevention or a pound of cure: bioeconomic risk analysis of invasive species. *Proceedings of the Royal Society of London Series B: Biological Sciences*, 269, 2407–2413.

Lindholm, A.K., Breden, F., Alexander, H.J., *et al.* (2005). Invasion success and genetic diversity of introduced populations of guppies *Poecilia reticulata* in Australia. *Molecular Ecology*, 14, 3671–3682.

Lockwood, J.L., Cassey, P. and Blackburn, T. (2005). The role of propagule pressure in explaining species invasions. *Trends in Ecology and Evolution*, 20, 223–228.

Lockwood, J.L., Cassey, P. and Blackburn, T.M. (2009). The more you introduce the more you get: the role of colonization pressure and propagule pressure in invasion ecology. *Diversity and Distributions*, 15, 904–910.

López-Sepulcre, A., Gordon, S.P., Paterson, I.G., Bentzen, P. and Reznick, D.N. (2013). Beyond lifetime reproductive success: the posthumous reproductive dynamics of male Trinidadian guppies. *Proceedings of the Royal Society B: Biological Sciences*, 280.

Lowe, S., Browne, M., Boudjelas, S. and De Poorter, M. (2000). *100 of the World's Worst Invasive Alien Species: A Selection from the Global Invasive Species Database*. The Invasive Species Specialist Group (ISSG) , a specialist group of the Species Survival Commission (SSC) of the World Conservation Union (IUCN), Auckland.

Luyten, P.H. and Liley, N.R. (1985). Geographic variation in the sexual behaviour of the guppy, *Poecilia reticulata* (Peters). *Behaviour*, 95, 164–179.

Luyten, P.H. and Liley, N.R. (1991). Sexual selection and competitive mating success of male guppies (*Poecilia reticulata*) from four Trinidad populations. *Behavioral Ecology and Sociobiology*, V, 28, 329–336.

Magellan, K., Pettersson, L.B. and Magurran, A.E. (2005). Quantifying male attractiveness and mating behaviour through phenotypic size manipulation in the Trinidadian guppy, *Poecilia reticulata*. *Behavioral Ecology and Sociobiology*, 58, 366–374.

Magurran, A.E. (1998). Population differentiation without speciation. *Philosophical Transactions of the Royal Society of London, B*, 353, 275–286.

Magurran, A.E. (2005). *Evolutionary Ecology: The Trinidadian Guppy*. Oxford, UK: Oxford University Press.

Magurran, A.E. and Seghers, B.H. (1990a). Population differences in the schooling behaviour of newborn guppies, *Poecilia reticulata*. *Ethology*, 84, 334–342.

Magurran, A.E. and Seghers, B.H. (1990b). Risk sensitive courtship in the guppy (*Poecilia reticulata*). *Behaviour*, 112, 194–201.

Magurran, A.E. and Seghers, B.H. (1991). Variation in schooling and aggression amongst guppy (*Poecilia reticulata*) populations in Trinidad. *Behaviour*, 118, 214–234.

Magurran, A.E. and Seghers, B.H. (1994). Sexual conflict as a consequence of ecology: evidence from guppy, *Poecilia reticulata*, populations in Trinidad. *Proceedings of the Royal Society B: Biological Sciences*, 225, 31–36.

Magurran, A.E., Seghers, B.H., Carvalho, G.R. and Shaw, P.W. (1992). Behavioral consequences of an artificial introduction of guppies (*Poecilia-reticulata*) in N-Trinidad: evidence for the evolution of antipredator behavior in the wild. *Proceedings of the Royal Society of London Series B: Biological Sciences*, 248, 117–122.

Magurran, A.E., Seghers, B.H., Shaw, P.W. and Carvalho, G.R. (1995). The behavioral diversity and evolution of guppy, *Poecilia reticulata*, populations in Trinidad. *Advances in the Study of Behavior*, Vol. 24. San Diego, CA: Academic Press Inc.

Marchetti, M.P., Moyle, P.B. and Levine, R. (2004a). Alien fishes in California watersheds: characteristics of successful and failed invaders. *Ecological Applications*, 14, 587–596.

Marchetti, M.P., Moyle, P.B. and Levine, R. (2004b). Invasive species profiling? Exploring the characteristics of non-native fishes across invasion stages in California. *Freshwater Biology*, 49, 646–661.

Marshall, M.C., Binderup, A.J., Zandonà, E., *et al.* (2012). Effects of consumer interactions on benthic resources and ecosystem processes in a neotropical stream. *PLoS ONE*, 7, e45230.

Mcdowall, R.M. (2004). Shoot first, and then ask questions: a look at aquarium fish imports and invasiveness in New Zealand. *New Zealand Journal of Marine and Freshwater Research*, 38, 503–510.

Mills, M.D., Rader, R.B. and Belk, M.C. (2004). Complex interactions between native and invasive fish: the simultaneous effects of multiple negative interactions. *Oecologia*, 141, 713–721.

Morgan, D.L., Gill, H.S., Maddern, M.G. and Beatty, S.J. (2004). Distribution and impacts of introduced freshwater fishes in Western Australia. *New Zealand Journal of Marine and Freshwater Research*, 38, 511–523.

Murdoch, W.W., Avery, S. and Smyth, M.E.B. (1975). Switching in predatory fish. *Ecology*, 56, 1094–1105.

Neff, B.D., Pitcher, T.E. and Ramnarine, I.W. (2008). Inter-population variation in multiple paternity and reproductive skew in the guppy. *Molecular Ecology*, 17, 2975–2984.

O'Steen, S., Cullum, A.J. and Bennett, A.F. (2002). Rapid evolution of escape ability in Trinidadian guppies (*Poecilia reticulata*). *Evolution*, 56, 776–784.

Ojanguren, A.F. and Magurran, A.E. (2004). Uncoupling the links between male mating tactics and female attractiveness. *Proceedings of the Royal Society of London. Series B: Biological Sciences*, 271, S427–S429.

Phillip, D.A.T. (1998). *Biodiversity of Freshwater Fishes of Trinidad and Tobago, West Indies*. St Andrews, UK: University of St Andrews.

Pitcher, T.J. (1986). *Functions of Shoaling Behaviour in Teleosts. The Behaviour of Teleost Fishes*. Berlin: Springer.

Pitcher, T.J., Magurran, A.E. and Winfield, I.J. (1982). Fish in larger shoals find food faster. *Behavioral Ecology and Sociobiology*, 10, 149–151.

Pitcher, T.J., Neff, B.D., Rodd, F.H. and Rowe, L. (2003). Multiple mating and sequential mate choice in guppies: females trade up. *Proceedings of the Royal Society B: Biological Sciences*, 270, 1623–1629.

Pollux, B., Meredith, R., Springer, M., Garland, T. and Reznick, D. (2014). The evolution of the placenta drives a shift in sexual selection in livebearing fish. *Nature*, 513, 233–236.

Rahel, F.J. and Olden, J.D. (2008). Assessing the effects of climate change on aquatic invasive species. *Conservation Biology*, 22, 521–533.

Reader, S.M. and Laland, K.N. (2000). Diffusion of foraging innovations in the guppy. *Animal Behaviour*, 60, 175–180.

Reader, S.M. and Laland, K.N. (2003). *Animal Innovation*. Oxford, UK: Oxford University Press.

Reader, S.M., Kendal, J.R. and Laland, K.N. (2003). Social learning of foraging sites and escape routes in wild Trinidadian guppies. *Animal Behaviour*, 66, 729–739.

Reeve, A.J., Ojanguren, A.F., Deacon, A.E., *et al.* (2014). Interplay of temperature and light influences wild guppy (*Poecilia reticulata*) daily reproductive activity. *Biological Journal of the Linnean Society*, 111(3), 511–520.

Rehage, J.S. and Sih, A. (2004). Dispersal behavior, boldness, and the link to invasiveness: a comparison of four Gambusia species. *Biological Invasions*, 6, 379–391.

Rejmánek, M. and Richardson, D.M. (1996). What attributes make some plant species more invasive? *Ecology*, 77(6), 1655–1661.

Reznick, D. and Endler, J.A. (1982). The impact of predation on life-history evolution in Trinidadian guppies (*Poecilia reticulata*). *Evolution*, 36, 160–177.

Reznick, D.A., Bryga, H. and Endler, J.A. (1990). Experimentally induced life-history evolution in a natural population. *Nature*, 346, 357–359.

Reznick, D.N., Shaw, F.H., Rodd, F.H. and Shaw, R.G. (1997). Evaluation of the rate of evolution in natural populations of guppies (*Poecilia reticulata*). *Science*, 275, 1934–1937.

Reznick, D., Butler, M.J. and Rodd, H. (2001). Life-history evolution in guppies. VII. The comparative ecology of high- and low-predation environments. *American Naturalist*, 157, 126–140.

Ringler, N.H. (1985). Individual and temporal variation in prey switching by brown trout, *Salmo trutta*. *Copeia*, 4, 918–926.

Rixon, C.A.M., Duggan, I.C., Bergeron, N.M.N., Ricciardi, A. and Macisaac, H.J. (2005). Invasion risks posed by the aquarium trade and live fish markets on the Laurentian Great Lakes. *Biodiversity and Conservation*, 14, 1365–1381.

Rodd, F.H. and Sokolowski, M.B. (1995). Complex origins of variation in the sexual behaviour of male Trinidadian guppies, *Poecilia reticulata*: interactions between social environment, heredity, body size and age. *Animal Behaviour*, 49, 1139–1159.

Rosenqvist, G. and Houde, A. (1997). Prior exposure to male phenotypes influences mate choice in the guppy. *Poecilia reticulata*. *Behavioral Ecology*, 8, 194–198.

Rowley, J.J.L., Rayner, T.S. and Pyke, G.H. (2005). New records and invasive potential of the poeciliid fish *Phalloceros caudimaculatus*. *New Zealand Journal of Marine and Freshwater Research*, 39, 1013–1022.

Sakai, A.K., Allendorf, F.W., Holt, J.S., *et al.* (2001). The population biology of invasive species. *Annual Review of Ecological Systems*, 32, 305–332.

Schlupp, I. and Ryan, M.J. (1996). Mixed-species shoals and the maintenance of a sexual–asexual mating system in mollies. *Animal Behaviour*, 52, 885–890.

Seghers, B.H. (1974). Schooling behavior in guppy (*Poecilia-reticulata*): evolutionary response to predation. *Evolution*, 28, 486–489.

Shaw, P., Carvalho, G., Seghers, B. and Magurran, A. (1992). Genetic consequences of an artificial introduction of guppies (*Poecilia reticulata*) in N. Trinidad. *Proceedings of the Royal Society of London. Series B: Biological Sciences*, 248, 111–116.

Sievers, C., Willing, E.-M., Hoffmann, M., *et al.* (2012). Reasons for the invasive success of a guppy (*Poecilia reticulata*) population in Trinidad. *PLoS ONE*, 7, e38404.

Sih, A., Bell, A. and Johnson, J.C. (2004). Behavioral syndromes: an ecological and evolutionary overview. *Trends in Ecology and Evolution*, 19, 372–378.

Simberloff, D. (2009). The role of propagule pressure in biological invasions. *Annual Review of Ecology, Evolution, and Systematics*, 40, 81–102.

Smith, B.R. and Blumstein, D.T. (2010). Behavioral types as predictors of survival in Trinidadian guppies (*Poecilia reticulata*). *Behavioral Ecology*, arq084.

Smith, B.R. and Blumstein, D.T. (2012). Structural consistency of behavioural syndromes: does predator training lead to multi-contextual behavioural change? *Behaviour*, 149, 187–213.

Snell-Rood, E.C. (2013). An overview of the evolutionary causes and consequences of behavioural plasticity. *Animal Behaviour*, 85, 1004–1011.

Soares, D. and Bierman, H.S. (2013). Aerial jumping in the Trinidadian guppy (*Poecilia reticulata*). *PLoS ONE*, 8, e61617.

Sol, D. (2003). Behavioural flexibility: a neglected issue in the ecological and evolutionary literature. In *Animal Innovation*, Reader, S.M. and Laland, K.N., eds. Oxford: Oxford University Press, pp. 63–82.

Sol, D., Timmermans, S. and Lefebvre, L. (2002). Behavioural flexibility and invasion success in birds. *Animal Behaviour*, 63, 495–516.

Sol, D., Maspons, J., Vall-Llosera, M., *et al.* (2012). Unraveling the life history of successful invaders. *Science*, 337, 580–583.

Sol, D., Lapiedra, O. and Gonzalez-Lagos, C. (2013). Behavioural adjustments for a life in the city. *Animal Behaviour*, 85, 1101–1112.

Templeton, C.N. and Shriner, W.M. (2004). Multiple selection pressures influence Trinidadian guppy (*Poecilia reticulata*) antipredator behavior. *Behavioral Ecology*, 15, 673–678.

Thibault, R.E. and Schultz, R.J. (1978). Reproductive adaptations among viviparous fishes (Cyprinodontiformes: Poeciliidae). *Evolution*, 32, 320–333.

Valero, A., Macias Garcia, C. and Magurran, A.E. (2008). Heterospecific harassment of native endangered goodeids by invasive guppies in Mexico. *Biology Letters*, 4, 149–152.

Van Oosterhout, C., Smith, A.M., Hanfling, B., *et al.* (2007). The guppy as a conservation model: implications of parasitism and inbreeding for reintroduction success. *Conservation Biology*, 21, 1573–1583.

Vila-Gispert, A., Alcaraz, C. and Garcia-Berthou, E. (2005). Life-history traits of invasive fish in small Mediterranean streams. *Biological Invasions*, 7, 107–116.

Walther, G.-R., Roques, A., Hulme, P.E., *et al.* (2009). Alien species in a warmer world: risks and opportunities. *Trends in Ecology and Evolution*, 24, 686–693.

Warburton, K. and Lees, N. (1996). Species discrimination in guppies: learned responses to visual cues. *Animal Behaviour*, 52, 371–378.

Wilbur, H.M. and Rudolf, V.H.W. (2006). Life-history evolution in uncertain environments: bet hedging in time. *The American Naturalist*, 168, 398–411.

Williamson, M.H. and Fitter, A. (1996). The varying success of invaders. *Ecology*, 77, 1661–1666.

Winge, O. (1937). Succession of broods in *Lebistes*. *Nature*, 140, 467.

Wright, T.F., Eberhard, J.R., Hobson, E.A., Avery, M.L. and Russello, M.A. (2010). Behavioral flexibility and species invasions: the adaptive flexibility hypothesis. *Ethology Ecology and Evolution*, 22, 393–404.

Zandonà, E., Auer, S.K., Kilham, S.S., *et al.* (2011). Diet quality and prey selectivity correlate with life histories and predation regime in Trinidadian guppies. *Functional Ecology*, 25, 964–973.

16 How Behaviour Has Helped Invasive Crayfish to Conquer Freshwater Ecosystems

Elena Tricarico and Laura Aquiloni

Crayfish, the largest freshwater macroinvertebrates, are considered keystone species and ecosystem engineers in aquatic habitats. They form a highly diversified group within the order of crustacean decapods (644 species), naturally distributed across all continents (except continental Africa and Antarctica) with two centres of diversity, one in the southeastern United States (northern hemisphere centre) and one in south-east Australia (southern hemisphere centre) (Crandall and Buhay, 2008; Richman *et al.*, 2015). Many introductions mediated by humans occurred within and between continents, altering the biogeography of the taxon (Gherardi, 2010). Since the end of the 1800s, a number of species, particularly from North America, have been introduced into Europe for aquaculture and restocking purposes (Souty-Grosset *et al.*, 2006). Once in the wild, the majority of the introduced species dramatically impacted the indigenous crayfish through either direct or indirect competition and exerted a strong impact on the invaded communities and ecosystems (Gherardi, 2007; Lodge *et al.*, 2012; Twardochleb *et al.*, 2013). It thus became pivotal to investigate the ethological traits of alien crayfish that have made them successful invaders to establish appropriate methods for their management. The present chapter will review these behaviours, particularly focusing on some American crayfish, and showing how these species were enabled to invade diverse habitats with consequent high impacts on native ecosystems. The majority of literature concern North American species introduced elsewhere within North America and in Europe (the Cambaridae: *Orconectes* spp., *Procambarus* spp. and the Astacidae: *Pacifastacus leniusculus*; Lodge *et al.*, 2012). The rusty crayfish *Orconectes rusticus*, the best-studied member of the genus *Orconectes*, is the most widespread alien invasive crayfish in the Laurentian Great Lakes, where it causes several impacts, and up to now it is not present in Europe (Kouba *et al.*, 2014). Due to its high invasiveness, the red swamp crayfish *Procambarus clarkii* is included in the list of the 100 worst alien species in Europe (DAISIE, 2011). The signal crayfish *P. leniusculus* is another well-known invader in North America, Europe and Asia. Crayfish belonging to the Australian Parastacidae (*Cherax* spp.) were introduced, among the other places, to Africa, Asia, the Caribbean, and Europe, but they are reported as having very few wild populations and their behaviour and impacts are less studied (Souty-Grosset *et al.*, 2006; Jaklič and Vrezec, 2011).

Biological Invasions and Animal Behaviour, eds J.S. Weis and D. Sol. Published by Cambridge University Press. © Cambridge University Press 2016.

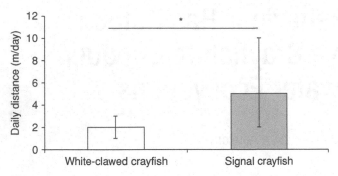

Figure 16.1 Range per daily distance for the native white-clawed crayfish ($N = 20$) and the alien signal crayfish ($N = 15$). Bars represent the median with first–third quartiles. The asterisk denotes a significant difference after the Mann–Whitney U test. Modified after Bubb *et al.* (2006).

Locomotory Behaviour

Alien crayfish occupy diverse habitats and tolerate a wide range of environmental conditions (Souty-Grosset *et al.*, 2006; Lodge *et al.*, 2012). Coupled with this plasticity, their capability to actively spread is crucial for conquering new habitats. Crayfish, both native and alien, usually display an 'ephemeral home range', alternating periods of intense locomotion with those of slow speed or no movement (see Gherardi, 2002; Gherardi *et al.*, 2002; Bubb *et al.*, 2004; Aquiloni *et al.*, 2005). Their locomotory behaviour generally increases with temperatures (*P. clarkii*: Gherardi *et al.*, 2002; Aquiloni *et al.*, 2005; *P. leniusculus*: Bubb *et al.*, 2002, 2004; Johnson *et al.*, 2014; but not in the spiny-cheek crayfish *Orconectes limosus* during the mating period: Buřič *et al.*, 2009a). Some alien species show nocturnal habits (e.g. *P. leniusculus*: Johnson *et al.*, 2014), while others are more adaptable and, depending on the environment and period of life cycle, can be active both by day or by night (e.g. *P. clarkii*: Gherardi *et al.*, 2002; Aquiloni *et al.*, 2005; *O. limosus*: Musil *et al.*, 2010), or more during the day (*O. limosus* in the breeding season; Buřič *et al.*, 2009a). Males can be more nomadic during the reproductive period, when they are searching for partners, while females may move less when they bear eggs and more after hatching (Light, 2003; Bubb *et al.*, 2004; Buřič *et al.*, 2009a; Hudina *et al.*, 2011; Wutz and Geist, 2013).

Alien crayfish rapidly cover greater distance than the native ones, confirming their invasive potential and the great threat they pose to the invaded ecosystems. *Procambarus clarkii* in Spain, Italy and Portugal (Barbaresi and Gherardi, 2000; Gherardi *et al.*, 2002; Aquiloni *et al.*, 2005; up to 4 km day^{-1}), and *O. limosus* in the Czech Republic show a high dispersal rate with no directional preference during summer (up to 15 m day^{-1}: Buřič *et al.*, 2009a, b). In the UK, *P. leniusculus* actively moves downstream, generally more than the co-occurring native crayfish *Austropotamobius pallipes* (up to 13 m day^{-1} versus up to 7 m day^{-1}: Bubb *et al.*, 2004, 2006; Figure 16.1). Even *Orconectes rusticus* in a long-term study in a North American invaded lake showed a high dispersal rate (0.68 km year^{-1}; Wilson *et al.*, 2004). The removal of small barriers for the conservation of native crayfish may favour the spread of alien species, as evidenced for *P. leniusculus* by Light (2003) in California (USA) and by Rosewarne *et al.* (2013) in the UK, where the presence of gauging weirs or artificial/natural barriers can reduce its dispersal. This measure may not be efficacious, however, for species such as *P. clarkii*, capable of active land dispersal from invaded waters for up to 24 h consecutively (Gherardi, 2002; Gherardi *et al.*, 2002). The presence of hydroelectric dams seems to be more efficacious to successfully preventing the expansion of *P. clarkii* in areas inhabited by the native *A. pallipes* (Dana *et al.*, 2011). Similarly, *O. limosus* and *O. rusticus* have some ability to walk overland to access a watercourse (Claussen *et al.*, 2000; Holdich and Black, 2007).

Shelters and burrows help crayfish to withstand environmental extremes (e.g. high or low temperatures, dehydration) and protect them from predators/conspecifics during sensitive phases of their life cycle (moulting, reproduction) (Gherardi, 2002). Overall, native European crayfish use shelters and do not construct burrows (but see *Astacus astacus*: Souty-Grosset *et al.*, 2006). *Pacifastacus leniusculus*, considered a tertiary burrower crayfish (i.e. it only burrows for reproduction or under extreme conditions as drought; Hogger, 1988), in North America, as an invasive in Europe, frequently digs burrows under rocks or river and lake banks, reaching a density of 14 per linear metre with severe impacts on bank morphology and stability (Souty-Grosset *et al.*, 2006). In many countries, damage to agricultural fields caused by *P. clarkii*'s burrowing activity (reported density up to 6.8 burrows m^{-2} and increasing with the amount of fine sediment in the soil; Correia and Ferreira, 1995; Barbaresi *et al.*, 2004) has made this species achieve pest status (Hobbs *et al.*, 1989). *Orconectes limosus* is capable of constructing burrows in England at least, while in other invaded European countries, this activity has been rarely observed (Holdich and Black, 2007). The Australian yabby, *Cherax destructor*, is a well-known burrowing species (with burrows connected to the water table, burrow Type 2: Horwitz and Richardson, 1986), but it is reported in the wild only in Spain without any impacts up to now (Souty-Grosset *et al.*, 2006).

Due to their ephemeral home range, crayfish usually do not show high shelter/burrow fidelity (but see *O. limosus*: Buřič *et al.*, 2009b). In the UK, *P. leniusculus* remained at one refuge for several days to weeks before moving to a different refuge (Bubb *et al.*, 2004). Similarly, in Italy, *P. clarkii* occupied and dug a burrow for a relatively short time (6 h on average; Barbaresi *et al.*, 2004): once abandoned, the old burrows collapsed and the crayfish dug new ones, increasing the density of burrows in a small area and the possibility of bank collapse. Ilhéu *et al.* (2003) found a low fidelity for the same species in Portugal, with individuals both actively digging burrows and simply occupying available natural shelters. Burrowing by *P. clarkii* in European coastal wetlands (Scalici *et al.*, 2010) could reduce coastal protection from severe storms and sea-level rise. *Procambarus clarkii* is considered a bioturbator: its intense burrowing activity increases water turbidity followed by reduced light penetration and plant production, and thus inhibits primary production (Angeler *et al.*, 2001; Rodríguez *et al.*, 2003). Similarly, *P. leniusculus* increases suspended sediments in invaded rivers (e.g. Harvey *et al.*, 2011).

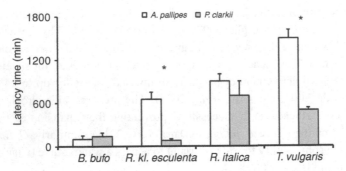

Figure 16.2 Time spent in capturing one live larva of either *Bufo bufo* or *Rana*. kl. *esculenta* or *Rana italica* or *Triturus vulgaris* by starved individuals of the indigenous *Austropotamobius pallipes* and the alien *Procambarus clarkii*. N = 20 per crayfish for each different food items. Bars represent the mean with standard error. The asterisk denotes a significant difference after the Student's *t*-test. Modified after Renai and Gherardi (2004).

Predatory and Anti-predatory Behaviour

Crayfish are omnivorous and opportunistic feeders, consuming a wide range of food items such as macrophytes, macroinvertebrates, fish, periphyton, detritus, moulted conspecifics and tadpoles (Gherardi, 2007; Lodge *et al.*, 2012). *Procambarus clarkii*, for example, feeds on the diverse items present in a given invaded habitat in proportion to their availability so that its diet can change with habitats (Gherardi, 2006). It quickly learns to feed on unknown prey: in laboratory *P. clarkii*, naïve individuals require less than 12 h to learn to maximize capture rate of larvae (Diptera, *Chaoborus* sp.), showing its predation capabilities in newly invaded habitats (Ramalho and Anástacio, 2011). Gray and Jackson (2012) found that this species, occupying the footprints of hippopotamus on the edge on Lake Naivasha (Kenya), consumed living terrestrial plants, pinpointing its new opportunistic feeding capability. Because of its intense consumption of young rice plants, it is considered an agricultural pest in some Mediterranean countries (Anastàcio *et al.*, 2005). Overall, alien crayfish are both effective and voracious predators, more so than native ones, posing an additional threat for species of conservation concern, such as amphibians and invertebrates (Banha and Anastàcio, 2011; Haddaway *et al.*, 2012; Lodge *et al.*, 2012; Chucoll, 2013). Crayfish presence is negatively related to the breeding probability for several salamander, frog and toad species (Cruz *et al.*, 2006; Ficetola *et al.*, 2011). Some alien species such as *O. rusticus*, *P. clarkii* and *P. leniusculus* can have more destructive effects than others such as *O. limosus* (Wilson *et al.*, 2004; Dunoyer *et al.*, 2014). In the laboratory, *P. clarkii* consumed amphibian larvae more efficiently than the native *A. pallipes* (Gherardi *et al.*, 2001; Renai and Gherardi, 2004) (Figure 16.2). Apparently, the poisons contained in several amphibians are not an effective deterrent for *P. clarkii*. In California, Gamradt and Kats (1996) found that the introduced *P. clarkii* were able to consume *Taricha torosa* larvae, notwithstanding their tetrodotoxin poison which is an effective defence against other predators.

In northern Europe, eggs of the common toad, *Bufo bufo*, are unpalatable to newts and predatory insects, but are readily consumed by the invasive *P. leniusculus* (Axelsson *et al.*, 1997). From laboratory experiments, *P. clarkii* seems to be able to consume trout alevins (Gherardi *et al.*, 2001), but its impact on fish populations in nature should be investigated. This species may increase the vulnerability of some fish to predators by evicting them from shelters (Guan and Wiles, 1997; Matsuzaki *et al.*, 2012). *Pacifastacus leniusculus* significantly reduces the number of Atlantic salmon, using shelters in artificial test arenas, by consuming their eggs and juveniles (Griffiths *et al.*, 2004). In its introduced North American range, *Orconectes virilis* competes with endemic fish for food (Carpenter, 2005), while in Canada *P. leniusculus* seems to have caused the collapse of a stickleback species pair (Taylor *et al.*, 2006). *Orconectes rusticus* significantly reduced trout egg abundance in Great Lakes, hampering trout rehabilitation efforts (Jonas *et al.*, 2005). In Italy, the intense predatory behaviour of *P. clarkii* eliminated a once locally abundant semiaquatic beetle, *Carabus clatratus* (Casale and Busato, 2008), and in Japan threatened an endangered odonate, *Libellula angelina* (Miyake and Miyashita, 2011). However, in Africa, predation by *P. clarkii* has a positive effect, reducing the populations of the snails that host the trematodes causing human schistosomiasis (Mkoji *et al.*, 1999).

The ability to form associations between odours and predators is well developed in invasive species: the introduction into new habitats may expose individuals to unknown predators. The faster they learn, the more efficient their predator-avoidance behaviours will be in decreasing the risk of predation, enhancing their success in new environments (Hazlett *et al.*, 2002). Several laboratory experiments showed the wide behavioural flexibility of *P. clarkii* when coping with new types of predators. This species uses a broader range of information about predation risk than native species, reacting more strongly to heterospecific alarm cues that elicit the typical alarm reaction (i.e. stop any movements to avoid being detected; Hazlett *et al.*, 2003). It is capable of learning and remembering associations between different predation-risk cues. When trained to associate a novel cue (i.e. goldfish odour) with predation risk following experimental pairing with conspecific alarm odour, individuals of this invasive species remembered the association longer than the native *A. pallipes* (Acquistapace *et al.*, 2003). *Procambarus clarkii* is also able to assess an unknown fish species as risky based on a single pairing of conspecific alarm and fish odours, and remembers this association without reinforcement for up to 3 weeks (Hazlett *et al.*, 2002), similarly to *O. rusticus* (Hazlett, 2000). However, 'alarm' substances stimulated feeding-related activities in *P. clarkii* cultivated in aquaculture ponds (Acquistapace *et al.*, 2004). This crayfish, once reared in an environment where predation risks are reduced, can respond differently to cues than those in more risky habitats, further underlining its extreme flexibility. *Orconectes virilis* responded similarly to alarm cues from conspecifics, sympatric heterospecifics, and novel heterospecifics, suggesting that it has the ability to respond adaptively to a wide range of predation risk cues in new habitats (Pecor *et al.*, 2010). As evidenced by Stebbing *et al.* (2010), *P. leniusculus* also utilizes both conspecific and heterospecific sources of info-chemicals to assess risk in the environment. Predator-odour conditioned water from

Perca fluviatilis (perch) elicited no response, but conditioned *Anguilla anguilla* (European eel) water stimulated significant responses in juveniles and adults of the species.

Aggressive Behaviour and Interspecific Competition

Crayfish are well-known model organisms for the study of agonistic behaviour and have been used since the 1950s (Gherardi, 2002). Levels of aggression differ among species, and alien species usually dominate the native ones, being more aggressive and able to better exploit resources as food and shelters. In Europe, alien crayfish can outcompete the native species even carrying the crayfish plague *Aphanomyces astaci*, lethal for the latter ones (Souty-Grosset *et al.*, 2006). In England, *P. leniusculus* shows preference for the same habitats of *A. pallipes* and, being more aggressive, has displaced the native crayfish, leading to some population extinctions within 5 years (Hiley, 2003; Bubb *et al.*, 2006). However, in Finland, *P. leniusculus* seemed to coexist for 30 years in a lake with the native *A. astacus*. Then, the latter started to decline. Westman *et al.* (2002) suggested that its disappearance was due to several interacting mechanisms, of which harvest and competition with *P. leniusculus* were initially the main reasons, with the ultimate reason being the almost complete cessation of successful reproduction, presumably due to reproductive interference between the two species (see section on reproductive behaviour). In the laboratory, invasive *P. leniusculus* is superior to the native *Cambaroides japonicus* for shelter occupation, and also preys on the latter (Nakata and Goshima, 2003, 2006); in fact, *P. leniusculus* has reduced the range of *C. japonicus* on the island of Hokkaido (Kawai and Hiruta, 1999). In North America, the sooty crayfish, *Pacifastacus nigrescens*, native to the western United States, has become extinct partly due to interspecific competition with *P. leniusculus*, which was introduced into its range (Bouchard, 1977). *Pacifastacus leniusculus* also had a relevant role in reducing the range of the already narrowly endemic shasta crayfish, *P. fortis*, in the western states of the United States (Light *et al.*, 1995). Similarly, *Procambarus clarkii* is more competitive than the indigenous North American *P. acutus acutus* (Gherardi and Daniels, 2004), in the presence or absence of a shelter, even when *P. clarkii* excluded the other species from the shelter but did not use it. *Procambarus clarkii* has greater chelar strength than the native *A. pallipes*, even dominating it (Gherardi and Cioni, 2004). Chela size also influences the invasion potential of *O. rusticus* (Garvey and Stein, 1993), which caused the local extinction of native American crayfish via competition (Szela and Perry, 2013) and hybridization (Perry *et al.*, 2001a, b). The pioneering studies on *O. rusticus* were made by the Stein and Lodge teams, who first revealed how the differential susceptibility of alien and native species to fish predators can interact with interspecific competition and be responsible for the displacement of native crayfish. Indeed, the native *O. virilis*, being excluded from shelters by the alien *O. rusticus* and *O. propinquus*, was more predated by largemouth bass *Micropterus salmoides* (Garvey *et al.*, 1994). Hill and Lodge (1999) confirmed the competitive superiority of *O. rusticus*, and found that in the presence of a predator, *O. virilis* growth declined substantially, *O. rusticus* growth declined slightly, and *O. propinquus* growth was unaffected. Mortality of all three crayfishes increased in the presence of *M. salmoides*, with *O. virilis* experiencing the greatest and *O. rusticus* the least mortality. Similarly, in the laboratory, *O. rusticus* adults were not susceptible to predation by *M. salmoides* and did not alter shelter use when fish were present. Even its juveniles were less susceptible to predation than the native *O. sanborni* by occupying shelters more often than the native juveniles (Butler and Stein, 1985).

There are some exceptions to the dominance of alien species over natives. The native European river crab, *Potamon fluviatile*, can outcompete both the native *A. pallipes* and the alien *P. clarkii* (Gherardi and Cioni, 2004). Even the crayfish *Astacopsis franklinii*, endemic to Tasmania, in the laboratory, dominated interactions by gaining first possession of a shelter and maintaining it, when opposed with equally sized specimens of the introduced *C. destructor* (Elvey *et al.*, 1996).

Sometimes, coexistence between alien and native species seems possible. Laboratory observations were conducted in Kenya on agonistic interactions between the introduced *P. clarkii* and the native river crab *Potamonautes loveni* in the absence or presence of a resource (food or shelter) and on their occupancy of different microhabitats. Crayfish showed higher competitive ability and plasticity compared to the crab (Tricarico *et al.*, 2012). Nevertheless, the crayfish and crabs seem to coexist in the field by occupying different microhabitats (Tricarico *et al.*, 2012), and further observations could better assess the stability of this cohabitation. Similarly, in the laboratory, introduced *Orconectes neglectus* juveniles were dominant over the endemic *O. eupunctus* juveniles in the presence of limited food. In the field, however, the alien juveniles did not inhibit growth or reduce the survival of native ones, suggesting alternative mechanisms responsible for the apparent displacement of *O. eupunctus* by *O. neglectus*, such as differential predation or reproductive interference (Larson and Magoulick, 2009). In the laboratory, the introduced *O. hylas* was unable to dominate two endemic species (*O. peruncus* and *O. quadruncus*); life history and ecological factors are the likely determinants of the displacement (Rahm *et al.*, 2005). Recently, Peters and Lodge (2013) observed that coexistence between the introduced *O. rusticus* and the native *O. virilis* is possible because, in the presence of *O. rusticus*, *O. virilis* alters its habitat use. Lakes in which native species persisted for many years had significantly less cobble and sand substrate habitats, and significantly more vegetated habitats compared to lakes from which native crayfish have been displaced. As evidenced by Twardochleb *et al.* (2013), in studies of aggression between native and alien crayfish, investigators most often use naïve individuals with no experience with invaders (e.g. Larson and Magoulick, 2009). Hayes *et al.* (2009) suggested that native crayfish with prior experience with the alien perform better in aggressive encounters than their naïve counterparts, representing a case of acclimation or rapid evolution.

Recent studies focus on interactions between invasive crayfish. In interspecific pairs formed by *P. leniusculus* and *O. limosus*, the former is more prone to fighting than the latter, which consistently retreats from staged bouts as fights became more intense (Hudina *et al.*, 2011; Hudina and Hock, 2012) (Figure 16.3). *Procambarus clarkii* dominates in size-matched heterospecific pairs with *P. leniusculus* males (Alonso and Martinez, 2006). However, the latter species can reach larger dimensions in the wild and potentially outcompete the former one. In addition, Hanshew and Garcia (2012)

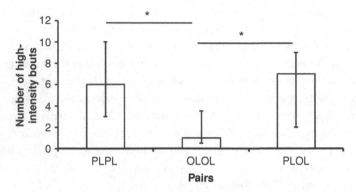

Figure 16.3 Number of high-intensity bouts in conspecific (PLPL = both opponents *Pacifastacus leniusculus*; OLOL = both opponents *Orconectes limosus*) and interspecific pairs (PLOL = *Pacifastacus leniusculus* vs. *Orconectes limosus*) of alien crayfish species. $N = 16$ per pair. Bars represent the median with first–third quartiles. The asterisk denotes a significant difference after Kruskal–Wallis ANOVA with post hoc multiple comparisons of the mean ranks, Mann–Whitney U test. Modified after Hudina and Hock (2012).

investigated shelter occupancy and resource competition between the same species pairs, with *P. clarkii* invading the native habitat of *P. leniusculus* in southern Oregon (USA). The two species appeared to be equal competitors for shelter, with *P. clarkii* modifying its shelter occupancy behaviour in the presence of *P. leniusculus* and having broader microhabitat preferences. Gherardi *et al.* (2013) analysed the agonistic behaviour of pairs composed of similarly sized males in combinations of three alien species: *O. limosus*, *P. leniusculus* and *P. clarkii*, at two different temperatures (20°C and 27°C) in the light of climate change. The first two species reduced their agonistic behaviour at the higher temperature, but *P. clarkii* showed the same aggressiveness, outcompeting both species, and suggesting that in the future – with climate warming – it will dominate European water basins, in accordance with the distribution forecast by Capinha *et al.* (2013). The competitive superiority of *P. clarkii* is also demonstrated by long-term field studies, still in progress, conducted in the wetlands of 'Parc Naturel Régional de la Brenne' (Indre department, Centre region, France). Here, both *P. clarkii* and *O. limosus* have coexisted since 2007, but the latter species is declining due to the higher invasiveness and aggressiveness of the former one (C. Souty-Grosset, pers. comm.).

Personality

The scientific interest towards animal personality, particularly in invertebrates, has recently increased. Studies on animal personality could contribute to identifying traits associated with the behaviour of alien species (Gherardi *et al.*, 2012; Hudina *et al.*, 2014). The dispersal process per se might select for bold/aggressive/active individuals, who then have a particularly strong tendency to disrupt their invaded communities.

In one study that has demonstrated this issue in alien crayfish, Pintor *et al.* (2008) showed positive correlations among aggression, activity and boldness in populations of *P. leniusculus* in both native and invaded ranges in Oregon and California. The overall aggressive behaviour of this crayfish, as opposed to the non-aggressive, inactive and shy *P. fortis*, might explain its success as an invasive species (against the endangered shasta crayfish), as the result of its ability to both outcompete native crayfish in highly productive habitats and establish itself in low-productivity streams.

Reproductive Behaviour

Alien crayfish commonly are *r*-selected species, reaching sexual maturity early, and producing a higher number of eggs (e.g. 570 for *O. rusticus*; 600 for *P. clarkii*; 400–500 for *P. leniusculus*; vs up to 260 for *A. astacus*; 200 for *A. pallipes*; Souty-Grosset *et al.*, 2006). Fecundity is indeed one of the strongest traits predicting invasiveness of crayfish species as reported by Larson and Olden (2010). The life span of alien species is short (usually 4 vs 10–20 years of native European crayfish; Souty-Grosset *et al.*, 2006). Crayfish usually have promiscuous mating systems, with both males and females potentially mating with more than one partner, although this has been demonstrated only for females of few species (e.g. *P. clarkii*: Yue *et al.*, 2010). Crayfish can select mates that allow them to maximize their reproductive success, and are capable of recognizing potential mates and breeding conspecifics. The reproductive behaviour of *P. clarkii* was investigated in depth for management purposes. As in other crayfish species (Gherardi, 2002), *P. clarkii* females select mates with large body size, with chelar asymmetry having no apparent effect on this choice (Aquiloni and Gherardi, 2008a). Mate choice is not only a prerogative of females: in *P. clarkii*, males are more attracted by partners with large body sizes and which are virgins (Aquiloni and Gherardi, 2008b). Some alien crayfish can also reproduce asexually via parthenogenesis in the new invaded habitats, for example, the marmorkreb *P. fallax* forma *virginalis* or in *O. limosus*, where this phenomenon has been reported in captivity (Buřič *et al.*, 2011).

A number of studies have reported evidence of sex recognition and mate attractants in several crayfish species (Stebbing *et al.*, 2003). The presence of sex pheromones released during breeding season by mature females – that stimulate courtship and mating behaviour – was clearly demonstrated for the first time in *P. leniusculus* by Stebbing *et al.* (2003). This could be particularly useful for the control of alien species, but the chemical purification of the female sex pheromones is still in progress (Breithaupt and Thiel, 2011). Some studies reveal that during the mating season *P. clarkii* shows unusual discrimination abilities for an invertebrate (Aquiloni *et al.*, 2008). In this species, the male is involved in mate search (Aquiloni *et al.*, 2009) using olfaction exclusively, unlike the native *A. pallipes* males that rely on both olfaction and vision (Acquistapace *et al.*, 2002). *Procambarus clarkii* males can detect potential mates at a longer distance than native species, even in turbid water (Gherardi, 2006). In addition, *P. clarkii* females can recognize potential mates as individuals (Aquiloni and Gherardi, 2008c; Aquiloni *et al.*, 2009), using a combination of visual and chemical stimuli, to

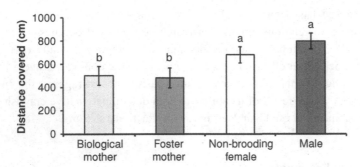

Figure 16.4 Distance covered (cm) by the alien *Procambarus clarkii* adults of four types in the presence of 30 third-stage juveniles of the same brood after 30 min from the start of the experiment. $N = 13$ per adult type. Bars represent the mean with standard error. Letters denote the hierarchy after a post hoc Tukey test. Modified after Aquiloni and Gherardi (2008d).

discriminate the best male in the crowded social context during the reproductive period (Gherardi *et al.*, 1999).

In Finland, interspecific mating was reported between the native *A. astacus* and the alien *P. leniusculus* (Westman *et al.*, 2002), resulting in females laying sterile eggs. Although both species suffer from the ensuing loss of recruitment, the consequences are less serious for *P. leniusculus*, which has a higher capacity for population increase than *A. astacus*. Hybridization with *O. rusticus* increased the displacement rate of *O. propinquus* (Perry *et al.*, 2001a). *Orconectes rusticus* mates with *O. propinquus* producing fertile offspring, able to outcompete both parental species for food and shelter (Perry *et al.*, 2001b). However, subsequent generations show a lower survival rate, leading to an increased rate of decline in *O. propinquus* (see Lodge *et al.*, 2012).

Finally, despite being *r*-selected, *P. clarkii* exhibits extended parental care in laboratory conditions. Aquiloni and Gherardi (2008d) investigated the return behaviour of its third-stage juveniles when offered four types of adults (biological mothers, foster mothers, non-brooding females and males), and the posture and the behaviours of these adults. Contrary to non-brooding individuals, both biological and foster mothers displayed relatively low locomotion, executed few cleaning and feeding acts, and never attempted to prey on juveniles (Figure 16.4). They often assumed a 'spoon-like telson posture' that seemed to facilitate offspring's approaches. Juveniles performed more tail-flips away in the presence of the non-brooding adults, while they accepted foster mothers, along with biological mothers, evidencing how this behaviour, poorly studied for crayfish, could be useful for this species to thrive in harsh environmental conditions.

Conclusions

It is clear how several behavioural traits favour the successful invasion of some alien crayfish. They are highly plastic and adaptable, displaying different and new behaviours – for example, burrowing or parthenogenesis – in the invaded areas; they actively and widely disperse, and can face predators, even new ones, due to their ability to respond adaptively to a wide range of predation risk cues. Omnivorous feeding habits, common in crayfish, are better developed in invasive species than in native ones: they are more voracious and can consume new – even unpalatable – prey, leading to species extinction and severe ecosystem changes. Aggressive behaviour, sometimes coupled with a lower susceptibility to predation, and hybridization potential, helps them to outcompete the native species, even if the latter in some cases can evolve some adaptive mechanisms. Alien crayfish can even display personality traits that further enhance their invasion abilities. They are highly prolific, better able to recognize mates at a distance and in turbid water than native species. Finally, some of them are better able to cope with future climate change, which may enhance their aggressiveness and aid in further expanding their invasions.

Behavioural research thus has helped us to identify the 'winning' traits of invasive species, but it is also essential to understand behaviours such as mating in *P. clarkii*, for developing new control methods in the light of integrative pest management. However, despite the growing number of studies on invasive crayfish, most of them concern North American species such as *O. rusticus*, *P. leniusculus* and *P. clarkii* in Europe and North America. It is important to have more in-depth studies from other countries where these species were introduced (in Asia or Africa), and on other species, for example others in the genera *Orconectes* and *Cherax*. Research is needed to compare native and introduced populations of alien species in order to find which behavioural traits evolve more rapidly – or are only seen – in the new habitats, as recently found by Sargent and Lodge (2014) in the United States for growth rate and reproductive outputs in different *O. rusticus* populations.

Among the topics in this review, we suggest further studies on the use of burrows, particularly for species considered 'non-burrowers' in their native area; on the differential susceptibility of alien/native species to predators; longer-term studies as the invasion process continues; reproductive behaviour, for example the identification of sexual pheromones and the mother–offspring relationship; and on personality in other alien crayfish species. Finally, in light of global climate change, gathering more information on the potential changes in behaviour of alien species under different environmental conditions is particularly crucial in order to understand their increasing threats and to conserve native species, ecosystems and biodiversity.

Acknowledgements

Our warmest thank to Dr Frances Lucy (Ireland) for English and draft revision, and to two anonymous referees who greatly improved the first version of the chapter with their comments.

ET thanks the Regional Council of Poitou Charentes (France) for having allowed her to write this chapter while she was hosted by Professor Catherine Souty-Grosset within the 'Programme Régional de Bourses de Chercheur Invité' at the University of Poitiers, Laboratory 'Ecologie, Evolution, Symbiose', UMR CNRS 6556.

References

Acquistapace, P., Aquiloni, L., Hazlett, B.A. and Gherardi, F. (2002). Multimodal communication in crayfish: sex recognition during mate search by male *Austropotamobius pallipes*. *Canadian Journal Zoology*, 80, 2041–2045.

Acquistapace, P., Hazlett, B.A. and Gherardi, F. (2003). Unsuccessful predation and learning of predator cues by crayfish. *Journal of Crustacean Biology*, 23, 364–370.

Acquistapace, P., Daniels, W.H. and Gherardi, F. (2004). Behavioral responses to 'alarm odors' in potentially invasive and noninvasive crayfish species from aquaculture ponds. *Behaviour*, 141, 691–702.

Alonso, F. and Martinez, R. (2006). Shelter competition between two invasive crayfish species: a laboratory study. *Bulletin française de la Pêcheet de la Pisciculture*, 380–381, 1121–1131.

Anastácio, P.M., Correia A.M. and Menino, J.P. (2005). Processes and patterns of plant destruction by crayfish: effects of crayfish size and development stages of rice. *Archiv für Hydrobiologie*, 162, 37–51.

Angeler, D.G., Sanchez-Carrillo, S., García, G. and Alvarez-Cobelas, M. (2001). The influence of *Procambarus clarkii* (Cambaridae, Decapoda) on water quality and sediment characteristics in a Spanish floodplain wetland. *Hydrobiology*, 464, 88–98.

Aquiloni, L. and Gherardi, F. (2008a). Evidence of female cryptic choice in crayfish. *Biology Letters*, 4, 163–165.

Aquiloni, L. and Gherardi, F. (2008b). Mutual mate choice in crayfish: large body size is selected by both sexes, virginity by males only. *Journal of Zoology London*, 274, 171–179.

Aquiloni, L. and Gherardi, F. (2008c). Assessing mate size in the red swamp crayfish *Procambarus clarkii*: effects of visual versus chemical stimuli. *Freshwater Biology*, 53, 461–469.

Aquiloni, L. and Gherardi, F. (2008d). Extended mother-offspring relationships in crayfish: the return behaviour of *Procambarus clarkii* juveniles. *Ethology*, 114, 946–954.

Aquiloni, L., Ilheu, M. and Gherardi, F. (2005). Habitat use and dispersal of the invasive crayfish *Procambarus clarkii* in ephemeral water bodies in Portugal. *Marine Freshwater Behaviour and Physiology*, 38, 225–236.

Aquiloni, L., Buřič, M. and Gherardi, F. (2008). Crayfish females eavesdrop on fighting males before choosing the dominant mate. *Current Biology*, 18, 462–463.

Aquiloni, L., Massolo, A. and Gherardi, F. (2009). Sex identification in female crayfish is bimodal. *Naturwissenschaften*, 9, 103–110.

Axelsson, E.P., Nyström, P., Didenmark, J. and Brönmark, C. (1997). Crayfish predation on amphibian eggs and larvae. *Amphibia-Reptilia*, 18, 217–228.

Banha, F. and Anastácio, P.M. (2011). Interactions between invasive crayfish and native river shrimp. *Knowledge and Management of Aquatic Ecosystems*, 401, 17.

Barbaresi, S. and Gherardi, F. (2000). Invasive crayfish: activity patterns of *Procambarus clarkii* in the rice fields of the Lower Guadalquivir (Spain). *Archiv für Hydrobiologie*, 150, 153–168.

Barbaresi, S., Tricarico, E. and Gherardi, F. (2004). Factors inducing the intense burrowing activity by the red-swamp crayfish, *Procambarus clarkii*, an invasive species. *Naturwissenschaften*, 91, 342–345.

Bouchard, R.W. (1977). Distribution, systematic status and ecological notes on five poorly known species of crayfishes in western North America (Decapoda: Astacidea and Cambaridae). *Freshwater Crayfish*, 3, 409–423.

Breithaupt, T. and Thiel, M. (eds.) (2011). *Chemical Communication in Crustaceans*. New York: Springer.

Bubb, D.H., Lucas M.C. and Thom T.J. (2002). Winter movements and activity of the signal cray-fish *Pacifastacus leniusculus* in an upland river, determined by radio telemetry. *Hydrobiologia*, 483, 111–119.

Bubb, D.H., Thom, T.J. and Lucas, M.C. (2004). Movement and dispersal of the invasive signal crayfish *Pacifastacus leniusculus* in upland rivers. *Freshwater Biology*, 49, 357–368.

Bubb, D.H., Thom, T.J. and Lucas, M.C. (2006). Movement, dispersal and refuge use of co-occurring introduced and native crayfish. *Freshwater Biology*, 51, 1359–1368.

Buřič, M., Kouba, A. and Kozák, P. (2009a). Spring mating period in *Orconectes limosus*: the reason for movement. *Aquatic Science*, 71, 473–477.

Buřič, M., Kozák, P. and Kouba, A. (2009b). Movement patterns and ranging behavior of the invasive spiny-cheek crayfish in a small reservoir tributary. *Fundamental Applied Limnology*, 174, 329–337.

Buřič, M., Hulák, M., Kouba, A., Petrusek, A. and Kozák, P. (2011). A successful crayfish invader is capable of facultative parthenogenesis: a novel reproduction mode in decapod crustaceans. *PloS ONE*, 6, e20281.

Butler, M.J. and Stein, R.A. (1985). An analysis of the mechanisms governing species replace-ments in crayfish. *Oecologia*, 66, 168–177.

Capinha, C., Larson, E.R., Tricarico, E., Olden J.D. and Gherardi, F. (2013). Climate change, species invasions, and diseases threaten European native crayfishes. *Conservation Biology*, 27, 731–740.

Carpenter, J. (2005). Competition for food between an introduced crayfish and two fishes endemic to the Colorado River basin. *Environmental Biology of Fishes*, 72, 335–342.

Casale, A. and Busato, E. (2008). A real time extinction: the case of *Carabus clatratus* in Italy (Coleoptera, Carabidae). In *Back to the Roots and Back to the Future. Towards a New Synthesis amongst Taxonomic, Ecological and Biogeographical Approaches in Carabidology*, ed. Penev, L., Erwin, T. and Hassmann, T. Sofia, Bulgaria: Pensoft, pp. 353–362.

Chucoll, C. (2013). Feeding ecology and ecological impact of an alien 'warm-water' omnivore in cold lakes. *Limnologica*, 43, 219–229.

Claussen, D.L., Hopper, R.A. and Sanker, A.M. (2000). The effects of temperature, body size, and hydration state on the terrestrial locomotion of the crayfish *Orconectes rusticus*. *Journal of Crustacean Biology*, 20, 218–223.

Correia, A.M. and Ferreira, O. (1995). Burrowing behavior of the introduced red swamp crayfish *Procambarus clarkii* (Decapoda: Cambaridae) in Portugal. *Journal of Crustacean Biology*, 15, 248–257.

Crandall, K.A. and Buhay, J.E. (2008). Global diversity of crayfish (Astacidae, Cambaridae, and Parastacidae–Decapoda) in freshwater. *Hydrobiologia*, 595, 295–301.

Cruz, M.J., Rebelo, R. and Crespo E.G. (2006). Effects of an introduced crayfish, *Procambarus clarkii*, on the distribution of south-western Iberian amphibians in their breeding habitats. *Ecography*, 29, 329–338.

Delivering Alien Invasive Species Inventory for Europe (DAISIE) (2011). 100 of The Worst. Available at: http://www.europe-aliens.org/speciesTheWorst.do, accessed 21 April 2016.

Dana, E.D., García De Lomas, J. González, R. and Ortega, F. (2011). Effectiveness of dam con-struction to contain the invasive crayfish *Procambarus clarkii* in a Mediterranean mountain stream. *Ecological Engineering*, 37, 1607–1613.

Dunoyer, L., Dijoux, L., Bollache, L. and Lagrue, C. (2014). Effects of crayfish on leaf lit-ter breakdown and shredder prey: are native and introduced species functionally redundant? *Biological Invasions*, 16, 1545–1555.

Elvey, W., Richardson, A.M.M. and Bermuta, L. (1996). Interactions between the introduced yabby, *Cherax destructor*, and the endemic crayfish, *Astacopsis franklinii*, in Tasmanian streams. *Freshwater Crayfish*, 11, 349–363.

Ficetola, G.F., Siesa, M.E., Manenti, R., *et al.* (2011). Early assessment of the impact of alien species: differential consequences of an invasive crayfish on adult and larval amphibians. *Diversity and Distributions*, 17, 1141–1151.

Gamradt, S.C. and Kats, L.B. (1996). Effect of introduced crayfish and mosquito fish on California newts (*Taricha torosa*). *Conservation Biology*, 10, 1155–1162.

Garvey, J.E. and Stein, R.A. (1993). Evaluating how chela size influences the invasion potential of an introduced crayfish (*Orconectes rusticus*). *American Midland Naturalist*, 129, 172–181.

Garvey, J.E., Stein, R.A. and Thomas, H.M. (1994). Assessing how fish predation and interspecific prey competition influence a crayfish assemblage. *Journal of Ecology*, 75, 532–547.

Gherardi, F. (2002). Behaviour. In *Biology of Freshwater Crayfish*, ed. Holdich, D.M. Oxford, UK: Blackwell Science, pp. 258–290.

Gherardi, F. (2006). Crayfish invading Europe: the case study of *Procambarus clarkii*. *Marine and Freshwater Behaviour and Physiology*, 39, 175–191.

Gherardi, F. (2007). Understanding the impact of invasive crayfish. In *Biological Invaders in Inland Waters: Profiles, Distribution, and Threats*, Springer Series in Invasion Ecology, ed. Gherardi, F. Invading Nature: Dordrecht, The Netherlands: Springer, pp. 507–542.

Gherardi, F. (2010). Invasive crayfish and freshwater fishes of the world. *Revue Scientifique et technique dell'O.I.E.* (*Office International des Épizooties*), 29, 241–254.

Gherardi, F. and Cioni, A. (2004). Agonism and interference competition in freshwater decapods. *Behaviour*, 141, 1297–1324.

Gherardi, F. and Daniels, W.H. (2004). Agonism and shelter competition between invasive and indigenous crayfish species. *Canadian Journal of Zoology*, 82, 1923–1932.

Gherardi, F., Barbaresi, S. and Raddi, A. (1999). The agonistic behaviour in the red swamp crayfish, *Procambarus clarkii*: functions of the chelae. *Freshwater Crayfish*, 12, 233–243.

Gherardi, F., Renai, B. and Corti, C. (2001). Crayfish predation on tadpoles: a comparison between a native (*Austropotamobius pallipes*) and an alien species (*Procambarus clarkii*). *Bulletin française de la Pêcheet de la Pisciculture*, 361, 659–668.

Gherardi, F., Tricarico, E. and Ilhéu, M. (2002). Movement patterns of an invasive crayfish, *Procambarus clarkii*, in a temporary stream of southern Portugal. *Ethology Ecology and Evolution*, 14, 183–197.

Gherardi, F., Aquiloni, L. and Tricarico, E. (2012). Behavioral plasticity, behavioral syndromes and animal personality in crustacean decapods: an imperfect map is better than no map. *Current Zoology*, 58, 567–579.

Gherardi, F., Coignet, A., Souty-Grosset, C., Spigoli, D. and Aquiloni, L. (2013). Global warming and the agonistic behavior of invasive crayfishes in Europe. *Freshwater Biology*, 58, 1958–1967.

Gray, J. and Jackson, M.C. (2012). 'Leaves and eats shoots': direct terrestrial feeding can supplement invasive red swamp crayfish in times of need. *PLoS ONE*, 7, e42575.

Griffiths, S.W., Collen, P. and Armstrong, J.D. (2004). Competition for shelter among overwintering signal crayfish and juvenile Atlantic salmon. *Journal of Fish Biology*, 65, 436–447.

Guan, R.-Z. and Wiles, P.R. (1997). Ecological impact of introduced crayfish on benthic fishes in a British lowland river. *Conservation Biology*, 11, 641–647.

Haddaway, N.R., Wilcox, R.H., Heptonstall, R.E.A., *et al.* (2012). Predatory functional response and prey choice identify predation differences between native/invasive and parasitised/unparasitised crayfish. *PLoS ONE*, 7, e32229.

Hanshew, B.A. and Garcia, T.S. (2012). Invasion of the shelter snatchers: behavioural plasticity in invasive red swamp crayfish, *Procambarus clarkii. Freshwater Biology*, 57, 2285–2296.

Harvey, G.L., Moorhouse, T.P., Clifford, N.J., *et al.* (2011). Evaluating the role of invasive aquatic species as drivers of fine sediment-related river management problems: the case of the signal crayfish (*Pacifastacus leniusculus*). *Progress in Physical Geography*, 35, 517–533.

Hayes, N.M., Butkas, K.J., Olden, J.D. and Vander Zanden, M.J. (2009). Behavioural and growth differences between experienced and naïve populations of a native crayfish in the presence of invasive rusty crayfish. *Freshwater Biology*, 54, 1876–1887.

Hazlett, B.A. (2000). Information use by an invading species: do invaders respond more to alarm odors than native species? *Biological Invasions*, 2, 289–294.

Hazlett, B.A., Acquistapace, P. and Gherardi, F. (2002). Differences in memory capabilities in invasive and native crayfish. *Journal of Crustacean Biology*, 22, 439–448.

Hazlett, B.A., Burba, A., Gherardi, F. and Acquistapace, P. (2003). Invasive species use a broader range of predation-risk cues than native species. *Biological Invasions*, 5, 223–228.

Hiley, P.D. (2003). The slow quiet invasion of signal crayfish (*Pacifastacus leniusculus*) in England – prospects for the white-clawed crayfish (*Austropotamobius pallipes*). In *Management and Conservation of Crayfish*. Proceedings of a conference held in Nottingham on 7th November (2002). Holdich, D.M. and Sibley, P.J. (eds). Bristol: Environment Agency, pp. 127–138.

Hill, A.M. and Lodge, D.M. (1999). Replacement of resident crayfishes by an exotic crayfish: the roles of competition and predation. *Ecological Applications*, 9, 678–690.

Hobbs, H.H., Jass, J.P. and Huner, J.V. (1989). A review of global crayfish introductions with particular emphasis on two North American species (Decapoda: Cambaridae). *Crustaceana*, 56, 299–316.

Hogger, J.B. (1988). Ecology, population biology and behaviour. In *Freshwater Crayfish. Biology, Management and Exploitation*, ed. Holdich, D.M. and Lowery, R.S. Portland, OR: Croom Helm and Timber Press, pp. 114–144.

Holdich, D.M. and Black, J. (2007). The spiny-cheek crayfish, *Orconectes limosus* (Rafinesque, 1817) [Crustacea: Decapoda: Cambaridae], digs into the UK. *Aquatic Invasions*, 2, 1–15.

Horwitz, P.H.J. and Richardson A.M.M. (1986). An ecological classification of the burrows of Australian freshwater crayfish. *Marine and Freshwater Research*, 37, 237–242.

Hudina, S. and Hock, K. (2012). Behavioural determinants of agonistic success in invasive crayfish. *Behavioural Processes*, 91, 77–81.

Hudina, S., Lucić, A., Žganec, K. and Janković, S. (2011). Characteristics and movement patterns of a recent established invasive *Pacifastacus leniusculus* population in the river Mura, Croatia. *Knowledge and Management of Aquatic Ecosystems*, 403, 07.

Hudina, S., Hock, K. and Žganec, K. (2014). The role of aggression in range expansion and biological invasions. *Current Zoology*, 60, 401–409.

Ilhéu, M., Acquistapace, P., Benvenuto, C. and Gherardi, F. (2003). Shelter use of the red-swamp crayfish (*Procambarus clarkii*) in dry-season stream pools. *Archiv für Hydrobiologie*, 157, 535–546.

Jaklič, M. and Vrezec, A. (2011). The first tropical alien crayfish species in European waters: the red claw *Cherax quadricarinatus* (von Martens, 1868) (Decapoda, Parastacidae). *Crustaceana*, 84, 651–665.

Johnson, F., Rice S.P. and Reid I. (2014). The activity of signal crayfish (*Pacifastacus leniusculus*) in relation to thermal and hydraulic dynamics of an alluvial stream, UK. *Hydrobiologia*, 724, 41–54.

Jonas, J.L., Claramunt, R.M., Fitzsimons, J. D., Marsden, J.E. and Ellrott, B.J. (2005). Estimates of egg deposition and effects of lake trout (*Salvelinus namaycush*) egg predators in three regions of the Great Lakes. *Canadian Journal of Fisheries and Aquatic Sciences*, 62, 2254–2264.

Kawai, T. and Hiruta, M. (1999). Distribution of crayfish (*Pacifastacus leniusculus* and *Cambaroides japonicus*) in Lake Shikaribetsu and Shihoro, Hokkaido, Japan. *Crayfish News*, 21, 11.

Kouba, A., Petrusek, A. and Kozák, P. (2014). Continental-wide distribution of crayfish species in Europe: update and maps. *Knowledge and Management of Aquatic Ecosyatems*, 413, 05.

Larson, E.R. and Magoulick, D.D. (2009). Does juvenile competition explain displacement of a native crayfish by an introduced crayfish? *Biological Invasions*, 11, 725–735.

Larson, E.R. and Olden, J.D. (2010). Latent extinction and invasion risk of crayfishes in the southeastern United States. *Conservation Biology*, 24, 1099–1110.

Light, T. (2003). Success and failure in a lotic crayfish invasion: the roles of hydrologic variability and habitat alteration. *Freshwater Biology*, 48, 1886–1897.

Light, T., Erman, D.C., Myrick, C. and Clark, J. (1995). Decline of the Shasta crayfish (*Pacifastacus fortis* Faxon) of northeastern California. *Conservation Biology*, 9, 1567–1577.

Lodge, D.M., Deines, A., Gherardi, F., *et al.* (2012). Global introductions of crayfishes: evaluating the impact of species invasions on ecosystem services. *Annual Review of Ecology, Evolution, and Systematics*, 43, 449–472.

Matsuzaki, S.S., Sakamoto, M., Kawabe, K. and Takamura, N. (2012). A laboratory study of the effects of shelter availability and invasive crayfish on the growth of native stream fish. *Freshwater Biology*, 57, 874–882.

Miyake, M. and Miyashita, T. (2011). Identification of alien predators that should not be removed for controlling invasive crayfish threatening endangered odonates. *Aquatic Conservation: Marine Freshwater Ecosystems*, 21, 292–298.

Mkoji, G.M., Hofkin, B.V., Kuris, A.M., *et al.* (1999). Impact of the crayfish *Procambarus clarkii* on *Schistosoma haematobium* transmission in Kenya. *The American Journal of Tropical Medicine and Hygiene*, 61, 751–759.

Musil, M., Buřič, M., Policar, T., Kouba, A. and Kozák, P. (2010). Comparison of day and night activity between noble (*Astacus astacus*) and spiny-cheek crayfish (*Orconectes limosus*). *Freshwater Crayfish*, 17, 189–193.

Nakata, K. and Goshima, S. (2003). Competition for shelter of preferred size between the native crayfish species *Cambaroides japonicus* and the alien crayfish species *Pacifastacus leniusculus* in Japan in relation to prior residence, sex difference, and body size. *Journal of Crustacean Biology*, 23, 897–907.

Nakata, K. and Goshima, S. (2006). Asymmetry in mutual predation between the endangered Japanese native crayfish *Cambaroides japonicus* and the North American invasive crayfish *Pacifastacus leniusculus*: a possible reason for species replacement. *Journal of Crustacean Biology*, 26, 134–140.

Pecor, K.W., Deering, C.M., Firnberg, M.T., Pastino, A.K. and Wolfson, S.J. (2010). The use of conspecific and heterospecific alarm cues by virile crayfish (*Orconectes virilis*) from an exotic population. *Marine and Freshwater Behaviour and Physiology*, 43, 37–44.

Perry, W.L., Feder, J.L., Dwyer, G. and Lodge, D.M. (2001a). Hybrid zone dynamics and species replacement between *Orconectes* crayfishes in a northern Wisconsin lake. *Evolution*, 55, 1153–1166.

Perry, W.L., Feder, J.L., Dwyer, G. and Lodge, D.M. (2001b). Implications of hybridization between introduced and resident *Orconectes* crayfishes. *Conservation Biology*, 15, 1656–1666.

Peters, J.A. and Lodge, D.M. (2013). Habitat, predation, and coexistence between invasive and native crayfishes: prioritizing lakes for invasion prevention. *Biological Invasions*, 15, 2489–2502.

Pintor, L.M., Sih, A. and Bauer, M.L. (2008). Differences in aggression, activity and boldness between native and introduced populations of an invasive crayfish. *Oikos*, 117, 1629–1636.

Rahm, E.J., Griffith, S.A., Noltie, D.B. and DiStefano, R.J. (2005). Laboratory agonistic interactions demonstrate failure of an introduced crayfish to dominate two imperiled endemic crayfishes. *Crustaceana*, 78, 437–456.

Renai, B. and Gherardi, F. (2004). Predatory efficiency of crayfish: comparison between indigenous and non-indigenous species. *Biological Invasions*, 6, 89–99.

Ramalho, R.O. and Anástacio, P.M. (2011). Crayfish learning abilities: how does familiarization period affect the capture rate of a new prey item? *Ecological Research*, 26, 53–58.

Richman, N.I., Böhm, M., Adams, S.B., *et al.* (2015). Multiple drivers of decline in the global status of freshwater crayfish (Decapoda: Astacidea). *Philosophical Transactions of the Royal Society B: Biological Sciences*, 370, 20140060.

Rodríguez, C.F., Bécares, E. and Fernández-Aláez, M. (2003). Shift from clear to turbid phase in Lake Chozas (NW Spain) due to the introduction of American red swamp crayfish (*Procambarus clarkii*). *Hydrobiologia*, 506–509, 421–426.

Rosewarne, P.J., Piper, A.T., Wright, R.M. and Dunn, A.M. (2013). Do low-head riverine structures hinder the spread of invasive crayfish? Case study of signal crayfish (*Pacifastacus leniusculus*) movements at a flow gauging weir. *Management of Biological Invasions*, 4, 273–282.

Sargent, L.W. and Lodge, D.M. (2014). Evolution of invasive traits in nonindigenous species: increased survival and faster growth in invasive populations of rusty crayfish (*Orconectes rusticus*). *Evolutionary Applications*, 7, 949–961.

Scalici, M., Chiesa, S., Scuderi, S., Celauro, D. and Gibertini, G. (2010). Population structure and dynamics of *Procambarus clarkii* (Girard, 1852) in a Mediterranean brackish wetland (Central Italy). *Biological Invasions*, 12, 1415–1425.

Souty-Grosset, C., Holdich, D.M., Noël, P.Y., Reynolds, J.D. and Haffner, P. (eds) (2006). *Atlas of Crayfish in Europe*. Paris: Muséum national d'Histoire naturelle (Patrimoines naturels, 64).

Stebbing, P.D., Bentley, M.G. and Watson, G.J. (2003). Mating behaviour and evidence for a female released courtship pheromone in the signal crayfish *Pacifastacus leniusculus*. *Journal of Chemical Ecology*, 29, 465–475.

Stebbing, P.D., Watson, G.J. and Bentley, M.G. (2010). The response to disturbance chemicals and predator odours of juvenile and adult signal crayfish *Pacifastacus leniusculus* (Dana). *Marine and Freshwater Behaviour and Physiology*, 43, 183–195.

Szela, K. and Perry, W.L. (2013). Laboratory competition hierarchies between potentially invasive rusty crayfish (*Orconectes rusticus*) and native crayfishes of conservation concern. *American Midland Naturalist*, 169, 345–353.

Taylor, E.B., Boughman, J.W., Groenenboom, M., *et al.* (2006). Speciation in reverse: morphological and genetic evidence of the collapse of a three-spined stickleback (*Gasterosteus aculeatus*) species pair. *Molecular Ecology*, 15, 343–355.

Tricarico, E., Gherardi, F., Giuliani, C., *et al.* (2012). How the invasive crayfish *Procambarus clarkii* and the native crab *Potamonautes loveni* can coexist in the Lake Naivasha catchment (Kenya). *Book of Abstracts*, The Crustacean Society Summer Meeting, 10th Colloquium Crustacea Decapoda Mediterranea, 3–7 June, Athens, p. 168.

Twardochleb, L.A., Olden, J.D. and Larson, E.R. (2013). A global meta-analysis of the ecological impacts of nonnative crayfish. *Freshwater Science*, 32, 1367–1382.

Westman, K., Savolainen, R. and Julkunen, M. (2002). Replacement of the native crayfish *Astacus astacus* by the introduced species *Pacifastacus leniusculus* in a small, enclosed Finnish lake: a 30-year study. *Ecography*, 25, 53–73.

Wilson, K.A., Magnuson, J.J., Lodge, D.M., *et al.* (2004). A long-term rusty crayfish (*Orconectes rusticus*) invasion: dispersal patterns and community change in a north temperate lake. *Canadian Journal of Fisheries and Aquatic Sciences*, 61, 2255–2266.

Wutz, S. and Geist, J. (2013). Sex- and size-specific migration patterns and habitat preferences of invasive signal crayfish (*Pacifastacus leniusculus* Dana). *Limnologica*, 43, 59–66.

Yue, G.H., Li, J.L., Wang, C.M., *et al.* (2010). High prevalence of multiple paternity in the invasive crayfish species, *Procambarus clarkii*. *International Journal of Biological Sciences*, 6, 107–115.

17 Behaviours of Pacific Lionfish Facilitate Invasion of the Atlantic

Mark A. Albins

Introduction

Behavioural characteristics can play a large role in determining invasion success. Behaviours of an invasive species can influence colonization and establishment, range expansion and the outcome of interactions with native species. A better understanding of how behavioural characteristics contribute to successful invasions is important for managing particular invasions. It can also help identify behavioural attributes that tend to be common among different invasive taxa, thus increasing our ability to predict which species might represent a high risk of becoming future invaders. Invasions are also essentially behavioural experiments, and can shed light on the plasticity of behaviours and the forces that maintain those behaviours.

Here I review what is currently known about the Pacific lionfish (*Pterois volitans*), a particularly successful marine invader, and examine a variety of morphological, physiological and behavioural traits that have likely contributed to their rapid and successful invasion of the Atlantic. While I have attempted to focus on behavioural traits, I present these in the context of associated morphological and physiological traits in order to provide a more comprehensive picture of what has made lionfish one of the most successful marine invaders to date.

Lionfish Invasion

The Pacific red lionfish, *P. volitans*, and the devil firefish, *P. miles* (hereafter collectively referred to as lionfish) were initially introduced into Atlantic waters as early as the mid-1980s, likely via intentional or unintentional releases from aquaria. They have since become established, spread rapidly and demonstrated the potential to substantially alter native communities by consuming large quantities of native prey (Albins and Hixon, 2013; Côté and Hixon, 2013).

Between 1985 and 1999, sporadic sightings of individuals and small groups of lionfish were reported from five sites in southeast Florida (Schofield, 2009). Similarly to other non-native tropical fish species released from aquaria in southeast Florida (Semmens *et al.*, 2004), lionfish did not appear to become established immediately

Biological Invasions and Animal Behaviour, eds J.S. Weis and D. Sol. Published by Cambridge University Press. © Cambridge University Press 2016.

upon introduction. However, it is clear that sometime between 1999 and 2004, lionfish became the first exotic marine fish to become reproductively established in the western Atlantic (Whitfield *et al.*, 2002; Ruiz-Carus *et al.*, 2006). By 2000, four additional lionfish had been sighted in southeast Florida, a single juvenile lionfish was found in a tide-pool in Bermuda, and several individuals were sighted at a number of deep reef sites off the coast of North and South Carolina (Whitfield *et al.*, 2002; Ruiz-Carus *et al.*, 2006). By 2004, lionfish sightings off the southeastern seaboard of the United States had become fairly common, and lionfish had been sighted in the Bahamas (Schofield, 2010). Since then, lionfish have spread rapidly throughout the tropical and subtropical western Atlantic, Caribbean and Gulf of Mexico, covering an area of approximately 7.3 million km^2 (Côté and Hixon, 2013), with significant numbers reported from sites as far north as Cape Hatteras, North Carolina, and as far south as the Caribbean coasts of Venezuela and Panama (Schofield, 2010).

In addition to their rapid spread, lionfish in some invaded areas have reached densities far exceeding those reported from their native range (Schiel *et al.*, 1986; Fishelson, 1997; Green and Côté, 2009; Green *et al.*, 2012; Kulbicki *et al.*, 2012). Lionfish are also reaching greater maximum sizes in the invaded range than in their native Pacific (Randall *et al.*, 1996; Whitfield *et al.*, 2007). Together, these contrasts suggest some form of ecological release (sensu Elton, 1958), whether due to escape from predators and/or parasites, increased access to food and/or other limiting resources, or some combination of these.

Invasive lionfish are generalist predators, consuming a broad diversity of invertebrates as well as native fishes from at least 25 different families including ecologically and economically important species such as herbivores and predators (Albins and Hixon, 2008; Morris and Akins, 2009; Muñoz *et al.*, 2011; Layman and Allgeier, 2012; Valdez-Moreno *et al.*, 2012; Côté *et al.*, 2013). Several experimental and observational studies conducted over a variety of spatial and temporal scales have indicated that lionfish substantially reduce the abundance, biomass and species richness of small native reef fishes, likely via predation on both small species and juveniles of larger-bodied species (Albins and Hixon, 2008; Green *et al.*, 2012, 2014; Albins, 2013, 2015).

The rapid spread of invasive lionfish across such a large area, combined with evidence of their strong negative effects on native reef fish populations and communities, has led to serious concerns that the lionfish invasion could turn out to be one of the most damaging marine invasions to date (Sutherland *et al.*, 2010; Albins and Hixon, 2013).

Traits of Lionfish Contribute to Invasion Success

Colonization, Establishment and Spread

The ornamental aquarium trade is the most likely vector for the lionfish invasion (Semmens *et al.*, 2004; Morris and Whitfield, 2009; Côté and Hixon, 2013). Lionfish are very popular and attractive ornamental aquarium fish. They can also cause substantial problems for aquarists. This combination of desirable and undesirable traits has led to a large number of lionfish being imported, purchased and kept in aquaria, and conversely, a large number of them being released into the environment. Therefore, in order to understand the factors that contributed to their colonization of the Atlantic, we must examine those characteristics that make lionfish both desirable and undesirable aquarium fish. The following excerpt from an on-line magazine for the marine aquarist illustrates some of the reasons why lionfish are both highly desirable and potentially problematic pets:

Nothing embodies both the beauty and danger of the oceans more than lionfish. Not only are they astonishingly beautiful with their gracefully flowing fins, dramatic colourations, cautious movements, and fish-gulping mouths, but they're equipped with venomous spines capable of delivering painful stings upon an unwary hobbyist. In spite of all this bravado, lionfish are peaceful, extremely hardy and disease-resistant tank inhabitants that are well suited for the intermediate saltwater hobbyist. Frank Marini, PhD (2002)

Additionally, lionfish are well-adapted to a variety of habitats, have a high tolerance to a variety of environmental conditions, and appear to be resilient to physical trauma. These characteristics make them robust to the rigours of the aquarium trade (capture, transport, confinement, handling, infection, etc.) and also confer obvious advantages for a species gaining a foothold in a new environment. Many of the above-mentioned characteristics, in combination with several aspects of lionfish reproductive behaviour and physiology and the fact that their early life stages have the potential for long-distance dispersal, have resulted in an unprecedented rate of spread for an invasive marine species (Côté and Hixon, 2013).

Venomous Spines

Lionfish are visually striking, exotic and literally bristling with venomous spines. These spines, located in their dorsal, pelvic and anal fins (Halstead et al., 1955), serve as natural protection from predators and have several consequences for their introduction. First, lionfish do not appear to exhibit a strong aversion to humans either in the wild or in captivity. This makes them easy to capture alive and probably contributes to their charismatic and reactive behaviour when held in aquaria. On the other side of this coin, many aquarists have themselves felt the sting of lionfish. In fact, most lionfish envenomations in the United States have occurred during fish handling or tank cleaning by aquarists (Kizer et al., 1985; Vetrano et al., 2002). Experiencing their painful sting could be a strong motivation to find another home for one's exotic pet.

High Consumption Rates and Resistance to Starvation

Lionfish are also voracious predators on small fishes and invertebrates. These gape-limited predators will consume almost any fish or invertebrate that they can fit into their mouth. Individual lionfish consume approximately 13 g of prey per day in the wild (Green et al., 2011) and as much as 14.6 g per day in captivity (Fishelson, 1997). As lionfish grow, their consumption rate and ability to consume larger prey increase substantially. However, when prey are not available, lionfish are capable of surviving without food for extended periods. In a starvation study conducted by Fishelson (1997), lionfish lost only 5–16% of their body weight over a period of 3 months without food. While this ability to tolerate long stretches without food likely helps lionfish to survive in captivity as well as in the wild, the appetite of a growing lionfish could easily wipe out an aquarium full of other prized and valuable fishes and invertebrates, providing aquarists with another potential motivation for releasing them.

Tolerance of Environmental Extremes

Lionfish are also capable of tolerating a variety of environmental extremes. While the upper thermal tolerance limits of lionfish have not been reported, their chronic lethal minimum temperature is approximately 10°C, and they are resistant to rapid drops in temperature (Kimball *et al.*, 2004). Lionfish also appear to survive in extremely low-salinity conditions. They have survived over extended periods (28 days) of exposure to salinities as low as 7 ppt, and withstood salinity fluctuations of approximately 28 ppt every 6 h for several days (Jud *et al.*, 2015). No studies have specifically looked at oxygen limitation in lionfish; however, during an attempted comparative feeding experiment in which lionfish, coney grouper (*Cephalopholis fulva*) and two goby species (*Coryphopterus glaucofraenum* and *Gnatholepis thompsoni*) were being held in small aquaria, a main salt water flow-through pump and all aerators failed to function overnight. By morning, all of the coney and most of the gobies were dead, presumably due to low oxygen conditions, but all lionfish survived (M. Cook, unpublished data). Tolerance to a variety of environmental extremes likely contributes to the popularity of lionfish in the ornamental aquarium trade and to their ability to survive in marginal conditions in the wild.

Resistance to Physical Trauma

Lionfish appear to be resistant to infections and to heal rapidly from substantial traumatic injury. During field observations (author's unpublished data), a lionfish that had been monitored for several days on a small artificial reef was apparently speared and subsequently escaped or was released by an unknown person. The lionfish had one large spear wound (approximately 2 cm in diameter at the skin surface and 1 cm in diameter at the narrowest point) which extended completely through the dorsal muscle tissue just beneath the dorsal fin. The fish was resting inside a hole in the reef, and aside from the obvious spear wound, seemed to be in good condition. The fish remained on the reef and was observed periodically over the next several weeks. Within 24 h of the first post-injury sighting, the fish appeared to be covered in thick mucus and the wound had begun to close; within a week, the mucous layer had mostly sloughed off, and the muscle tissue had healed almost completely; within two weeks there was no apparent excess mucous layer, the wound had closed completely and was covered with a thin layer of pigmented skin tissue, and by the third week, the fish was observed hunting in the sea grass near the reef with almost no sign of the wound, except for a scar and an interruption in the skin pigmentation pattern at the old wound site. There was never overt evidence of an infection in the wound of this fish. It is possible that some form of bacterial resistance was conferred by the excess production and sloughing of skin mucus. However, without knowing the rate of healing of similar injuries to other species, it is not known whether this represents a remarkable recovery or not.

Habitat Generalists

Considering their high tolerance for sub-optimal environmental conditions (e.g. low temperatures, low salinities), it is not surprising that lionfish appear to be habitat generalists. In the invaded range, lionfish have been found in a variety of habitats including tidepools, temperate hard-bottom and artificial reefs (Whitfield et al., 2002), shallow and mesophotic coral reefs (Albins and Hixon, 2008; Lesser and Slattery, 2011), seagrass beds (Claydon et al., 2012), mangroves (Barbour et al., 2010) and even up to 5.5 km from the ocean in brackish estuaries (Jud et al., 2011). They can also be found from the surface to depths of 300 m (R.G. Gilmore, personal communication). In their native range, higher densities of lionfish are found in reef-associated habitats including sandy slopes, reef channels and artificial reefs, as well as areas of high turbidity and low salinity, than on coral reefs themselves (Cure et al., 2014).

Reproduction and Larval Dispersal

One of the most striking aspects of the lionfish invasion has been their rapid spread over a large area (Albins and Hixon, 2013). Hurricanes may have facilitated the spread of lionfish across the Florida Current barrier (from Florida to the Bahamas) and increased their spread through the Bahamian archipelago (Johnston and Purkis, 2015). However, there is no evidence that such weather features facilitate the spread of lionfish more than any other marine organism with pelagic larvae. Estimates of the rate of advance of lionfish southward across the Caribbean (250 to 300 km yr^{-1}) (Côté and Green, 2012), are much higher than average rate of spread across a variety of introduced marine organisms (52 km yr^{-1}), and as high as or higher than the most mobile marine invaders (Kinlan and Hastings, 2005: a polychaete worm [247 km yr^{-1}] and a mussel [235 km yr^{-1}]). Lionfish have also exhibited high population growth rates, at least during the initial stages of the invasion (Claydon et al., 2009; Albins and Hixon, 2013). The rapid spread and high population growth rates of invasive lionfish are likely facilitated by a number of life history, morphological and behavioural characteristics including fast growth and early maturation, high fecundity, high fertilization success, high survival of eggs and larvae, and a high capacity for those propagules to travel long distances.

The capacity of a species to increase its population size is tightly linked with its rate of maturation, fecundity and longevity. Lionfish appear to grow more rapidly and mature earlier (before 1 year of age) than similarly sized native mesopredators (Morris, 2009; Albins, 2013; Côté and Hixon, 2013). Female lionfish release between 10 000 and 40 000 eggs per spawning event, and are asynchronous, indeterminate batch spawners capable of sustained reproduction throughout the year, likely spawning every 3 to 4 days under favourable conditions, resulting in an annual fecundity of approximately 2 million eggs (Morris, 2009; Morris *et al.*, 2011). Lionfish are known to live up to 30 years in captivity and individuals up to 8 years of age have been collected in the Atlantic (Potts *et al.*, 2011).

Lionfish spawning occurs in pairs, with the females releasing two buoyant, gelatinous egg masses into the water column (Fishelson, 1975; Morris, 2009) which are then fertilized by the male. Encapsulation of eggs in a gelatinous floating mass contrasts with most external spawning fishes, which typically release free-floating eggs or deposit eggs on the seafloor. These egg masses may increase fertilization rates by preventing sperm dilution (Morris *et al.*, 2011). In a related species, *Dendrochirus zebra*, it has been proposed that the gelatinous material surrounding the eggs decreases post-fertilization predation via some form of chemical deterrent (Moyer and Zaiser, 1981). While the buoyant, gelatinous mass may serve to increase fertilization rates, and protect the eggs from predators, it may also facilitate dispersal by keeping the eggs near the surface where wind-driven currents move faster than those at depth (Freshwater *et al.*, 2009).

After approximately 2–3 days, the gelatinous mass breaks up and the lionfish embryos hatch and become pelagic larvae, which spend approximately 26 days living in the water column (Ahrenholz and Morris, 2010). During this pelagic larval period, which is similar in length to that of many other marine fish species, lionfish propagules are subjected to the movement of ocean currents. It is during this dispersive egg/larval stage that lionfish are most likely to move long distances and colonize new areas. There have been no studies investigating lionfish behaviour during the larval stage. However, larval behaviour in other marine fishes, including sustained directional swimming and vertical positioning in the water column (Leis *et al.*, 2007), has the potential to enhance or modify passive transport mechanisms. Passive dispersal modelling for lionfish has resulted in slower range expansion predictions than those observed (Freshwater *et al.*, 2009; Côté and Hixon, 2013). Therefore, it is possible that lionfish larval behaviours could serve to increase potential dispersal rate and/or distance in the species.

Post-Settlement Movement

Compared with their capacity to move over long distances during early life stages (eggs and larvae), juvenile and adult lionfish are relatively site-attached. In one study of lionfish movements within a Florida estuary, Jud and Layman (2012) found juvenile and adult lionfish to be extremely site-attached, with average movement of approximately 28 m and with 74% of recaptures occurring within 10 m of the previous capture site after weeks to months at liberty. In a separate study conducted on patch and continuous coral reefs in the Bahamas, Tamburello and Côté (2014) reported that while most lionfish moved less than 50 m, some moved as far as 800 m between resightings, with a maximum movement over the course of the study of approximately 1.4 km over 15 days at liberty. Additionally, lionfish movement among habitat patches appeared to be density dependent, with most movements occurring from patches with high lionfish abundance to patches with low abundance (Tamburello and Côté, 2014). Another study conducted off North Carolina reported that adult lionfish displayed high site fidelity to areas no broader than 400 m in diameter, with daily movements that never exceeded 150 m (Bacheler *et al.*, 2015). While juvenile and adult lionfish movements appear to occur over distances too small to contribute to their rapid large-scale spread across the invaded range, these distances are more than sufficient to contribute to their ability to spread over local areas, while maximizing resource availability and minimizing intraspecific competition. Adult movement of lionfish, combined with their ability to survive in a variety of habitats and tolerate a broad range of salinities and temperatures, could also potentially facilitate future movement across certain barriers to larval dispersal like the Amazon–Orinoco river plume (Luiz *et al.*, 2013). If lionfish do successfully cross the Amazon–Orinoco river plume, researchers have speculated – based on thermal tolerance limits, and average annual sea surface temperatures – that the invasion range could eventually extend along the eastern seaboard of South America as far south as Uruguay (Morris and Whitfield, 2009).

Reduced Intraspecific Competition

Often, invasive species experience reduced intraspecific competition due to low local abundance during the initial stages and along 'fronts' of the invasion. However, this comes at a cost. If local densities are too low, it may be difficult for individuals to find mates, thereby stifling reproduction. Lionfish in their native range are reported to be primarily solitary, but are known to aggregate into small groups during spawning periods (Fishelson, 1975). This tendency to aggregate during spawning could allow lionfish to enjoy the benefits of reduced competition, while moderating the reproductive costs of low densities early in the invasion process and along the expanding margins of the invaded range. However, lionfish in core areas within the invaded range exhibit densities more than an order of magnitude greater than those observed in their native range (Green and Côté, 2009; Kulbicki *et al.*, 2012), and are commonly found in large, densely packed groups throughout the day and across all seasons, particularly on isolated patches of habitat (author's personal observations). If reports from the native range are correct about the solitary nature of lionfish except during spawning periods (Fishelson, 1975), invasive lionfish appear to have modified their social behaviours, either becoming more gregarious in the invaded range, or maintaining a continual state of spawning readiness. While invasive lionfish appear to maintain high densities at some locations in the invaded range, there is evidence of density-dependent growth (Benkwitt, 2014) and movement (Tamburello and Côté, 2014) in the species. There is also evidence that density-dependent mechanisms are resulting in limited lionfish population growth, with their numbers levelling off over time in some core areas (Green *et al.*, 2012).

Community Interactions

Interactions between invasive and native species can not only determine the degree of success of the invasive species, but can also determine the magnitude and nature of their effects on native species and communities. Predatory lionfish appear to have strong interactions with native prey and weak interactions with native predators. However, the ability of invasive lionfish to compete with native mesopredators for food resources, while suggested by their strong negative effects on prey densities and rapid growth rates, has not been thoroughly studied.

Lionfish as Predators

Lionfish consume a broad diversity of native prey (from at least 25 different families) including ecologically and economically important species such as herbivores and predators (Albins and Hixon, 2008; Morris and Akins, 2009; Muñoz et al., 2011; Layman and Allgeier, 2012; Côté et al., 2013). They have stronger effects on native prey, and grow faster than similarly sized native predators (Albins, 2013). Additionally, while lionfish allocate a similar amount of time to hunting activities, they consume larger prey and a wider diversity of prey in the invaded range than they do in the Pacific (Cure et al., 2012). These facts suggest that lionfish are particularly effective predators in the invaded range.

The apparent success of invasive lionfish may be due, in part, to the fact that they represent a novel predator type in the invaded system, and demonstrate a suite of predatory behaviours and morphological characteristics that native prey have not encountered during their evolutionary history (Albins and Hixon, 2013). Unlike many native mesopredators, which are susceptible to predation by larger apex predators and so must maintain a high level of vigilance while hunting, lionfish are well defended by venomous spines and are thus free to hover above the reef, stalking their prey with apparent impunity. Lionfish have a combination of spiny and fleshy projections on the head and face, a zebra-like barred colouration pattern, and elongated feather-like fin rays. They also have large fan-like pectoral fins, which they extend when stalking prey. These large fins may help to herd prey, and may also mask the movement of the propulsive caudal and soft dorsal/anal fins. Lionfish also appear to undulate their dorsal spines, in a manner reminiscent of a stalking cat's tail, while approaching potential prey. This combination of external morphology and behaviour may allow lionfish to blend with the reef background (crypsis) or appear to be a harmless invertebrate or plant (mimicry) (Albins and Hixon, 2013). Lionfish approach their prey slowly, flaring their large fan-like pectoral fins, and herd the prey into a corner of the reef or against the seafloor. During this process, lionfish sometimes blow a jet of water directed towards the head of the prey. This water jet likely disorients the prey, reducing their capacity to take evasive action, and may increase the probability of head-first capture (Albins and Lyons, 2012). Lionfish are known to hunt cooperatively with conspecifics in both their native range (Kendall, 1990) and the invaded range (author's personal observations). In their native range, they also form foraging associations with other predator species (Naumann and Wild, 2013). Such cooperative hunting behaviours presumably increase predation efficiency. Lionfish also have a specialized set of muscles attached to their gas bladder, which allows them to modify the pitch of their bodies while minimizing the use of their fins and any resulting disruption to surrounding water (Hornstra et al., 2004). Lionfish are thus able to stalk prey from a variety of angles without giving away their position to potential prey. They also quickly learn to discern between well-defended and undefended prey (Fishelson, 1997).

Lionfish represent a novel predator archetype in the invaded community, with a unique suite of morphological characteristics and behaviours. Evolutionary naïveté of native prey to lionfish may reduce prey vigilance and the effectiveness of prey avoidance behaviours (Marsh-Hunkin *et al.*, 2013; Black *et al.*, 2014; Kindinger, 2015), conferring a high degree of predatory efficiency and contributing to the remarkable success of the lionfish invasion.

Lionfish as Prey

One characteristic shared by many successful invaders is defence from predation (García-Berthou, 2007). For example, highly toxic cane toads have left a trail of dead predators in the wake of their invasion across Australia (Burnett, 1997). Lionfish are also well defended from predation. A series of venomous spines located in their dorsal, pelvic and anal fins contain a powerful neurotoxin (Cohen and Olek, 1989), which likely deters predators. In fact, rather than fleeing from potential predators, lionfish boldly approach threats (e.g. SCUBA divers, large predators) in a head-down posture with their long dorsal spines directed forward (Whitfield *et al.*, 2007; Green *et al.*, 2011). A single predator of lionfish, the cornetfish, *Fistularia commersonii*, has been documented from their native Pacific (Bernadsky and Goulet, 1991), and only isolated predation events have been reported from the invaded range (Maljković *et al.*, 2008). In fact, experimental feeding trials conducted in the invaded range suggest that few, if any, native predators regularly consume even the smallest lionfish (Morris, 2009; Raymond *et al.*, 2015; M. Cook, unpublished data). Native predators do consume lionfish that have been wounded or killed by spearfishers, and unlike predators of the invasive cane toad, appear to do so without suffering adverse effects from their venom. However, documented cases of native predators consuming healthy lionfish are rare. Additionally, parasites – which are effectively small predators – appear to infect lionfish at lower rates than other reef fishes (Sikkel *et al.*, 2014). While one study conducted in the Bahamas suggested a negative relationship between native apex predator density and lionfish density (Mumby *et al.*, 2011), other researchers were unable to find a relationship between lionfish and potential lionfish predators at a larger number of sites across the invaded range (Hackerott *et al.*, 2013). The preponderance of evidence suggests that direct interactions between invasive lionfish and native predators are weak. However, the potential for native predators (including parasites) to provide natural biotic resistance to invasive lionfish may prove a fruitful avenue for future research efforts.

Interspecific Competition

Competitive interactions between lionfish and native fishes have not been thoroughly studied. However, their high consumption rates and strong negative effects on prey, combined with evidence of substantial diet overlap with a number of native mesopredators, as indicated by isotopic analyses (Layman and Allgeier, 2012), suggest that lionfish may be strong exploitative competitors. An experiment designed to compare the effects of lionfish to those of a similarly sized native predator, the coney grouper *Cephalopholis fulva*, found that while lionfish had a stronger effect on prey abundance (2.5 times stronger), and grew more than six-times faster, they had no effect on the growth of this native grouper (Albins, 2013). Albins (2013) suggested that the grouper in the experiment may have compensated for the reduced availability of reef-associated prey fish by consuming alternative prey of lower quality or associated with higher risk, and that the lack of evidence for short-term competition from the experiment should not be taken as evidence against the likelihood of long-term competition between invasive lionfish and native mesopredators. Competition for food may not be the only arena in which lionfish prove challenging for native species. An experiment examining shelter competition between invasive lionfish and endangered Nassau grouper in outdoor tanks suggested that small lionfish may exclude Nassau grouper from shelter (Raymond *et al.*, 2015).

While this evidence suggests that lionfish may act as exploitative competitors in the invaded system, no overt interspecific aggression or evidence of interference competition has been documented between lionfish and other species in the wild. The role of competitive interactions in determining both the success of lionfish in the invaded system and the effects of lionfish on potential native competitors, is an area of research that deserves further attention.

Conclusion

The lionfish invasion represents one of the most successful and potentially damaging marine fish invasions to date. It is clear that the suite of morphological, physiological and behavioural traits exhibited by Pacific lionfish has made them well suited to colonize and become firmly established in Atlantic waters. It has also allowed them to rapidly expand their range, greatly increase their abundance, and cause substantial deleterious changes in the invaded system.

Somewhat ironically, one of the behaviours that has contributed to the lionfish invasion may also contribute to our ability to control their populations and mitigate their impacts. The bold behaviour of threatened lionfish likely contributes to their attractiveness for aquarists and may confer some protection from natural predators. This behaviour also makes it very easy for human divers to approach lionfish close enough to catch them using hand nets or spear them. This makes lionfish particularly susceptible to focused removal efforts by human divers. Even so, a vast army of SCUBA-equipped, spear-toting lionfish hunters would be incapable of eradicating invasive lionfish from the invaded range due to their sheer numbers, the vast area over which they have spread, and their ability to persist in depths beyond the limits of normal SCUBA equipment and training. Unfortunately, more efficient fisheries methods such as hook-and-line, longline, trawl nets, gill nets and traps are not amenable to lionfish control because they tend to catch many more native species than lionfish. Nonetheless, efforts to reduce their numbers via hand nets and spearfishing in particularly valuable and/or sensitive sites appear to be at least temporarily effective, and efforts are underway to encourage and expand manual lionfish control programmes. While early lionfish control efforts have had some success (Frazer *et al.*, 2012; Green *et al.*, 2014), Côté *et al.* (2014) found that lionfish on reefs where repeated culling had occurred became less active and hid deeper in the reef than those on unculled reefs, suggesting that long-term lionfish control by divers could become less effective over time. Technologies, such as lionfish-specific traps, are also under development to increase the efficiency and expand the reach of control efforts.

Many of the traits of lionfish discussed here are known risk factors for invasive species in general, and invasive fishes in particular. These include affiliation with humans (via the aquarium trade), defence from predation, broad diet, high environmental tolerances, fast growth, high propagule pressure, long life span, high fecundity, long reproductive season, early age at maturity, and novel archetype (García-Berthou, 2007; Morris, 2009). While not all of these traits are strictly behavioural, they are all tightly linked with behaviours. It is important to use such invasive species risk factors as guidelines for management and legislation to reduce future species invasions.

References

Ahrenholz, D.W. and Morris, J.A. (2010). Larval duration of the lionfish, *Pterois volitans* along the Bahamian Archipelago. *Environmental Biology of Fishes*, 88, 305–309.

Albins, M.A. (2013). Effects of invasive Pacific red lionfish *Pterois volitans* versus a native predator on Bahamian coral-reef fish communities. *Biological Invasions*, 15, 29–43.

Albins, M.A. (2015). Invasive Pacific lionfish *Pterois volitans* reduce abundance and species richness of native Bahamian coral-reef fishes. *Marine Ecology Progress Series*, 522, 231–243.

Albins, M.A. and Hixon, M.A. (2008). Invasive Indo-Pacific lionfish *Pterois volitans* reduce recruitment of Atlantic coral-reef fishes. *Marine Ecology Progress Series*, 367, 233–238.

Albins, M.A. and Hixon, M.A. (2013). Worst case scenario: potential long-term effects of invasive predatory lionfish (*Pterois volitans*) on Atlantic and Caribbean coral-reef communities. *Environmental Biology of Fishes*, 96, 1151–1157.

Albins, M.A. and Lyons, P.J. (2012). Invasive red lionfish *Pterois volitans* blow directed jets of water at prey fish. *Marine Ecology Progress Series*, 448, 1–5.

Bacheler, N.M., Whitfield, P.E., Muñoz, R.C., *et al.* (2015). Movement of invasive adult lionfish *Pterois volitans* using telemetry: importance of controls to estimate and explain variable detection probabilities. *Marine Ecology Progress Series*, 527, 205–220.

Barbour, A., Montgomery, M., Adamson, A., Díaz-Ferguson, E. and Silliman, B. (2010). Mangrove use by the invasive lionfish *Pterois volitans*. *Marine Ecology Progress Series*, 401, 291–294.

Benkwitt, C.E. (2014). Non-linear effects of invasive lionfish density on native coral-reef fish communities. *Biological Invasions*, 17, 1383–1395.

Bernadsky, G. and Goulet, D. (1991). A natural predator of the lionfish, *Pterois miles*. *Copeia*, 1, 231–234.

Black, A. N., Weimann, S.R., Imhoff, V.E., Richter, M.L. and Itzkowitz, M. (2014). A differential prey response to invasive lionfish, *Pterois volitans*: prey naiveté and risk-sensitive courtship. *Journal of Experimental Marine Biology and Ecology*, 460, 1–7.

Burnett, S. (1997). Colonizing cane toads cause population declines in native predators: reliable anecdotal information and management implications. *Pacific Conservation Biology*, 3, 65–72.

Claydon, J.A.B., Calosso, M.C. and Jacob, S.E. (2009). The red lionfish invasion of South Caicos, Turks and Caicos Islands. *Proceedings of the Gulf and Caribbean Fisheries Institute*, 400–402.

Claydon, J., Calosso, M. and Traiger, S. (2012). Progression of invasive lionfish in seagrass, mangrove and reef habitats. *Marine Ecology Progress Series*, 448, 119–129.

Cohen, A.S. and Olek, A.J. (1989). An extract of lionfish (*Pterois volitans*) spine tissue contains acetylcholine and a toxin that affects neuromuscular transmission. *Toxicon*, 27, 1367–1376.

Côté, I.M. and Green, S.J. (2012). Potential effects of climate change on a marine invasion: the importance of current context. *Current Zoology*, 58, 1–8.

Côté, I.M., and Hixon, M.A. (2013). Predatory fish invaders: Insights from Indo-Pacific lionfish in the western Atlantic and Caribbean. *Biological Conservation*, 164, 50–61.

Côté, I., Green, S., Morris, J., Akins, J. and Steinke, D. (2013). Diet richness of invasive Indo-Pacific lionfish revealed by DNA barcoding. *Marine Ecology Progress Series*, 472, 249–256.

Côté, I.M., Darling, E.S., Malpica-Cruz, L., *et al.* (2014). What doesn't kill you makes you wary? Effect of repeated culling on the behavior of an invasive predator. *PLoS ONE*, 9, e94248.

Cure, K., Benkwitt, C.E., Kindinger, T.L., *et al.* (2012). Comparative behavior of red lionfish *Pterois volitans* on native Pacific versus invaded Atlantic coral reefs. *Marine Ecology Progress Series*, 467, 181–192.

Cure, K., McIlwain, J. and Hixon, M. (2014). Habitat plasticity in native Pacific red lionfish *Pterois volitans* facilitates successful invasion of the Atlantic. *Marine Ecology Progress Series*, 506, 243–253.

Elton, C.S. (1958). *The Ecology of Invasions by Animals and Plants*. New York: Wiley.

Fishelson, L. (1975). Ethology and reproduction of pteroid fishes found in the Gulf of Aqaba (Red Sea), especially *Dendrochirus brachypterus* (Cuvier), (Pteroidae, Teleostei). *Pubblicazionidella Stazione Zoologica di Napoli*, 39 (Suppl.), 635–656.

Fishelson, L. (1997). Experiments and observations on food consumption, growth and starvation in *Dendrochirus brachypterus* and *Pterois volitans* (Pteroinae, Scorpaenidae). *Environmental Biology of Fishes*, 50, 391–403.

Frazer, T.K., Jacoby, C.A., Edwards, M.A., Barry, S.C. and Manfrino, C.M. (2012). Coping with the lionfish invasion: can targeted removals yield beneficial effects? *Reviews in Fisheries Science*, 20, 185–191.

Freshwater, D.W., Hines, A., Parham, S., *et al.* (2009). Mitochondrial control region sequence analyses indicate dispersal from the US East Coast as the source of the invasive Indo-Pacific lionfish *Pterois volitans* in the Bahamas. *Marine Biology*, 156, 1213–1221.

García-Berthou, E. (2007). The characteristics of invasive fishes: what has been learned so far? *Journal of Fish Biology*, 71, 33–55.

Green, S.J. and Côté, I.M. (2009). Record densities of Indo-Pacific lionfish on Bahamian coral reefs. *Coral Reefs*, 28, 107.

Green, S.J., Akins, J.L. and Côté, I.M. (2011). Foraging behavior and prey consumption in the Indo-Pacific lionfish on Bahamian coral reefs. *Marine Ecology Progress Series*, 433, 159–167.

Green, S.J., Akins, J.L., Maljković, A. and Côté, I.M. (2012). Invasive lionfish drive Atlantic coral reef fish declines. *PLoS ONE*, 7, e32596.

Green, S.J., Dulvy, N.K., Brooks, A.M., *et al.* (2014). Linking removal targets to the ecological effects of invaders: a predictive model and field test. *Ecological Applications*, 24, 1311–1322.

Hackerott, S., Valdivia, A., Green, S.J., *et al.* (2013). Native predators do not influence invasion success of Pacific lionfish on Caribbean reefs. *PLoS ONE*, 8, e68259.

Halstead, B.W., Chitwood, M.J. and Modglin, F.R. (1955). The anatomy of the venom apparatus of the zebrafish, *Pterois volitans* (Linnaeus). *The Anatomical Record*, 122, 317–333.

Hornstra, H.M., Herrel, A. and Montgomery, W.L. (2004). Gas bladder movement in lionfishes: a novel mechanism for control of pitch. *Journal of Morphology Special Issue: Seventh International Congress of Vertebrate Morphology.*

Johnston, M.W. and Purkis, S.J. (2015). Hurricanes accelerated the Florida-Bahamas lionfish invasion. *Global Change Biology*, 21, 2249–2260.

Jud, Z.R. and Layman, C.A. (2012). Site fidelity and movement patterns of invasive lionfish, *Pterois* spp., in a Florida estuary. *Journal of Experimental Marine Biology and Ecology*, 414–415, 69–74.

Jud, Z., Layman, C., Lee, J. and Arrington, D. (2011). Recent invasion of a Florida (USA) estuarine system by lionfish *Pterois volitans/P. miles. Aquatic Biology*, 13, 21–26.

Jud, Z.R., Nichols, P.K. and Layman, C.A. (2015). Broad salinity tolerance in the invasive lionfish *Pterois* spp. may facilitate estuarine colonization. *Environmental Biology of Fishes*, 98, 135–143.

Kendall, J.J. (1990). Further evidence of cooperative foraging by the turkeyfish, *Pterois miles* in the Gulf of Aqaba, Red Sea with comments on safety and first aid. In *Proceedings of the American Academy of Underwater Sciences Tenth Annual Scientific Diving Symposium.* St. Petersburg, FL: University of South Florida, pp. 209–223.

Kimball, M.E., Miller, J.M. Whitfield, P.E and Hare, J.A. (2004). Thermal tolerance and potential distribution of invasive lionfish (*Pterois volitans/miles* complex) on the east coast of the United States. *Marine Ecology Progress Series*, 283, 269–278.

Kindinger, T.L. (2015). Behavioral response of native Atlantic territorial three spot damselfish (*Stegastes planifrons*) toward invasive Pacific red lionfish (*Pterois volitans*). *Environmental Biology of Fishes*, 98, 487–498.

Kinlan, B.P. and Hastings, A. (2005). Rates of population spread and geographic range expansion: what exotic species tell us. In *Species Invasions: Insights into Ecology, Evolution, and Biogeography*, ed. Sax, D.F., Stachowicz, J.J. and Gaines, S.D. Sunderland, MA: Sinauer Associates, pp. 381–419.

Kizer, K.W., McKinney, H.E. and Auerbach, P.S. (1985). Scorpaenidae envenomation: a five-year poison center experience. *The Journal of the American Medical Association*, 253, 807–810.

Kulbicki, M., Beets, J., Chabanet, P., *et al.* (2012). Distributions of Indo-Pacific lionfishes *Pterois* spp. in their native ranges: implications for the Atlantic invasion. *Marine Ecology Progress Series*, 446, 189–205.

Layman, C.A. and Allgeier, J.E. (2012). Characterizing trophic ecology of generalist consumers: a case study of the invasive lionfish in the Bahamas. *Marine Ecology Progress Series*, 448, 131–141.

Leis, J.M., Wright, K.J. and Johnson, R.N. (2007). Behavior that influences dispersal and connectivity in the small, young larvae of a reef fish. *Marine Biology*, 153, 103–117.

Lesser, M.P. and Slattery, M. (2011). Phase shift to algal dominated communities at mesophotic depths associated with lionfish (*Pterois volitans*) invasion on a Bahamian coral reef. *Biological Invasions*, 13, 1855–1868.

Luiz, O., Floeter, S., Rocha, L. and Ferreira, C. (2013). Perspectives for the lionfish invasion in the South Atlantic: are Brazilian reefs protected by the currents? *Marine Ecology Progress Series*, 485, 1–7.

Maljković, A., Van Leeuwen, T.E. and Cove, S.N. (2008). Predation on the invasive red lionfish *Pterois volitans* (Pisces: Scorpaenidae), by native groupers in the Bahamas. *Coral Reefs*, 27, 501.

Marini, F. (2002). The lionfish info sheet: captive care and home husbandry. *Reefkeeping*. Available at: http://www.reefkeeping.com/issues/2002-11/fm/feature/, accessed 21 April 2016.

Marsh-Hunkin, K.E., Gochfeld, D.J. and Slattery, M. (2013). Antipredator responses to invasive lionfish, *Pterois volitans*: interspecific differences in cue utilization by two coral reef gobies. *Marine Biology*, 160, 1029–1040.

Morris, J.A.J. (2009). The biology and ecology of the invasive Indo-Pacific lionfish. PhD thesis. Raleigh, NC: North Carolina State University.

Morris, J.A. and Akins, J.L. (2009). Feeding ecology of invasive lionfish (*Pterois volitans*) in the Bahamian archipelago. *Environmental Biology of Fishes*, 86, 389–398.

Morris, J.A.J. and Whitfield, P.E. (2009). Biology, ecology, control and management of the invasive Indo-Pacific lionfish: an updated integrated assessment. NOAA Technical Memorandum.

Morris, J.A.J., Sullivan, C.V. and Govoni, J.J. (2011). Oogenesis and spawn formation in the invasive lionfish, *Pterois miles* and *Pterois volitans*. *Scientia Marina*, 75, 147–154.

Moyer, J.T. and Zaiser, M.J. (1981). Social organization and spawning behaviour of the Pteroine fish *Dendrochirus zebra* at Miyake-jima, Japan. *Japanese Journal of Ichthyology*, 28, 52–69.

Mumby, P.J., Harborne, A.R. and Brumbaugh, D.R. (2011). Grouper as a natural biocontrol of invasive lionfish. *PLoS ONE*, 6, e21510.

Muñoz, R.C., Currin, C.A. and Whitfield, P.E. (2011). Diet of invasive lionfish on hard bottom reefs of the Southeast USA: insights from stomach contents and stable isotopes. *Marine Ecology Progress Series*, 432, 181–193.

Naumann, M.S. and Wild, C. (2013). Foraging association of lionfish and moray eels in a Red Sea seagrass meadow. *Coral Reefs*, 32, 1111.

Potts, J.C., Berrane, D. and Morris, J.A.J. (2011). Age and growth of lionfish from the western North Atlantic. *Proceedings of the 63rd Gulf and Caribbean Fisheries Institute*, 63, 314.

Randall, J.E., Allen, G.R. and Steene, R.C. (1996). *Fishes of the Great Barrier Reef and Coral Sea. Revised and Expanded*. Honolulu, Hawaii: University of Hawaii Press.

Raymond, W.W., Albins, M.A. and Pusack, T.J. (2015). Competitive interactions for shelter between invasive Pacific red lionfish and native Nassau grouper. *Environmental Biology of Fishes*, 98, 57–65.

Ruiz-Carus, R., Matheson, R.E., Roberts, D.E. and Whitfield, P.E. (2006). The western Pacific red lionfish, *Pterois volitans* (Scorpaenidae), in Florida: evidence for reproduction and parasitism in the first exotic marine fish established in state waters. *Biological Conservation*, 128, 384–390.

Schiel, D.R., Kingsford, M.J. and Choat, J.H. (1986). Depth distribution and abundance of benthic organisms and fishes at the subtropical Kermadec Islands. *New Zealand Journal of Marine and Freshwater Research*, 20, 521–535.

Schofield, P.J. (2009). Geographic extent and chronology of the invasion of non-native lionfish (*Pterois volitans* [Linnaeus 1758] and *P. miles* [Bennett 1828]) in the Western North Atlantic and Caribbean Sea. *Aquatic Invasions*, 4, 473–479.

Schofield, P.J. (2010). Update on geographic spread of invasive lionfishes (*Pterois volitans* [Linnaeus, 1758] and *P. miles* [Bennett, 1828]) in the Western North Atlantic Ocean, Caribbean Sea and Gulf of Mexico. *Aquatic Invasions*, 5, S117–S122.

Semmens, B.X., Buhle, E.R., Salomon, A.K. and Pattengill-Semmens, C.V. (2004). A hotspot of non-native marine fishes: evidence for the aquarium trade as an invasion pathway. *Marine Ecology Progress Series*, 266, 239–244.

Sikkel, P.C., Tuttle, L.J., Cure, K., Coile, A.M. and Hixon, M.A. (2014). Low susceptibility of invasive red lionfish (*Pterois volitans*) to a generalist ectoparasite in both its introduced and native ranges. *PLoS ONE*, 9, e95854.

Sutherland, W.J., Clout, M., Côté, I.M., *et al.* (2010). A horizon scan of global conservation issues for 2010. *Trends in Ecology and Evolution*, 25, 1–7.

Tamburello, N. and Côté, I.M. (2014). Movement ecology of Indo-Pacific lionfish on Caribbean coral reefs and its implications for invasion dynamics. *Biological Invasions*, 17, 1639–1653.

Valdez-Moreno, M., Quintal-Lizama, C., Gómez-Lozano, R. and del C. García-Rivas, M. (2012). Monitoring an alien invasion: DNA barcoding and the identification of lionfish and their prey on coral reefs of the Mexican Caribbean. *PLoS ONE*, 7, e36636.

Vetrano, S.J., Lebowitz, J.B. and Marcus, S. (2002). Lionfish envenomation. *The Journal of Emergency Medicine*, 23, 379–382.

Whitfield, P.E., Gardner, T., Vives, S.P., *et al.* (2002). Biological invasion of the Indo-Pacific lionfish *Pterois volitans* along the Atlantic coast of North America. *Marine Ecology Progress Series*, 235, 289–297.

Whitfield, P.E., Hare, J.A., David, A.W., *et al.* (2007). Abundance estimates of the Indo-Pacific lionfish *Pterois volitans/miles* complex in the Western North Atlantic. *Biological Invasions*, 9, 53–64.

18 Wildlife Trade, Behaviour and Avian Invasions

Martina Carrete and José L. Tella

Concerns about the potential negative effects of biological invasions has promoted a large number of studies aimed to identify factors related to whether introduced species establish non-native populations and eventually become invasive, to develop preventative measures (Kolar and Lodge, 2001; Catford *et al.*, 2009). In this sense, the recognition that the invasion process is a succession of stages (uptake, transport, introduction, establishment and expansion), separated by barriers that act as selective filters preventing or allowing species to move on to the next invasion stage (Kolar and Lodge, 2001; Blackburn *et al.*, 2011) has been important in acknowledging that the most effective management option is to tackle invaders on their initial stages (Puth and Post, 2005): species that may not be taken up or transported do not have the opportunity to establish and become invasive (e.g. Blackburn and Duncan, 2001; Cassey *et al.*, 2004; Theoharides and Dukes, 2007; van Kleunen *et al.*, 2010). In spite of that, scientific attention has been mainly addressed to patterns and processes in the new range (establishment and expansion stages), with comparatively little effort devoted to understand pre-establishment stages such as the uptake and transport of species (Puth and Post, 2005).

In this chapter, we deal with trade as an important determinant of biological invasions, paying particular attention to the role played by behaviour in trade-related avian invasions.

Wildlife Trade and Biological Invasions

Increases in international travel and trade have escalated the extent and frequency of species transferred around the world, and this shows no sign of abating (Mack, 2003; Lockwood *et al.*, 2005; Alpert, 2006). The volume of international trade is illustrated by exporting quotas reported by the Convention on International Trade in Endangered Species of Wild Flora and Fauna (CITES); as an example, in 2008, nearly 4 million individual animals, belonging to over 400 species and originating in 30 countries, were traded worldwide for the pet market (Carrete and Tella, 2008a; Figure 18.1). The actual volume of trade is much higher, since CITES only reports quotas for those species

Biological Invasions and Animal Behaviour, eds J.S. Weis and D. Sol. Published by Cambridge University Press. © Cambridge University Press 2016.

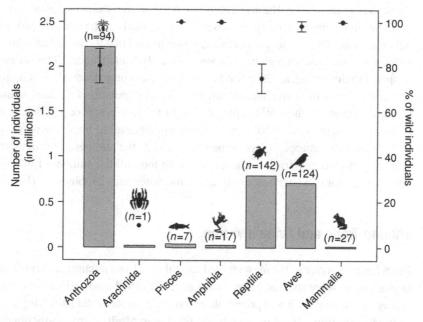

Figure 18.1 International trade of species included in CITES in 2008. The left axis represents the total number of individuals per taxa, while the right axis shows the within-taxa average proportion (± SD) of wild-caught traded individuals. Number of species within taxa is in brackets (redrawn from Carrete and Tella, 2008a).

potentially threatened by international trade and listed in its Appendixes. Using as study model US trade data, Romagosa *et al.* (2009) found that vertebrate families that were more frequently traded were also more likely to become established outside their native ranges, thus confirming that trade in live vertebrates is an important introduction pathway. The European Invasive Alien Species Gateway (DAISIE) lists 80 alien terrestrial vertebrate species known to have become established in Europe as a direct consequence of the trade in wild pets (Westphal *et al.*, 2008). Many other invasive non-vertebrate species also originate from the wild pet trade, resulting in devastating economic costs to agriculture and natural resource industries (Pimentel *et al.*, 2005; Shine *et al.*, 2009; Williams *et al.*, 2010). For example, the invasion of the protected marshlands of the Ebro Delta in Cataluña (Spain) by the apple snail (*Pomacea insularum*) introduced through the drains from a seller of wild pets has caused millions of euros of damage to rice crops, with an estimated cost of removal rounding €6 million (ENDCAP, 2012). Besides, costs associated with treating injuries or infections caused by wild pets have not been quantified, although examples may range from €250 per consultation to €2500 per day of hospitalization (ENDCAP, 2012). In the UK alone, it is estimated that there may be around 5600 cases of reptile-related salmonellosis annually (Toland *et al.*, 2012), and this is only one of the approximately 70 diseases that may be or can be attributable to wild pets (Warwick *et al.*, 2012).

Although outside the focus of this chapter, it is interesting to note that there is also a clear link between increasing amounts of trade and abundance of invasive plant species

(Westphal *et al.*, 2008; Hulme, 2009). In the UK, for example, plants more widely available in twentieth-century nurseries are more likely to be invasive today (Dehnen-Schmutz *et al.*, 2007). More concerning, it is expected that the introduction of new plant species will increase with global climate change (Hellmann *et al.*, 2008) as earlier onset of spring (Schwartz *et al.*, 2006) and warmer temperatures decrease the requirement for winter-hardiness in ornamental plants, and human population increases in the arid and semiarid regions of the world demand drought-tolerant plants (e.g. Seager and Vecchi, 2010). Thus, economic globalization offers opportunities to import new types of plants from previously untapped parts of the world, and as the number of horticultural trading partners between countries continues to rise, so too will the number of introductions of non-natives that can later become invasive and potentially problematic (Hulme, 2009).

Wildlife Trade and Avian Invasions

Birds have been one of the most studied animal taxa in an attempt to identify the characteristics of invasive species. Many works have identified associations between the probability of establishment and propagule pressure, environmental suitability and a variety of life history traits (Blackburn *et al.*, 2009). These results were mostly obtained taking advantage of the large number of deliberate introductions of avian species conducted by Acclimatization Societies during the so-called Great European diaspora, which still constitute the most accurate historical reports of introductions in animals (Blackburn *et al.*, 2009). However, the temporal window for the transport and introduction of bird species across the world is much wider.

Humans historically transported bird species across different areas for a variety of motives: as a source of food, for the use of their plumage, for sporting purposes, as an aesthetic amenity, as companion pets, as a form of biological control of pest species, for sentimental and nostalgic reasons, and even as scavengers (Lever, 2005). The earliest translocations of wild and early-domesticated birds for their consumption may have been several thousands of years ago, resulting in the first avian introductions (Blackburn *et al.*, 2009; Tella, 2011). Given the difficulties of compiling evidence of ancient bird introductions, we should expect a number of still unravelled historical introductions worldwide (Tella, 2011; see also Clavero *et al.*, 2015 for other taxa). In fact, different ancient cultures showed a great fascination for exotic birds and intensively traded them. The best known cases come from the Inca, Maya, Aztec and Paquimé civilizations in South and Central America, whose emperors obtained exotic birds through taxation of conquered provinces and the active trade of live individuals of species with native distributions hundreds and perhaps thousands of kilometres away (Haemig, 1978; Somerville *et al.*, 2010). Aztec emperors kept in their elaborate zoos a variety of exotic species, such as rheas, condors, eagles, waterbirds, parrots, quetzals and passerines, but lesser nobles also raised birds in their gardens and exotic birds were widely available in the marketplace where common people bought them as household pets (Haemig, 1978). This intensive trade of wild birds resulted in the Pre-Columbian introduction of the great-tailed grackle, *Quiscalus mexicanus*, and probably of other traded species which are now showing surprising disjunct distributions in the Neotropical region (Haemig, 1978, 2011).

Although the actual extent of ancient introductions is unknown, there is no doubt that transport of birds and their establishment in non-native regions increased in the last two centuries. A review of 1243 introduction events showed that most avian introductions took place between 1850 and 1950, coinciding with the migration of Europeans to the Nearctic and southern hemisphere (Blackburn and Duncan, 2001; Hulme *et al.*, 2008). It was the time when Acclimatization and Wildlife Societies were founded to introduce a number of Palearctic bird species in very distant sites such as New Zealand, Australia or Hawaii (Blackburn *et al.*, 2009). However, trade in birds has dramatically increased in recent decades, due to the increasing numbers of transport facilities and the large demand of exotic cage and pet birds in the developed world (Blackburn *et al.*, 2015). Estimates of the annual numbers of birds involved in international trade range from 2 to 7.5 million individuals in the 1970s and 1980s (Beissinger, 2001), and trade continues today, although at lower numbers (Figure 18.1). This trade reversed previous geographic trends, now translocating species from the southern hemisphere – mostly from South American and African countries – to the northern, richer countries (Beissinger, 2001). As a direct consequence, an unprecedented number of non-native bird species became established in the United States and European countries in recent years (Butler, 2005; Carrete and Tella, 2008b; Chiron *et al.*, 2009).

Taxonomic Biases and the Role of Behaviour on Bird Trade

The first stage of the invasion process is of particular importance because it acts as the primary filter for species entering the later stages: species only get the chance to be introduced, established and spread if they are first transported. Thus, there is a growing interest in understanding which species are entrained onto the biological invasion pathways (Costello and Solow, 2003; Kark and Sol, 2005; Blackburn and Cassey, 2007; Hulme *et al.*, 2008; van Kleunen *et al.*, 2010) as well as whether the transition between these stages would have an effect on their subsequent establishment success.

Trade has always been linked with human demands. Historically introduced bird species belonged significantly more often than expected by chance to the families Anatidae, Phasianidae, Columbidae, Psittacidae and Passeridae because they are most prized for hunting, ornamentation and pets (Blackburn and Duncan, 2001).

The prevalence of some families over others in international bird trade suggests that the possession of particular traits may have also contributed to their likelihood of being traded. For example, trade in some bird taxa such as swifts and hummingbirds could have been constrained by their ecology (i.e. flight and feeding habits), which makes their adaptation to captivity difficult. Behavioural traits may also play a major role in the non-random selection of traded species. Songs and vocalizations, for example, seem to have favoured the trade of some species over others across different civilizations. Although the Aztecs looked for exotic birds to use their colourful feathers for ritual purposes, several species were traded and kept in captivity solely for their aesthetic beauty and song (Haemig, 1978). Using a sample of North American and European passerines, Blackburn *et al.* (2014) found that bird species whose songs are known to have inspired classical music have been more represented in trade because they are more attractive for humans than others, and song attractiveness is positively related to number of birds available in Asian markets (Su *et al.*, 2014). It is thus not surprising that a number of songbird species (mainly Turdidae, Fringillidae and Emberizidae) were imported and introduced in the Antipodes by homesick European settlers, under the auspices of local Acclimatization Societies (Lever, 2005). Exotic songbirds have been kept in captivity by royalty for centuries, but only fairly recently have birds been available to the general public thanks to improvements in international transport facilities, and the contemporary movement of birds may be driven by different influences compared with that of previous centuries (Blackburn *et al.*, 2009). Beissinger (2001) compiled trade data recorded by CITES, showing that 70 species of songbirds (passerines) comprised 70% out of the nearly 5 million birds globally traded between 1991 and 1996. Parrots from 259 species made up 25% of the traded birds, while other species from 36 families just accounted for 5% of the global trade. Therefore, songbirds seem to remain the most in-demand group of birds, not only by Western societies but also by Eastern ones. There is a large trade in birds in Eastern countries such as Vietnam, Indonesia, Malaysia, China and Taiwan (Lau *et al.*, 1996; Jepson and Ladle, 2005; Shepherd, 2006; Shepherd *et al.*, 2012; Su *et al.*, 2015), where birds are used for bird-keeping and prayer animal release in religious ceremonies. In particular, bird-keeping is rooted deeply in Asian culture, associated with competitions where scoring factors include bird song quality and a hobby of 'bird-walking', where owners expose their cage birds to outdoors. Through a large survey of pet shops in Taiwan, Su *et al.* (2014) found that 160 out of 247 traded bird species were non-native, and that parrots and up to eight families of songbirds were over-represented in the pet trade, thus responding to the large demand of songbirds for bird competitions.

Parrots are highly appreciated as pets given their bright colours, playful behaviours and cognitive abilities, being thus the second most traded group of birds (Beissinger, 2001; Su *et al.*, 2014). Just between 1991 and 1996, three-quarters of the world's parrot species were internationally traded (contrasting with only 5% of the world species of passerines; Beisinger, 2001). However, as in songbirds, it also seems that some parrot species are preferred over others based on their behaviour. Tella and Hiraldo (2014) have shown that the 22 species of parrots native to Mexico differ widely in attractiveness to people (as reflected by their combined measures of body size, colouration and ability to imitate human speech), and that their attractiveness strongly correlated with their prices both in Mexican and US markets. Moreover, the most attractive and valuable species (Amazons and macaws) were caught more often than expected by their relative availability and accessibility in the wild. The same patterns were found for parrots illegally sold in domestic Mexican markets, for those smuggled to the United States, and for those legally exported before or after 1992, when the US ban led parrot exports to be mostly directed to European countries. These results support the idea that there is a long-term, cross-cultural preference for the most attractive parrot species and that behaviour is part of the appeal (Tella and Hiraldo, 2014).

Pre-Establishment Filters, Behaviour and Invasiveness

Given the human preferences for the behaviour of particular pet and cage bird species, it is not surprising that traded parrots and passerines are among those groups over-represented in past and current avian introductions worldwide (Blackburn *et al.*, 2010). However, not all traded species become invasive, and only a small proportion of trans-ported species are able to move across the successive invasion stages, i.e. introduc-tion, establishment and spread (Rodríguez-Cabal *et al.*, 2010). While propagule pres-sure (Cassey *et al.*, 2004), species-specific life history traits (Sol *et al.*, 2012a) and niche conservatism (Strubbe *et al.*, 2013) are all involved in the succesful establishment of introduced birds, behaviour may also play a role. Using a global database of avian introductions, Sol *et al.* (2005) found that species with a relatively larger brain (con-sidered as a proxy of behavioural flexibility) had higher probabilities of establishing in novel environments as a consequence of their higher innovation propensity. While the assumption here is that some species have higher propensity than others to change their behaviour during their life, behavioural flexibility can also be expressed as differences in behaviour among individuals from the population (Bolnick *et al.*, 2003; Carrete and Tella, 2011). Some credence for this possibility comes from a study of urban avian inva-sions showing that although larger brained species exhibit a higher propensity to invade urban areas, the effect turns out non-significant when intraspecific variability in fear of humans is included in the model (Carrete and Tella, 2011). The authors propose that behavioural flexibility can also be regarded as a specific trait encompassing variability in personality traits among individuals rather than the ability of individuals to change behaviours during their lifetime (Carrete and Tella, 2011, 2013).

Most comparative studies on avian invasions are based on the analysis of deliber-ate introductions conducted in the past (Blackburn *et al.*, 2009). Nowadays, however, intentional introductions are rather exceptional and current avian invasions are mostly accidental, often coming from birds escaping from cages (Carrete and Tella, 2008b; Romagosa *et al.*, 2009; Blackburn *et al.*, 2010; Abellán *et al.*, 2016). Identifying poten-tial invaders within this new human-driven invasion pathway is an urgent task, given the increasing social demand for pets in developed countries and the increasing number of traded species that have escaped into the wild (Carrete and Tella, 2008b; Hulme *et al.*, 2008; Romagosa *et al.*, 2009; Blackburn *et al.*, 2010; Abellán *et al.*, 2016). There are a number of differences between historical deliberate introductions and recent accidental introductions of birds. For example, large numbers of individuals were often introduced to assure establishment success (Lever, 2005), and thus large propagule sizes could have overcome demographic and genetic bottlenecks and, probably, behavioural-based Allee effects during deliberate introductions. For example, the establishment success of highly social or colonial breeding species could have been compromised if they were introduced in low numbers. However, current escapes of cage birds usually occur at low numbers and in scattered places, thus leading to multiple introduction events with small local propagule sizes (e.g. Edelaar *et al.*, 2015; Saavedra *et al.*, 2015), where the likeli-hoods for Allee effects (i.e. the positive correlation between population size or density and mean individual fitness) are higher.

Large differences also exist in the number of filters arising across the successive invasion stages, and in the way we should look at how exotic species are able to cross them. Hypotheses proposed to understand biological invasions usually deal with species- or population-average traits, while typically ignoring the role of inter-individual variation among potential invaders. Recently, however, it has been proposed that conditions faced by individuals during the invasion stages can impose strong selective pressures that progressively eliminate those with particular phenotypes and/or genotypes from the pool of potential invaders (Chapple et al., 2012; see also Chapter 2). This may be the case for bird introductions arising from international trade (Carrete et al., 2012). Contrary to deliberate introductions, which implied the capture, transport and rapid release of usually large numbers of individuals, trade-related accidental introductions have to cross a larger number of filters which also expand over longer periods of time. Traded birds are first captured in the wild and stored until large enough numbers have been acquired (Figure 18.2) to be transported to the – often distant – exporter companies. Birds are there kept in captivity in crowded conditions for a variable period of time until they are exported to distant countries, then entering into international flight transports. Immediately after arriving at importing countries, they enter sanitary quarantines. After quarantine, birds are again moved by terrestrial or flight transport to pet shops, which are usually scattered across the whole importing countries, or even are re-exported to others. After a variable time of exposition in pet shops, birds are bought and are kept as household pets or cage birds. Just a few of these birds will accidentally escape or be released by unsatisfied owners, several months or years after they were captured.

There is a number of ways in which individual behaviour may influence the transition across the above pre-establishment filters (Carrete et al., 2012). First, there is evidence for personality-related sampling bias in birds, showing that capture is often non-random as active, more exploratory or bold individuals are more likely to be captured than inactive or shy ones (Biro and Digenmanse, 2009; Stuber et al., 2013). Therefore, some behavioural types are less likely to enter the trade-invasion pathway. Thereafter, captured individuals must cope with new food sources and crowding, as well as with stressful captive and transport conditions to transit all the successive filters. Mortality estimates for wild-caught birds range within 7–62% between capture and acclimatization to captivity, 3–7% during international transport, 11–15% during quarantine, and up to 30% during the first months after quarantine (Thomsen et al., 1992). This high mortality could further select for particular personality traits and stress-tolerant individuals (Mason et al., 2013), likely representing a non-random sample with respect to activity, aggression and boldness, and making them better potential invaders (Carrete et al., 2012). This appealing hypothesis still requires empirical support. There is, however, recent evidence suggesting that selection during pre-establishment is a likely mechanism. The fact that the genetic diversity of different populations of the highly traded monk parakeet (Myiopsitta monachus) independently introduced in three continents is much lower than in its South American native range, together with a high genetic similarity among introduced populations, may suggest a common source but also that selection of particular genotype-linked behaviours may confer them the ability to successfully transit all the pre-establishment filters (Edelaar et al., 2015). Further research would help not only to advance our knowledge on the behavioural mechanisms

Figure 18.2 Traded passerines are typically baited with grain (a) to be caught with nets (b), and kept in crowded conditions (c) for a long time before survivors arrive at a pet shop. Images are from the capture of red-billed queleas (*Quelea quelea*) in Senegal, September 2014, for the international trade (photos: J. Blas).

involved in current trade-mediated avian invasions, but also on contemporary micro-evolutionary processes meaning that the correlated genotypes and personality traits of invasive populations may differ from those of the native populations where they originated (Mueller *et al.*, 2014).

Wild-Caught and Captive-Bred Cage Birds Differ in Their Invasiveness

While trade may be filtering individuals with a higher invasive potential, captive breeding may select against behaviours that are required to invade new environments. Some

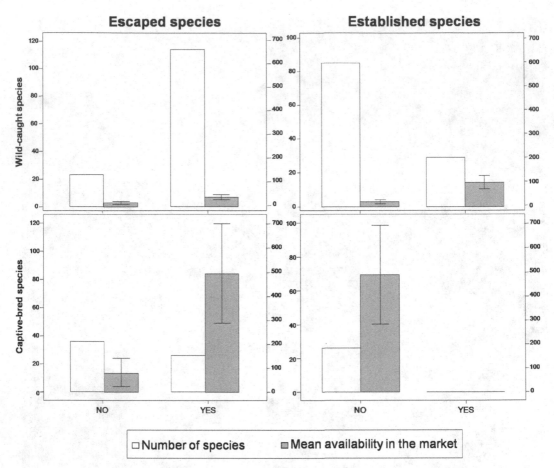

Figure 18.3 Fate of wild-caught and captive-bred cage bird species reaching the two main transition stages of the invasion process. Left panels: number of species escaped or not escaped from cages and reported free-living in the wild. Right panels: from the subset of escaped species, numbers recorded as breeders or non-breeders in the wild. The mean availability of species in the pet market is shown in both cases (redrawn from Carrete and Tella, 2008b).

originally wild-traded bird species have been bred in captivity for decades or even centuries, and we could expect that this process of domestication would have reduced the individual's skills for establishing in the wild (Mason *et al.*, 2013). Carrete and Tella (2008b) examined the differential invasiveness of wild-caught and captive-bred cage bird species by comparing their availability in markets with their invasion status in Spain. As for global international trade (Beissinger, 2001), parrots (Psittaciformes) and songbirds (Passeriformes) represented 99% of the >21 000 individuals recorded in the Spanish markets during 2004–2005, from 202 species. Although the invasion success of the species was positively correlated to their availability in markets (a measure of propagule pressure), and captive-bred individuals outnumbered wild-caught ones, the main determinant of invasiveness was the origin of traded birds: all species breeding and established in the wild were wild-caught, but none captive-bred (Figure 18.3).

These results suggest that wild-caught species are more likely to escape from cages and establish breeding populations, thereby successfully completing two main transition stages (release and establishment) of the invasion pathway. Individuals belonging to captive-bred species were also recorded in the wild, although at a lower proportion, but they were not able successfully to reproduce and, thus, to establish long-term viable populations. Several differences between individuals could be responsible for such differences in establishment success. Unlike wild-caught individuals, captive-bred birds often descend from a small pool of individuals, so the detrimental effects of inbreeding depression, loss of genetic diversity, and accumulation of deleterious mutations can be expected (Charlesworth and Charlesworth, 1987; Keller and Walter, 2002). Moreover, breeders often promote mating between relatives to generate and maintain new plumage mutations in captivity. Therefore, the eroded genetic diversity of captive individuals may have contributed to their lower fitness parameters (i.e. fecundity, mating success, survival), slower development and increased susceptibility to pathogens and environmental stress (Charlesworth and Charlesworth, 1987; Keller and Walter, 2002) compared to their wild-caught counterparts.

Added to genetic aspects, behavioural and physiological traits that are important to survive in the wild may also have been eroded in captivity (McDougall et al., 2006). The secretion of glucocorticoids, mainly corticosterone in birds, constitutes a major mechanism allowing individuals to cope with stressful situations. However, chronic exposure to stress hormones, as may typically occur during the pre-establishment stages of the invasion process, may result in numerous negative impacts on growth, immune function, cognitive ability and reproduction (Bortolotti et al., 2008, 2009). Considerable unplanned selection for more stress-tolerant individuals is expected to occur through captive breeding. Using an experimental approach, Cabezas et al. (2013) found that wild-caught pet parrots have longer corticosterone responses to acute stress than captive-bred ones. This pattern held both when comparing species of different origin and wild birds to the first generation born in captivity from the same species. Captive-bred birds could show an attenuated corticosterone response due to habituation, while high mortality rates during the international trade of wild-caught birds could have selected for those individuals that are better able to cope with stress (Carrete et al., 2012; Cabezas et al., 2013). Although more research is needed, the longer acute response found in wild-caught birds could help them to escape from cages and survive better when facing challenges in the new wild environments, so helping to explain their higher invasiveness.

Other behaviours related to cultural knowledge, anti-predator and foraging abilities, which constitute important skills in the wild, can also be eroded in captivity, as captive animals generally receive ample food and water as well as protection from predation and aggressive interactions (McDougall et al., 2006; Mason, 2010). Moreover, aggressiveness and activity could be counter-selected, as highly active or aggressive animals are more likely to be poorly adapted to captive conditions (De Boer et al., 2003; Künzl et al., 2003; Archard and Braithwaite, 2010), thus reducing their likelihood of reproducing in captivity compared to less active and aggressive individuals. Captivity may at the same time relax selection pressures, promoting the persistence of behaviours and genes that would otherwise be selected against in the wild. Artificial selection of certain phenotypes and/or genotypes in captivity can also promote sensory alterations, such as auditory and song abnormalities, that may interfere with both natural and sexual selection. For example, one captive-bred strain of the canary *Serinus canarius*, the Belgian Waterslager, has less sensitive hearing at high frequencies, and produces songs with more energy at low frequencies (Dooling *et al.*, 1971; Guttinger, 1985), which leads to a shift in their hearing range towards lower frequencies than wild-type canaries. All of these problems make it difficult for captive-bred birds to survive after escaping and to establish breeding populations in the wild. The wild-caught origin of cage birds may therefore predict their ability as invaders, thus constituting a simple indicator of invasion potential (Carrete and Tella, 2008b).

Behaviour and Impact of Invasive Traded Birds

Many internationally traded birds have established and become invasive in different regions of the world. They may cause a variety of impacts on their recipient environments, through predation, competition, hybridization and the spread of diseases to native fauna, or their negative effects as crop pests (Lever, 2005). Recent evaluations suggest that some non-native bird species have as severe impacts as non-native mammals in Europe (Kumschick and Nentwig, 2010). However, at a global scale, most information on the impact of non-native bird species comes from islands rather than from continental environments, and the actual impact of most introduced bird species remains largely unknown (Lever, 2005; Edelaar and Tella, 2012). As for other introduced nonnative taxa (Simberloff *et al.*, 2013), identifying the impacts of invasive bird species is a difficult task given the variety of impacts they may cause and the typical temporal gap between introduction and the upsurge of impacts (Edelaar and Tella, 2012).

Invasive birds may affect native ones through competition for nesting and food resources. Some studies have tried indirectly to identify these impacts through changes in fitness (Freed and Cann, 2009) or in the distribution and abundance of native species (Newson *et al.*, 2011), or even experimentally by assessing nest site competition (Strubbe and Mathyssen, 2009). Less attention has been paid, however, to the study of the behavioural mechanisms driving these interspecific competition patterns. The study of behavioural interactions is needed to fully understand the potential impact of invasive birds, and gain importance given that interactions between invasive and native birds may be context dependent, i.e. depend on the recipient habitats and communities. While some invasive species may impact native ones by exploiting more aggressively or efficiently the same food or nesting resources (competition hypothesis), others may simply take advantage of resources unexploited by natives (opportunism hypothesis) causing no or little impact, as may be the case for bird species introduced in anthropogenic habitats (Sol *et al.*, 2012b). A few studies recently examined these hypotheses by relying on behavioural interactions between invasive and natives. Sol *et al.* (2012b) combined field experiments and observations focusing on five bird species introduced in Australia, including the common myna (*Acridotheres tristis*), one of the invasive species for which more negative impacts have been recorded worldwide

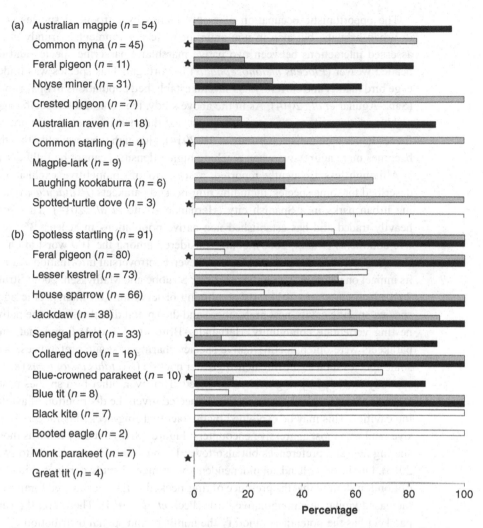

(a) Australian magpie (n = 54)
Common myna (n = 45) ★
Feral pigeon (n = 11) ★
Noyse miner (n = 31)
Crested pigeon (n = 7)
Australian raven (n = 18)
Common starling (n = 4) ★
Magpie-lark (n = 9)
Laughing kookaburra (n = 6)
Spotted-turtle dove (n = 3) ★

(b) Spotless startling (n = 91)
Feral pigeon (n = 80) ★
Lesser kestrel (n = 73)
House sparrow (n = 66)
Jackdaw (n = 38)
Senegal parrot (n = 33) ★
Collared dove (n = 16)
Blue-crowned parakeet (n = 10) ★
Blue tit (n = 8)
Black kite (n = 7)
Booted eagle (n = 2)
Monk parakeet (n = 7) ★
Great tit (n = 4)

0   20   40   60   80   100
Percentage

Figure 18.4. Aggressive interactions between (a) exotic and native species in an Australian city, and (b) between the exotic ring-necked parakeet and both exotic and native species in a Spanish city. Bars indicate the percentage of encounters that ended in aggressions (white bars), and percentage of aggressions initiated (grey bars) and won by the focal species (black bars). Asterisks indicate alien species, and the number of recorded encounters is shown in brackets (redrawn from Sol *et al.*, 2012b and Hernández-Brito *et al.*, 2014).

(Lever, 2005). This study showed that with the only exception of the feral pigeon (*Columba livia*), all other invasive species were less aggressive than natives (Figure 18.4a) and were generally excluded from artificially created food patches. These exotic species were also more abundant in urban environments where the diversity and abundance of natives were lower than in the natural surrounding environments. Therefore, invasive species exploited ecological opportunities that most natives rarely used, supporting the opportunism hypothesis (Sol *et al.*, 2012b).

The opportunistic occupation of empty niches by introduced species could occur in urban habitats, but also in non-anthropogenic environments. Grundy *et al.* (2014) assessed interactions between two native marshland passerine species and the black-headed weaver (*Ploceus melanocephalus*) in Portugal. This species was traded for the cage bird market and currently has several established populations in Spain and Portugal (Sanz-Aguilar *et al.*, 2014). As in the above study, the lack of interspecific aggressions and breeding territory segregation suggest that the invasive species was not competing for resources with natives (Grundy *et al.*, 2014), although the possibility that the species becomes more aggressive when reaching higher densities cannot be fully discarded.

Although less frequently reported, a scenario of competitive exclusion has been described for some species, including ring-necked parakeets (*Psittacula krameri*) invading urban parks in a Spanish city (Hernández-Brito *et al.*, 2014). This species was heavily traded and has established non-native populations in at least 35 countries on five continents (Butler, 2003), being considered among the 100 worst alien species in Europe (DAISIE, 2011) despite little and even controversial information available about its impact on native avifauna (Lever, 2005; Strubbe and Mathyssen, 2009; Strubbe *et al.*, 2010; Newson *et al.*, 2011). By combining observations on interspecific aggressions, species-specific cavity-nest preferences and the spatial distribution of the native cavity-nesting vertebrate community, Hernández-Brito *et al.* (2014) found that ring-necked parakeets were outcompeting native species sharing nest site preferences. As a result, some species such as the threatened greater noctule (*Nyctalus lasiopterus*) seemed to be spatially excluded by ring-necked parakeets. However, other bird species nested closer to nests of ring-necked parakeets than expected given the distribution of available nesting cavities. This may be explained by the fact that ring-necked parakeets were aggressive (and won most aggressive encounters; Figure 18.4b) not only towards those species sharing nest site preferences but also towards predators (Hernández-Brito *et al.*, 2014, 2015). On the other hand, an independent experimental study conducted in city gardens of London showed that the presence of ring-necked parakeets reduced foraging rates and increased vigilance among native birds (Peck *et al.*, 2014). Therefore, the ring-necked parakeet has the potential to modify the numbers and spatial distribution of coexisting foraging and breeding species through behavioural-mediated exclusion of both competitors and predators (Hernández-Brito *et al.*, 2014; Peck *et al.*, 2014).

The above results show complex interspecific interactions where some native species may be excluded while other may be favoured by the spread of an aggressive exotic species. But complexity may increase further with the coexistence of several exotic species competing with natives for nest cavities, as shown by Orchan *et al.* (2013). These authors investigated the network of aggressive and nest-occupancy interactions among three invasive species and the native species breeding in an urban park in Israel. Cavity enlarging by the early breeding ring-necked parakeet enhanced breeding in otherwise unavailable cavities for the common myna. This in turn excluded the smaller invasive vinous-breasted starling (*Sturnus burmannicus*), which is the main competitor of the native Syrian woodpecker (*Dendrocopos syriacus*). Therefore, the study of behavioural interactions among multiple invasive species is crucial for understanding their differential invasion success and impact on native communities.

Another example of behaviours of exotic species that can impact native biota is hybridization with natives. A remarkable case is that of the invasive ruddy duck (*Oxyura jamaicensis*) in Europe. After seven individuals were introduced to the UK in 1948, escaped birds rapidly reproduced to reach a population of tens of thousands and spreading over 22 Western Palearctic countries. This invasion caused no major concern until ruddy ducks colonized Spain in the 1990s, where they began frequently to hybridize with the endangered native white-headed duck (*Oxyura leucocephala*). A transboundary, costly eradication plan was needed to eradicate ruddy ducks to avoid genetic introgression and further endangerment of the native species (Muñoz-Fuentes *et al.*, 2013). Hybridization between other more cryptic species such as quails can be easily overlooked. The Japanese quail (*Coturnix japonica*) was first introduced in Europe in the 1950s primarily for meat and egg production, but it was also used for game restocking. The release of Japanese quails for hunting was banned in several European countries after it was realized that it was hybridizing with the phenotypically similar, native European quail (*Coturnix coturnix*) (Sánchez-Donoso, 2014): restocking was only allowed with farm-reared European quails. However, recent genetic studies showed that the majority of game farm quails are actually not European quails, but hybrids between the two species, raising concerns about this practice (Sánchez-Donoso *et al.*, 2012). As has been suggested for captive-bred parrots and passerines (Carrete and Tella, 2008b), the survival in the wild of released captive-bred hybrid quails is low (Puigcerver *et al.*, 2014). However, massive releases of hybrids provide an unusually large propagule pressure, making it possible that some individuals survive and reproduce in the wild (Puigcerver *et al.*, 2014). More worryingly, field experiments showed no behavioural/reproductive barriers between native and hybrid quails, and even that farmed hybrid females attract more wild European quail males than European quail females are able to attract for mating (Figure 18.5; Puigcerver *et al.*, 2014). Moreover, experiments in captivity revealed a higher fertilization success in hybrids than in European male quails and that the sperm of hybrids outcompetes the sperm of European males (Sánchez-Donoso *et al.*, 2016). Therefore, the higher mating and fertilization success of hybrids may explain the significant genetic introgression in wild European quail populations (Sánchez-Donoso *et al.*, 2014) despite high mortality and nest predation rates in released hybrids (Puigcerver *et al.*, 2014).

Behaviour and the Prevention of Invasions by Traded Birds

The number of recent avian introductions is increasing worldwide as a result of changes in societal demands and a higher availability of species due to transport facilities (Blackburn *et al.*, 2010). Although trading and keeping cage birds is a long-standing and widespread tradition among cultures (Jepson and Ladle, 2005; Tella, 2011; Pires, 2012; Su *et al.*, 2014; Tella and Hiraldo, 2014), today birds are probably more popular than ever, and are likely only to continue to increase as the pet of choice for more and more people. People keep millions of birds in captivity because they bring colour, song and amusement into their lives, and accidental escapes or unplanned releases of cage birds

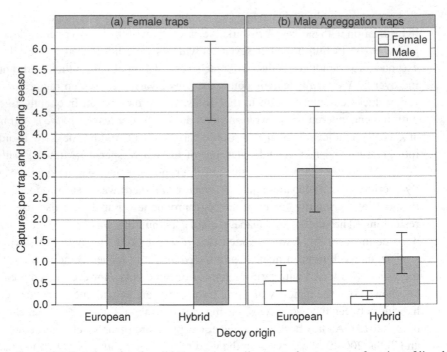

Figure 18.5 Female and male wild European quails attracted to traps as a function of live decoy type (European quail or hybrid originating from game farms) for two types of traps: (a) one female used as decoy, and (b) a group of males used as decoy (redrawn from Puigcerver *et al.*, 2014).

will continue to seed new avian invasions. Although prevention measures are much more effective and less costly than the control and eradication of the established and often spreading introduced populations (Edelaar and Tella, 2012), nowadays it is unfeasible to prohibit bird-keeping at global or national scales. However, the behavioural differences between captive-bred and wild-caught cage birds that results in a lower risk of invasion by the former species (Carrete and Tella, 2008b), constitutes a straightforward management action to halt avian invasions. Building on these results, Spain created a law in 2011 (remodelled in 2013: Real Decreto 6013/2013) to act against biological invasions, whereby the importation of wild-caught birds is forbidden. This Spanish law continued the trade ban in wild birds initiated in 2005 by the European Union in response to the avian influenza crisis, but with the explicit argument of halting further avian invasions. Despite concerns and heated debate arising from the blanket ban on the wild-bird trade in the EU (Cooney and Jepson, 2006; Gilardi, 2006; Rivalan *et al.*, 2007), this management action has been effective. Bird markets and suppliers rapidly switched to increase captive breeding to satisfy the demand for cage birds, illegal trade remains today anecdotal, and at least in Spain the annual rate of exotic bird species newly introduced into the wild was drastically reduced (Tella and Carrete, unpublished data).

Spanish and European bans may represent a successful management action against avian invasions at a regional but not at a global scale; international trade in birds continues to concern considerable numbers, mostly involving wild-caught individuals (Figure 18.1), and it is expected to be redirected to other countries after the EU ban. Moreover, demand for pets is increasing not only in developed countries, but also in developing Asian, Middle Eastern and American countries. Mexico constitutes a striking example. While thousands of native parrots are annually poached and sold at low prices in the domestic market (Tella and Hiraldo, 2014), the recent economic upsurge of the country allowed an eight-fold increase in the importations of exotic parrot species during the past decade, including 10 wild-caught and potentially invasive species among those most imported (Cantú *et al.*, 2007). In fact, one of the traded species – the monk parakeet – has already established non-native populations in several Mexican localities (MacGregor-Fors *et al.*, 2011). Applying the precautionary principle (although bearing in mind the need for more studies to generalize the lower invasiveness of captive-bred cage birds), a worldwide ban on wild-bird trading should be seriously considered with the aim of preventing further avian invasions.

References

Abellán, P., Carrete, M., Anadón, J.D., Cardador, L. and Tella, J.L. (2016). Non-random patterns and temporal trends (1912–2012) in the transport, introduction and establishment of exotic birds in Spain and Portugal. *Diversity and Distributions*, 22(3), 263–373.

Alpert, P. (2006). The advantages and disadvantages of being introduced. *Biological Invasions*, 8, 1523–1534.

Archard, G.A. and Braithwaite, V.A. (2010). The importance of wild populations in studies of animal temperament. *Journal of Zoology*, 281, 149–160.

Beissinger, S.R. (2001). Trade in live wild birds: potentials, principles and practices of sustainable use. In *Conservation of Exploited Species*, ed. Reynolds, J.D., Mace, G.M., Redford, K.H. and Robinson, J.G. Cambridge: Cambridge University Press, pp. 182–202.

Biro, P.A. and Dingemanse, N.J. (2009). Sampling bias resulting from animal personality. *Trends in Ecology and Evolution*, 24, 66–67.

Blackburn, T.M. and Cassey, P. (2007). Patterns of non-randomness in the exotic avifauna of Florida. *Diversity and Distributions*, 13, 519–526.

Blackburn, T.M. and Duncan, R.P. (2001). Establishment patterns of exotic birds are constrained by non-random patterns in introduction. *Journal of Biogeography*, 28, 927–939.

Blackburn, T.M., Lockwood, J.L. and Cassey, P.B. (2009). *Avian Invasions: The Ecology and Evolution of Exotic Birds*. Oxford, UK: Oxford University Press.

Blackburn, T.M., Gaston, K.J. and Parnell, M. (2010). Changes in non-randomness in the expanding introduced avifauna of the world. *Ecography*, 33, 168–174.

Blackburn, T.M. Essl, F., Evans, T., *et al.* (2011). A unified classification of alien species based on the magnitude of their environmental impacts. *PLoS Biology*, 12, e1001850.

Blackburn, T.M., Su, S. and Cassey, P. (2014). A potential metric of the attractiveness of bird song to humans. *Ethology*, 120, 305–312.

Blackburn, T.M., Dyer, E., Su, S. and Cassey, P. (2015). Long after the event, or four things we (should) know about bird invasions. *Journal of Ornithology*, doi 10.1007/s10336-015-1155-z.

Bolnick, D.I., Svanbäck, R., Fordyce, J.A., *et al.* (2003). The ecology of individuals: incidence and implications of individual specialization. *American Naturalist*, 161, 1–28.

Bortolotti, G.R., Marchant, T., Blas, J. and Cabezas, S. (2009). Tracking stress: localisation, deposition and stability of corticosterone in feathers. *Journal of Experimental Biology*, 212, 1477–1482.

Bortolotti, G.R., Marchant, T.A., Blas, J. and German, T. (2008). Corticosterone in feathers is a long-term, integrated measure of avian stress physiology. *Functional Ecology*, 22, 494–500.

Butler, C.J. (2003). Population Biology of the Introduced Rose Ringed Parakeet *Psittacula krameri* in the UK. PhD thesis. Oxford, UK: University of Oxford.

Butler, C.J. (2005). Feral parrots in the continental United States and United Kingdom: past, present, and future. *Journal of Avian Medicine and Surgery*, 19, 142–149.

Cabezas, S., Carrete, M., Tella, J.L., Marchant, T.A. and Bortolotti, G.R. (2013). Differences in acute stress responses between wild-caught and captive-bred birds: a physiological mechanism contributing to current avian invasions? *Biological Invasions*, 15, 521–527.

Cantú J.C., Sánchez, M.E., Groselet, M. and Silva, J. (2007). *The Illegal Parrot Trade in Mexico: A Comprehensive Assessment*. Washington DC: Defenders of Wildlife.

Carrete, M. and Tella, J.L. (2008a). Non-native wildlife risk assessment: a call for scientific inquiry response. *Frontiers in Ecology and Environment*, 10, 466–467.

Carrete, M. and Tella, J.L. (2008b). Wild-bird trade and exotic invasions: a new link of conservation concern? *Frontiers in Ecology and Environment*, 6, 207–211.

Carrete, M. and Tella, J.L. (2011). Inter-individual variability in fear of humans and relative brain size of the species are related to contemporary urban invasion in birds. *PLoS ONE*, 6, e18859.

Carrete, M. and Tella, J.L. (2013). High individual consistency in fear of humans throughout the adult lifespan of rural and urban burrowing owls. *Scientific Reports*, 3, 3524.

Carrete, M., Edelaar, P., Blas, J., *et al.* (2012). Don't neglect pre-establishment individual selection in deliberate introductions. *Trends in Ecology and Evolution*, 27, 67–68.

Cassey, P., Blackburn, T.M., Sol, D., *et al.* (2004). Global patterns of introduction effort and establishment success in birds. *Proceedings of the Royal Society of London B: Biological Sciences* (Suppl.), 271, S405–S408.

Catford, J.A., Jansson, R. and Nilsson, C. (2009). Reducing redundancy in invasion ecology by integrating hypotheses into a single theoretical framework. *Diversity and Distributions*, 15, 22–40.

Chapple, D.G., Simmonds, S.M. and Wong, B.B.M. (2012). Can behavioral and personality traits influence the success of unintentional species introductions? *Trends in Ecology and Evolution*, 27, 57–64.

Charlesworth, D. and Charlesworth, B. (1987). Inbreeding depression and its evolutionary consequences. *Annual Review in Ecology and Systematics*, 18, 237–268.

Chiron, F., Shirley, S. and Kark, S. (2009). Human-related processes drive the richness of exotic birds in Europe. *Proceedings of the Royal Society B: Biological Sciences*, 276, 47–53.

Clavero, M., Nores, C., Kubersky-Piredda, S. and Centeno-Cuadros, A. (2015). Interdisciplinarity to reconstruct historical introductions: solving the status of cryptogenic crayfish. *Biological Review*, doi: 10.1111/brv.12205.

Cooney, R. and Jepson P. (2006). The international wild bird trade: what's wrong with blanket bans? *Oryx*, 40, 18–23.

Costello, C.J. and Solow, A.R. (2003). On the pattern of discovery of introduced species. *Proceedings of the National Academy of Sciences, USA*, 100, 3321–3323.

Delivering Alien Invasive Species Inventory for Europe (DAISIE) (2011). 100 of The Worst. Available at: http://www.europe-aliens.org/speciesTheWorst.do, accessed 21 April 2016.

De Boer, S.F., Van Der Vegt, B.J. and Koolhaas, J.M. (2003). Individual variation in aggression of feral rodent strains: a standard for the genetics of aggression and violence? *Behavior Genetics*, 33, 485–501.

Dehnen-Schmutz, K., Touza, J., Perrings, C. and Williamson, M. (2007). A century of the ornamental plant trade and its impact on invasion success. *Diversity and Distributions*, 13, 527–534.

Dooling, R.J., Mulligan, J.A. and Miller, J.D. (1971). Auditory sensitivity and song spectrum of the common canary (*Serinus canarius*). *Journal of Acoustical Society of America*, 50, 700–708.

Edelaar, P. and Tella, J.L. (2012). Managing non-native species: don't wait until their impacts are proven. *Ibis*, 154, 635–637.

Edelaar, P., Roques, S., Hobson, E.A., *et al.* (2015). Shared genetic diversity across the global invasive range of the Monk Parakeet suggests a common restricted geographic origin and the possibility of convergent selection. *Molecular Ecology*, 24(9), 2164–2176.

ENDCAP (2012). Wild pets in the European Union. Available at: http://endcap.eu/wp-content/uploads/2013/02/Report-Wild-Pets-in-the-European-Union.pdf, accessed 21 April 2016.

Freed, L.A. and Cann, R.L. (2009). Negative effects of an introduced bird species on growth and survival in a native bird community. *Current Biology*, 19, 1736–1740.

Gilardi, J.D. (2006). Captured for conservation: will cages save wild birds? A response to Cooney and Jepson. *Oryx*, 40, 24–26.

Grundy, J. P., Franco, A. and Sullivan, M.J. (2014). Testing multiple pathways for impacts of the non-native Black-headed Weaver *Ploceus melanocephalus* on native birds in Iberia in the early phase of invasion. *Ibis*, 156, 355–365.

Guttinger, H.R. (1985). Consequences of domestication on the song structures in the canary. *Behaviour*, 92, 255–278.

Haemig, P.D. (1978). Aztec Emperor Auitzotl and the great-tailed grackle. *Biotropica*, 10, 11–17.

Haemig, P.D. (2011). Introduction of the great-tailed grackle by Aztec Emperor Auitzotl: four-stage analysis with new information. *Ardeola*, 58, 387–397.

Hellmann, J.J., Byers, J.E., Bierwagen, B.G. and Dukes, J.S. (2008). Five potential consequences of climate change for invasive species. *Conservation Biology*, 22, 534–543.

Hernández-Brito, D., Carrete, M., Popa-Lisseanu, A., Ibáñez, C. and Tella, J.L. (2014). Crowding in the city: losing and winning competitors of an invasive bird. *PLoS ONE*, 9, e100593.

Hernández-Brito, D., Luna, A., Carrete, M. and Tella, J.L. (2015). Alien rose-ringed parakeets (*Psittacula krameri*) often attack and even cause the death of black rats (*Rattus rattus*). *Hystrix*, 25, 121–123.

Hulme, P.E. (2009). Trade, transport and trouble: managing invasive species pathways in an era of globalization. *Journal of Applied Ecology*, 46, 10–18.

Hulme, P.E., Bacher, S., Kenis, M., *et al.* (2008). Grasping at the routes of biological invasions: a framework for integrating pathways into policy. *Journal of Applied Ecology*, 45, 403–414.

Jepson, P. and Ladle, R. (2005). Bird-keeping in Indonesia. Conservation impacts and the potential for substitution-based conservation responses. *Oryx*, 39, 442–449.

Kark, S. and Sol, D. (2005). Establishment success across convergent Mediterranean ecosystems: an analysis of bird introductions. *Conservation Biology*, 19, 1519–1527.

Keller, L.F. and Walter, D.M. (2002). Inbreeding effects in wild populations. *Trends in Ecology and Evolution*, 17, 236–241.

Kolar, C.S. and Lodge, D.M. (2001). Progress in invasion biology: predicting invaders. *Trends in Ecology and Evolution*, 16, 199–204.

Kumschick, S. and Nentwig, W. (2010). Some alien birds have as severe an impact as the most effectual alien mammals in Europe. *Biological Conservation*, 143, 2757–2762.

Künzl, C., Kaiser, S., Meier, E. and Sachser, N. (2003). Is a wild mammal kept and reared in captivity still a wild animal? *Hormones and Behavior*, 43, 187–196.

Lau, M.W.N., Ades, G., Goodyer, N. and Zou, F. (1996). Wildlife trade in Southern China including Hong Kong and Macao. Biodiversity Working Group of the China Council for International Cooperation on Environment and Development Project, Hong Kong.

Lever, C. (2005). *Naturalised Birds of the World*. London: T and AD Poyser.

Lockwood, J.L., Cassey, P. and Blackburn, T. (2005). The role of propagule pressure in explaining species invasions. *Trends in Ecology and Evolution*, 20, 223–228.

MacGregor-Fors, I., Calderón-Parra, R., Meléndez-Herrada, A., López-López, S. and Schondube, J.E. (2011). Pretty, but dangerous! Records of non-native Monk Parakeets (*Myiopsitta monachus*) in Mexico. *Revista Mexicana de Biodiversidad*, 82, 1053–1056.

Mack, M.C. (2003). Phylogenetic constraint, absent life forms, and preadapted alien plants: a prescription for biological invasions. *International Journal of Plant Sciences*, 164, S185–S196.

Mason, G.J. (2010). Species differences in responses to captivity: stress, welfare and the comparative method. *Trends in Ecology and Evolution*, 25, 713–721.

Mason, G.J., Burn, C.C., Dallaire, J.A., *et al.* (2013). Plastic animals in cages: behavioural flexibility and responses to captivity. *Animal Behaviour*, 85, 1113–1126.

McDougall, P.T., Réale, D., Sol, D. and Reader, S.M. (2006). Wildlife conservation and animal temperament: causes and consequences of evolutionary change for captive, reintroduced and wild populations. *Animal Conservation*, 9, 39–48.

Mueller, J.C., Edelaar, P., Carrete, M., *et al.* (2014). Behaviour-related DRD4 polymorphisms in invasive bird populations. *Molecular Ecology*, 23, 2876–2885.

Muñoz-Fuentes, V., Green, A.J. and Negro, J.J. (2013). Genetic studies facilitated management decisions on the invasion of the ruddy duck in Europe. *Biological Invasions*, 15, 723–728.

Newson, S.E., Johnston, A., Parrott, D. and Leech, D.I. (2011). Evaluating the population-level impact of an invasive species, ring-necked parakeet *Psittacula krameri*, on native avifauna. *Ibis*, 153, 509–516.

Orchan, Y., Chiron, F., Shwartz, A. and Kark, S. (2013). The complex interaction network among multiple invasive bird species in a cavity-nesting community. *Biological Invasions*, 15, 429–445.

Peck, H.L., Pringle, H.E., Marshall, H.H., Owens, I.P.F. and Lord, A.M. (2014). Experimental evidence of impacts of an invasive parakeet on foraging behavior of native birds. *Behavioral Ecology*, 25, 582–590.

Pimentel D., Zuniga R. and Morrison D. (2005). Update on the environmental and economic costs associated with alien-invasive species in the United States. *Ecological Economics*, 52, 273–288.

Pires, S. (2012). The illegal parrot trade: a literature review. *Global Crime*, 13, 176–190.

Puigcerver, M., Sanchez-Donoso, I., Vilà, C., *et al.* (2014). Decreased fitness of restocked hybrid quails prevents fast admixture with wild European quails. *Biological Conservation*, 171, 74–81.

Puth, L.M. and Post, D.M. (2005). Studying invasion: have we missed the boat? *Ecology Letters*, 8, 715–721.

Rivalan, P., Delmas, V., Angulo, E., *et al.* (2007). Can bans stimulate wildlife trade? *Nature*, 447, 529–530.

Rodriguez-Cabal, M.A., Williamson, M. and Simberloff, D. (2013). Overestimation of establishment success of non-native birds in Hawaii and Britain. *Biological Invasions*, 15, 249–252.

Romagosa, C.M., Guyer, C. and Wooten, M.C. (2009). Contribution of the live-vertebrate trade toward taxonomic homogenization. *Conservation Biology*, 23, 1001–1007.

Saavedra, S., Maraver, A., Anadón, J.D. and Tella, J.L. (2015). Multiple recent introduction events and the control and spread of mynas (*Acridotheres* sp.) in Spain and Portugal. *Animal Biodiversity and Conservation*, 38, 121–127.

Sánchez-Donoso, I. (2014). Impact of game restocking on common quail populations. PhD Dissertation. Barcelona, Spain: University of Barcelona.

Sánchez-Donoso, I., Vilà, C., Puigcerver, M., *et al.* (2012). Are farm-reared quails for game restocking really common quails (*Coturnix coturnix*)? A genetic approach. *PloS ONE*, 7, e39031.

Sánchez-Donoso, I., Huisman, J., Echegaray, J., *et al.* (2014). Detecting slow introgression of invasive alleles in an extensively restocked game bird. *Frontiers in Ecology and Evolution*, 2, 15.

Sánchez-Donoso, I., Morales-Rodriguez, P.A., Puigcerver, M., *et al.* (2016). Postcopulatory sexual selection favors fertilization success of restocking hybrid quails over native common quails (*Coturnix coturnix*). *Journal of Ornithology*, 157, 33–42.

Sanz-Aguilar, A., Anadón, J.D., Edelaar, P., Carrete, M., Tella, J.L. (2014). Can establishment success be determined through demographic parameters? A case study on five introduced bird species. *PloS ONE*, 9, e110019.

Schwartz, M.D., Ahas, R. and Aasa, A. (2006). Onset of spring starting earlier across the northern hemisphere. *Global Change Biology*, 12, 343–351.

Seager, R. and Vecchi, G.A. (2010). Greenhouse warming and the 21st century hydroclimate of southwestern North America. *Proceedings of the National Academy of Sciences USA*, 107, 21277–21282.

Shepherd, C.R. (2006). The bird trade in Medan, north Sumatra: an overview. *Birding ASIA*, 5, 16–24.

Shepherd, C.R., Stengel, C.J. and Nijman, V. (2012). The export and re-export of CITES-listed birds from the Solomon Islands. *TRAFFIC*, Southeast Asia, Petaling Jaya, Selangor, Malaysia.

Shine C., Kettunen, M., ten Brink, P., Genovesi, P. and Gollasch, S. (2009). Technical support to EU strategy on invasive species (IAS) – Recommendations on policy options to control the negative impacts of IAS on biodiversity in Europe and the EU. Final report for the European Commission. Institute for European Environmental Policy (IEEP), Brussels, Belgium. 35 pp. Available at http://ec.europa.eu/environment/nature/invasivealien/docs/Shine2009_IAS_Final%20report.pdf, accessed 21 April 2016.

Simberloff, D., Martin, J.-L., Genovesi, P., *et al.* (2013). Impacts of biological invasions: what's what and the way forward. *Trends in Ecology and Evolution*, 28, 58–66.

Sol, D., Duncan, R.P., Blackburn, T.M., *et al.* (2005). Big brains, enhanced cognition, and response of birds to novel environments. *Proceedings of the National Academy of Sciences, USA*, 102, 5460–5465.

Sol, D., Maspons, J., Vall-llosera, M., *et al.* (2012a). Unraveling the life history of successful invaders. *Science*, 337, 580–583.

Sol, D., Bartomeus, I. and Griffin, A.S. (2012b). The paradox of invasion in birds: competitive superiority or ecological opportunism? *Oecologia*, 169, 553–564.

Somerville, A.D., Nelson, B.A. and Knudson, K.J. (2010). Isotopic investigation of pre-Hispanic macaw breeding in Northwest Mexico. *Journal of Anthropological Archaeology*, 29, 125–135.

Strubbe, D. and Matthysen, E. (2009). Experimental evidence for nest-site competition between invasive ring-necked parakeets (*Psittacula krameri*) and native nuthatches (*Sitta europaea*). *Biological Conservation*, 142, 1588–1594.

Strubbe, D., Matthysen, E. and Graham, C.H. (2010). Assessing the potential impact of invasive ring-necked parakeets *Psittacula krameri* on native nuthatches *Sitta europeae* in Belgium. *Journal of Applied Ecology*, 47, 549–557.

Strubbe, D., Broennimann, O., Chiron, F. and Matthysen, E. (2013). Niche conservatism in non-native birds in Europe: niche unfilling rather than niche expansion. *Global Ecology and Biogeography*, 22, 962–970.

Stuber, E.F., Araya-Ajoy, Y.G., Mathot, K.J., *et al.* (2013). Slow explorers take less risk: a problem of sampling bias in ecological studies. *Behavioral Ecology*, 24, 1092–1098.

Su, S., Cassey, P. and Blackburn, T.M. (2014). Patterns of non-randomness in the composition and characteristics of the Taiwanese bird trade. *Biological Invasions*, 16, 2563–2575.

Su, S., Cassey, P., Vall-llosera, M. and Blackburn, T.M. (2015). Going cheap: determinants of bird price in the Taiwanese pet market. *PloS ONE*, 10, e0127482.

Tella, J.L. (2011). The unknown extent of ancient bird introductions. *Ardeola*, 58, 399–404.

Tella, J.L. and Hiraldo, F. (2014). Illegal and legal parrot trade shows a long-term, cross-cultural preference for the most attractive species increasing their risk of extinction. *PloS ONE*, 9, e107546.

Theoharides, K.A. and Dukes, J.S. (2007). Plant invasion across space and time: factors affecting nonindigenous species success during four stages of invasion. *New Phytologist*, 176, 256–273.

Thomsen, J.B., Edwards, S.R. and Muliken, T.A. (1992). *Perceptions, Conservation and Management of Wild Birds in Trade*. Cambridge: TRAFFIC International, WWF and IUCN.

Toland, E., Warwick, C. and Arena P. (2012). The exotic pet trade: pet hate. *The Biologist*, 59, 14–18.

van Kleunen, M., Dawson, W., Schlaepfer, D., Jeschke, J.M. and Fischer, M. (2010). Are invaders different? A conceptual framework of comparative approaches for assessing determinants of invasiveness. *Ecology Letters*, 13, 947–958.

Warwick, C., Arena, P.C., Steedman, C. and Jessop, M. (2012). A review of captive exotic animal-linked zoonoses. *Journal of Environmental Health Research*, 12, 9–24.

Westphal, M.I., Browne, M., MacKinnon, K. and Noble, I. (2008). The link between international trade and the global distribution of invasive alien species. *Biological Invasions*, 10, 391–398.

Williams, F., Eschen, R., Harris, A., *et al.* (2010). *The Economic Cost of Invasive Non-Native Species on Great Britain*. Wallingford, UK: CABI.

Index

Printed in the United States
By Bookmasters